# Current and Future Developments in Physiology

# (Volume 2)

# Plant Physiology: From Historical Roots to Future Frontiers

Edited by

**Ergun Kaya**
*Muğla Sıtkı Koçman University*
*Faculty of Science, Molecular Biology and Genetics*
*Department*
*48000 Menteşe, Muğla, Türkiye*

# Current and Future Developments in Physiology

*(Volume 2)*

*Plant Physiology: From Historical Roots to Future Frontiers*

Editor: Ergun Kaya

ISSN (Online): 2468-7537

ISSN (Print):  2468-7545

ISBN (Online): 978-981-5305-84-5

ISBN (Print): 978-981-5305-85-2

ISBN (Paperback): 978-981-5305-86-9

Published by Bentham Science Publishers Pte. Ltd. Singapore. All Rights Reserved.

First published in 2024.

need for a court order if at any point you breach any terms of this License Agreement. In no event will any delay or failure by Bentham Science Publishers in enforcing your compliance with this License Agreement constitute a waiver of any of its rights.

3. You acknowledge that you have read this License Agreement, and agree to be bound by its terms and conditions. To the extent that any other terms and conditions presented on any website of Bentham Science Publishers conflict with, or are inconsistent with, the terms and conditions set out in this License Agreement, you acknowledge that the terms and conditions set out in this License Agreement shall prevail.

**Bentham Science Publishers Pte. Ltd.**
80 Robinson Road #02-00
Singapore 068898
Singapore
Email: subscriptions@benthamscience.net

**BENTHAM SCIENCE**

# CONTENTS

# FOREWORD

Current and Future Developments in Physiology (Volume 2) is a book that is published for the benefit of botany. The people spearheading the initial steps of this service hope that this book will serve as a valuable resource for all plant physiologists in publishing important, fundamental discoveries that further our understanding of plant growth, development, and metabolism. The editor and authors of the chapters see their work as providing committed support to the study of plant physiology as a whole.

With the rapid advancement of technology, science has begun to occupy an increasingly larger place in the lives of humanity. Research in plant physiology is increasingly evolving into the fields of artificial intelligence and bioinformatics, which include computerized applications. Concurrently, it constantly investigates in more depth the problems of developmental metabolism under the leadership of physiologists well-trained in molecular approaches, biophysics, and biochemistry methods. In this context, this book combines previous knowledge in the field of basic plant physiology with artificial intelligence and bioinformatics, including biotechnology and molecular biology approaches.

Research is uncovering actual issues that are pressing the greatest intellectual pursuits of humanity to address. Therefore, this book is available for all plant sciences, where physiological approaches must be employed to solve encountered difficulties in order to improve this major field of research overall. In addition to serving as a method for bringing all plant physiologists together into a cohesive, effective working group, my hope is that it will serve as a resource for plant physiologists across all fields of study by offering a central setting in which we can collaborate in the advancement of plant physiology without interfering with other groups' scheduled activities.

I think that this book, which consists of eleven chapters, each containing useful information, will be an even more useful resource with the support and constructive criticism of basic science plant physiologists and applied physiologists from all nations. I would like to express my endless thanks to the editor, book authors, publication editors, graphic designers, and all those who contributed to the preparation and publication of the book. I hope you will read the book as a useful resource.

**Murat Turan**
Molecular Biology and Genetics Department
Erzurum Technical University
25100 Yakutiye, Erzurum, Türkiye

# PREFACE

Plant physiology is a science that studies the symptoms and causes of various vital events that occur throughout the life of plants. The vital events occurring in plants are the result of chemical and physical changes in the living matter of the cell. So, in more general terms, we can define the events that occur as a result of the physical and chemical changes that occur in living things as physiological. Plant physiology tries to answer the question of how and why these physiological events occur in plants, and thus plant physiology reveals the laws and principles in force for physiological events. While presenting these laws and principles, the laws of physics and chemistry are undoubtedly used to a large extent. This reveals that physiology has a close relationship with physics and chemistry. It should also be noted that plant physiology has a special importance in biology because it is a science based on quantitative results, just like chemistry and physical sciences. Because plant physiologists must not only provide descriptive explanations but also explain events with quantitative values.

This book of eleven chapters on classical plant physiology and basic mechanisms aims to present to the reader, by combining many approaches based on the molecular basis of the development of plant physiology, the elucidation of basic metabolic pathways, and the events taking place at the cellular level, starting from the history of plant physiology to current molecular approaches and artificial intelligence-supported applications.

I would like to thank all the book authors who made valuable contributions to the preparation of this book, my advisors Prof. Dr. Fusün GÜMÜŞEL, Prof. Dr. Yelda ÖZDEN ÇİFTÇİ, Dr. Fernanda VİDİGAL DUERTA SOUZA and Dr. Dave ELLIS, who helped me excel in this field and provide useful products, and our families who always supported me.

**Ergun Kaya**
Muğla Sıtkı Koçman University
Faculty of Science, Molecular Biology and Genetics Department
48000 Menteşe, Muğla, Türkiye

*To the world you are a mother, but to your children you are the world,*
*Dedicated to my mother and all mothers…*

# List of Contributors

| | |
|---|---|
| **Aysel Uğur** | Gazi University, Faculty of Dentistry, Basic Science Department, Ankara, Türkiye |
| **Damla Ekin Özkaya** | Muğla Sıtkı Koçman University, Faculty of Science, Molecular Biology and Genetics Department, 48000, Menteşe, Muğla, Türkiye<br>Okan University, Vocational School of Health Services, Medical Laboratory Techniques Department, 34959, Tuzla, İstanbul, Türkiye |
| **Ergun Kaya** | Muğla Sıtkı Koçman University, Faculty of Science, Molecular Biology and Genetics Department, 48000, Menteşe, Muğla, Türkiye |
| **Hacer Ağar** | Muğla Sıtkı Koçman University, Faculty of Science, Molecular Biology and Genetics Department, 48000, Menteşe, Muğla, Türkiye |
| **İrem Demir** | Mugla Sitki Kocman University, Institute of Science, Biology Department, Menteşe, Muğla, Türkiye |
| **Mohammad Mehdi Habibi** | University of Tsukuba, Faculty of Life and Environmental Sciences, Tennodai, Tsukuba, Ibaraki, Japan |
| **Mehmet Emin Uras** | Haliç University, Faculty of Arts and Sciences, Department of Molecular Biology and Genetics, Eyupsultan, Istanbul, Türkiye |
| **Mehmet Ali Balcı** | Muğla Sıtkı Koçman University, Faculty of Science, Department of Mathematics, Menteşe, Muğla, Türkiye |
| **Nurdan Saraç** | Muğla Sıtkı Koçman University, Faculty of Science, Biology Department, Menteşe, Muğla, Türkiye |
| **Ömer Akgüller1** | Muğla Sıtkı Koçman University, Faculty of Science, Department of Mathematics, Menteşe, Muğla, Türkiye |
| **Selin Galatalı** | Muğla Sıtkı Koçman University, Faculty of Science, Molecular Biology and Genetics Department, 48000, Menteşe, Muğla, Türkiye |
| **Sedat Çiçek** | Muğla Sıtkı Koçman University, Faculty of Science, Molecular Biology and Genetics Department, 48000, Menteşe, Muğla, Türkiye |
| **Selin Galatalı** | Muğla Sıtkı Koçman University, Faculty of Science, Molecular Biology and Genetics Department, 48000, Menteşe, Muğla, Türkiye |
| **Tuba Baygar** | Mugla Sitki Kocman University, Research Laboratories Center, Menteşe, Mugla, Türkiye |

**CHAPTER 1**

# Historical Development of Plant Physiology

**Ergun Kaya**[1,*]

[1] *Muğla Sıtkı Koçman University, Faculty of Science, Molecular Biology and Genetics Department, 48000, Menteşe, Muğla, Türkiye*

**Abstract:** Although the basis of plant science is identified with the history of humanity, studies in the field of plant physiology based on both the development process of science and technological developments date back to the very recent past. In 1727, English physiologist, inventor, and chemist Stephen Hales published a book called 'Vegetable Statick' and in his book, Hales explained how water is mobilized in plants and laid the foundations of plant physiology. Since then, great developments in technology and biotechnology have allowed plant physiology to grow in a logarithmic manner. Today, many metabolisms have been enlightened both at the cellular level and at the tissue and organ level, and new studies are being added to these studies every day. In addition to the significant advances brought about by technological advancement, research in the fields of nutrition, plant chemistry, particularly in the agricultural sector, and genetics and molecular biology, though often fraught with ethical issues, has produced some truly groundbreaking discoveries. Within this framework, the goal of this chapter is to elucidate the features of the development processes by examining the history of plant biotechnology development, how technological advancements have accelerated this process, and what key studies were conducted during these phases.

**Keywords:** Biotechnology, Cellular events, Plant metabolism, Plant physiology.

## INTRODUCTION

Plant physiology investigates the biological events that occur during the life process of a plant, from the beginning as a seed to the stage of producing seeds again. Plant functions are basically based on the principles of physics and chemistry. Plant physiology is studied with the application of modern physics and chemistry techniques to the plants studied, so research in plant physiology also deepens with the advancements in these branches of science [1, 2].

The development of tissues in plants is closely related to the environment in which the plant is located and the physiological events that occur accordingly.

---

[*] **Corresponding author Ergun Kaya:** Muğla Sıtkı Koçman University, Faculty of Science, Molecular Biology and Genetics Department, 48000, Menteşe, Muğla, Türkiye; E-mail: ergunkaya@mu.edu.tr

Light, humidity, temperature, water, and gravity are important environmental factors affecting plant growth. Structure and function are closely related; that is, living things come into question because of the coexistence of genes, enzymes, other molecules, organelles, cells, tissues, and organs. For this reason, studies on plant physiology are closely related to plant anatomy, cell biology, and structural and functional chemistry [3, 4].

The target principle of plant physiology is how plants reproduce, grow, and develop. From ancient times to the present, when people started collecting seeds to grow nutritional plants, they saw that plants needed optimum environmental conditions such as sunlight, warmth, and moist soil and that the best-quality seeds produced the most productive plants [5]. They did, however, observe the advantages of practices like fertilization, irrigation, and hoeing. Agricultural activities, which spanned a very long time, enabled the development of new varieties and the cultivation of various species (Fig. **1**). The information obtained from basic plant growth and morphological analyses, which are among all these preliminary study activities, laid the foundations of plant physiology [5, 6].

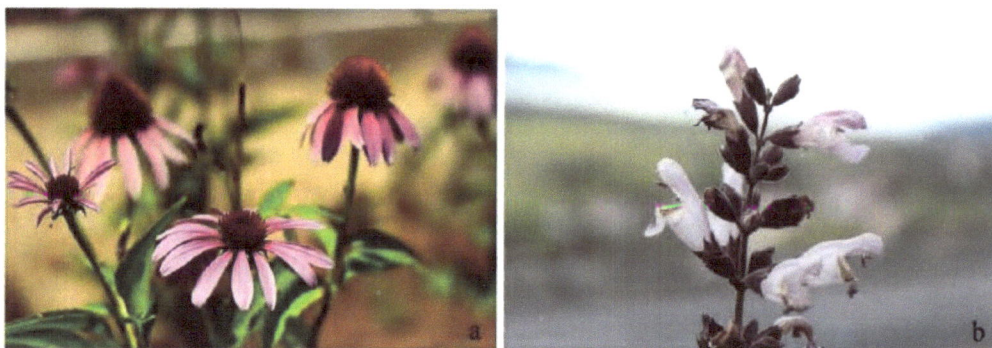

**Fig. (1).**  Medicinal-aromatic plant species cultivated in the collection garden of Muğla Metropolitan Municipality, Department of Agricultural Services, Garden for Local Seed Center. **a.** *Echinacea purpurea* (L.) Moench; **b.** *Salvia fruticosa* Mill.

## Early Experiments on Growth and Development

The first physiological approach to growth was directed at answering the question of where a plant gets the components necessary for its growth. Jean Baptista Van Helmont [7], who lived in the early 1600s, made a suggestion for this approach. According to Helmont, a Belgian doctor, the only source for the growth and development of a plant was water. The researcher irrigated the willow sapling he planted in a pot with only rainwater and grew a huge tree on a very small piece of soil. The researcher knew about $CO_2$ at the time but never anticipated that it could be one of the key growth drivers [7]. Some studies that have provided significant developments in plant physiology are listed chronologically in Table **1**, and are briefly summarized. In the 1700s, Antoine Lavoisier discovered that the matter resulting from organic synthesis was composed of oxygen and carbon on a large scale [8].

**Table 1. Some of the studies that pioneered the development of plant physiology [7 - 42].**

| Researcher | The Purpose of the Work | Date |
|---|---|---|
| Jean Baptista Van Helmont | Finding the answer to the question of where the plant gets the components necessary for its growth on *Salix* sp. | 1600s |
| Antoine Lavoisier | He found that the substance formed as a result of organic synthesis consists of oxygen and carbon on a large scale. | 1700s |
| Stephen Hales | He published his work "Vegetative Statics" as a result of his studies on transpiration, plant growth and gas exchange in plants. | 1727 |
| Joseph Priestley | He showed that plant leaves take in carbon dioxide in light and emit an equivalent amount of oxygen. | 1770 |
| Jan Ingenhousz | He repeated Priestley's experiments and investigated the effect of sunlight on living things. | 1778 |
| Jean Senebier | He carried out studies on the effect of sunlight on nature and photosynthesis. | 1790 |
| Nicholas de Saussure | He noted that water is involved in the photosynthesis process. Additionally, his findings on plant physiology have been reported in detail in many sources. | 1807 |
| Justus von Liebig | He introduced the "Law of the Minimum", which is the view that the growth and development of plants is limited by the least amount of nutrients in the soil. | 1830s |
| Julius Robert von Mayer | observed that the process converts light energy into the chemical energy of organic carbon | 1842 |
| Julius Sachs | He carried out the first studies on the uptake of nutrients and mineral substances from the soil by plants. | 1865 |
| Henry Dixen and John Joly | They studied the mechanism by which water and mineral substances are drawn up from the roots of plants with the help of open vessels of the xylem and tracheids. | 1895 |
| Gottlieb Haberlandt | He cultured the isolated plant cells in a simple nutrient solution, and this attempt was important in that it gave an idea about the potential of the cell and demonstrated a new technology as plant tissue culture. | 1902 |
| Wightman W. Garner and Harry A. Allard | They discovered photoperiodism, the regulation of flowering according to day length. | 1920 |
| Cornelis Bernardus van Niel | He used radioactive water to show that water, not carbon dioxide, was the source of the oxygen released during photosynthesis | 1929 |
| Erwin Bünning | He found that sleep movements, such as the drooping of bean leaves in the evening, are controlled not by the onset of darkness but by a biological "clock," a circadian rhythm. | 1930s |

(Table 1) cont.....

| Researcher | The Purpose of the Work | Date |
|---|---|---|
| Richard Snow | He demonstrated the role of indole acetic acid (IAA) in promoting cambial activity for rapid cell proliferation. | 1935 |
| Philip R. White | He produced a synthetic plant tissue culture medium composition (White medium-WM) containing vitamin B. | 1943 |
| Friedrich Laibach | Identification of Arabidopsis thaliana as a model organism for developmental and physiological analyses. | 1943 |
| Harry Borthwick | He discovered phytochrome and it was found to be the pigment at the center of photoperiodism. | 1952 |
| Toshio Murashige and Folke K. Skoog | They reported the composition of MS nutrient media, which is still widely used today. | 1962 |
| Melvin Calvin and Andrew Benson | They discovered that sugar is synthesized in the stroma of the chloroplast and solved the molecular details of this formation. | 1963 |
| Douglas E. Berg | He produced recombinant DNA for the first time using restriction enzymes. | 1972 |
| Patrick H. O'Farrell | He developed a high-resolution two-dimensional gel electrophoresis system. | 1975 |
| Frederick Sanger, Allan Maxam, and Walter Gilbert | They developed technologies for DNA sequencing. | 1977 |
| Kary Mullis | He discovered the polymerase chain reaction (PCR) for DNA amplification. | 1983 |
| Patricia C. Zambryski | The first use of genetic modification of *Agrobacterium tumefaciens* has been reported. | 1983 |
| Mike Bevan *et al.* | *Arabidopsis thaliana* full genome has been sequenced | 2001 |

Joseph Priestley, Jan Ingenhousz, and Jean Senebier showed that plant leaves take up carbon dioxide in light and emit an equivalent amount of oxygen [9, 10]. Later, Nicholas de Saussure noted the involvement of water in the process. In the dark, the opposite happened: plants breathed like animals, taking in oxygen and releasing carbon dioxide [11, 12]. Julius Robert von Mayer observed that the process converts light energy into the chemical energy of organic carbon. Thus, the growth of seedlings in the dark or roots in the soil came at the expense of this energy. Therefore, in the 19th century, photosynthesis, although not understood biochemically, was considered the primary and fundamental synthetic process in plant growth [13].

## Previous Studies in Plant Nutrition and Transport

In the 1830s, Justus von Liebig proposed the Law of Minimums. In this law, it is emphasized that the growth and development of plants are limited by the minimum amount of nutrients in the soil, and this view has been confirmed by experiments conducted around the world [14, 15].

One of the first studies on the uptake of nutrients and mineral substances from the soil in plants was carried out by Julius Sachs and his colleagues in order to determine that nitrogen, potassium, phosphate, sulfur and other elements in quantitative terms are of great importance in small soil components for plant growth, and they used various chemical analyses in these studies [14]. Due to the presence of inorganic nutrients, particularly nitrogen, in the soil, fertilization has long been recognized as being important. They can be introduced to the soil as inorganic salts, like potassium nitrate, . The organic matter of manure, or the residue of its decay, has been shown to contribute to improving the soil or soil structure. From these discoveries, different approaches to the modern agricultural use of chemical fertilizers have been introduced [14, 16].

One of the pioneering studies on water uptake from the soil in plants was carried out by Van Helmont on the willow plant. In this study, metabolic processes in the plant were observed by continuously watering it with much more water than the tree contains [17]. In 1727, Stephen Hales, an English clergyman, and amateur physiologist, published his work titled "Vegetative Statics" as a result of his studies on transpiration [18], plant growth, and gas exchange in plants. Hales suggested that water coming from the soil moves towards the leaves, where it is lost as water vapor through a process called transpiration. Subsequent research in the nineteenth and early twentieth centuries showed that water diffuses through stomata, and pores in the leaf epidermis [17, 18].

The opening and closing mechanism of stomata is based on the principle of stomata taking in and giving out water, depending on the presence or absence of water. Capillary forces occurring in the vascular bundles along the mesophyll layer in leaves, with some contribution from osmosis, pull water columns up through the open vessels and tracheids of the xylem that carry nutrients from the roots. The harmony between water molecules and their adhesion to cell walls prevents the tense water columns from breaking even in very tall trees (adhesion-cohesion force of water). This mechanism was first proposed by Henry Dixen and John Joly in 1895, and in the twentieth century, numerous researchers confirmed and developed this "adaptation-tension" theory of transportation [19, 20].

Stephen Hales also measured the root pressure of plants whose apical meristematic region had been cut. Subsequent studies have shown that under conditions of quality soil, appropriate moisture, and aeration, roots actively secrete high concentrations of salt into the root xylem, creating a high osmotic pressure (guttation) that forces water up the stem and out through pores at the tips of the leaves. In 1926, E. Munch proposed a similar mechanism for translocation, which is the transfer of sugars from leaves to roots and other plant parts *via* the phloem. This mechanism is now known as the pressure flow model [21, 22].

## Cellular and Molecular Plant Physiology

As the twentieth century approached, plant physiologists increasingly turned to chemistry and physics to answer fundamental questions. In addition, they created their groups and published new journals with the study's findings. This has had a significant impact on increasing the level and amount of research. It has been found that many of the results obtained from more comprehensive medical, animal, and microbiological research on the basic biochemistry of cell growth and function also apply to plant cells. Anatomical studies have yielded structural details to support physiological findings, and microscopic cell structure has been revealed by electron microscopy [23, 24].

Environmental, hormonal, cellular and genetic factors on growth and development have been studied in detail, but there are many mechanisms yet to be learned. For instance, plants create ethylene, a basic two-carbon gas that controls seed germination and begins the ripening of fruit. It has been discovered that auxin or cell growth hormone translocation causes phototropisms and geotropisms [25]. In some cases, auxins can also stimulate cell division. Other plant growth regulators, gibberellins, regulate cell division in the root tip and activate enzyme formation in seed germination [25, 26].

The development of plant tissue culture technology in the early 1900s led to the discovery of cytokinins, which are growth regulators effective in the division of plant cells [27]. In 1902, Gottlieb Haberlandt cultured the plant cells he isolated in a simple nutrient solution and presented this initiative as a classic article at the meeting of the Vienna Academy of Sciences in Germany [28]. In 1935, Snow demonstrated the role of indole acetic acid (IAA) in promoting cambial activity for rapid cell proliferation [29]. In 1943, Philip R. White introduced a synthetic plant tissue culture medium composition (White medium-WM) containing vitamin B, and tomato root cultures could be preserved in this environment for as long as 30 years [28]. In 1962, Toshio Murashige and Folke K. Skoog reported the plant tissue culture medium composition (Murashige-Skoog medium-MS medium) that has been widely used until today. This initiative is important in that it gives insight into the potential of the cell and demonstrates a new technology as plant tissue culture. Abscisic acid, another plant growth regulator, initiates the senescence and shedding of leaves in autumn and causes stomata to close under water stress. Today, studies on growth regulatory compounds are continuing, but there are still points that need to be clarified on the basic mechanisms of the interaction of these molecules with each other [28, 30].

Photoperiodism, the regulation of flowering according to day length, was discovered by Wightman W. Garner and Harry A. Allard in 1920. Further studies were found in the 1930s by German biologist Erwin Bünning that sleep movements, such as the drooping of bean leaves in the evening, were controlled not by the onset of darkness but by a biological "clock", a circadian rhythm. Phytochrome was discovered by Harry Borthwick in 1952 and was found to be the pigment central to photoperiodism [31, 32].

In recent years, molecular biology studies have gained great momentum in attempts to find genes responsible for physiological processes and metabolic pathways. The structure of chlorophyll in photosynthesis has been determined and it has been revealed that it is localized in the inner membranes of the chloroplasts of mesophyll cells. The finding that the red and blue portions of the light spectrum are effective led to the discovery that two light reactions are required [33]. In 1929, Cornelis Bernardus van Niel used radioactive water to show that water, not carbon dioxide, was the source of the oxygen released during photosynthesis [34, 35]. Sugar was found to be synthesized in the stroma of the chloroplast, and the molecular details of its formation were solved by Melvin Calvin and Andrew Benson in 1963. All plant cells have been found to respire. This is understood to be an energy-yielding process that involves a different membranous organelle, the mitochondria, which is essentially the same as in animals and yields metabolic energy available for transport reactions and synthesis of cell material [36].

The formation of fats and oils from carbohydrates was found to be similar to that in animals, but plants had the additional ability to convert fats in germinating seeds into carbohydrates such as glucose, which are used in cellulose wall formation. The symbiotic relationships of plants and microorganisms were investigated, especially in cases of reduced nitrogen formation from atmospheric nitrogen by nodule bacteria. In this way, various types of energy metabolism that occur in prokaryotic organisms were discovered, and inferences could thus be made about the role that specific microorganisms play in the cycling of elements. In addition, it became possible to elucidate many details about metabolic pathways and their regulation [37, 38].

Described by Friedrich Laibach in 1943 as a model organism for developmental and physiological analyses, Arabidopsis thaliana became central to scientists' efforts to understand plant genomes at the end of the twentieth century. The complete sequence of this genome was elucidated in 2000 by an international consortium of plant geneticists. Following the introduction of recombinant DNA technology into the scientific world, the acceleration of technological developments and the spread of DNA/Protein-based techniques after the twentieth century allowed studies to elucidate many cellular metabolic pathways to continue rapidly [39, 40].

## Programmed Cell Death or Apoptosis

The process of planned cell death is called apoptosis. It is applied in the early stages of development to get rid of undesirable cells, like the ones that grow between a developing hand's fingers. Adults employ apoptosis to get rid of cells that are too damaged to be repaired [41]. In plants, as in animals, programmed cell death or apoptosis is the final stage of genetically controlled cell differentiation. In some cases, they acquire special functions as they die (such as vascular tissues, and fibers) or, conversely, cells die after completing their duties. This type of programmed cell death is called developmental cell death and is controlled by an endogenous program. The other type of cell death occurs as a result of different environmental signals or pathogen attacks, including biotic and abiotic stimuli, and they cause changes in the original cell program. Senescence, also referred to as biological aging, is the progressive loss of functional traits in living things. In the latter stages of an organism's life cycle, at least, whole organism senescence refers to a rise in mortality or a fall in fecundity with advancing age. Senescence is often associated with apoptosis. Programmed cell death in plants occurs in response to specialized conditions involving the hypersensitivity response to pathogens and the development of tracheal elements. It may differ from programmed cell death because senescence can be delayed and reversible. This indicates that programmed cell death is a non-essential part of the senescence process [41, 42].

## The Age of Artificial Intelligence in Plant Physiology

Today, artificial intelligence technologies (Fig. **2**). operate in many areas, from automatic editing applications on mobile phones to facial recognition, using satellite systems and precision agriculture applications. Although the use of these technologies does not have a long history, they have recently enabled serious developments in many fields. Increasing computing power, increasing ability to collect, manage and store huge amounts of data, and algorithmic advances are the most obvious examples. Multiple applications of artificial intelligence have been developed, each with their own techniques, strengths, and weaknesses, making certain approaches a better match for certain problems than others [43, 44].

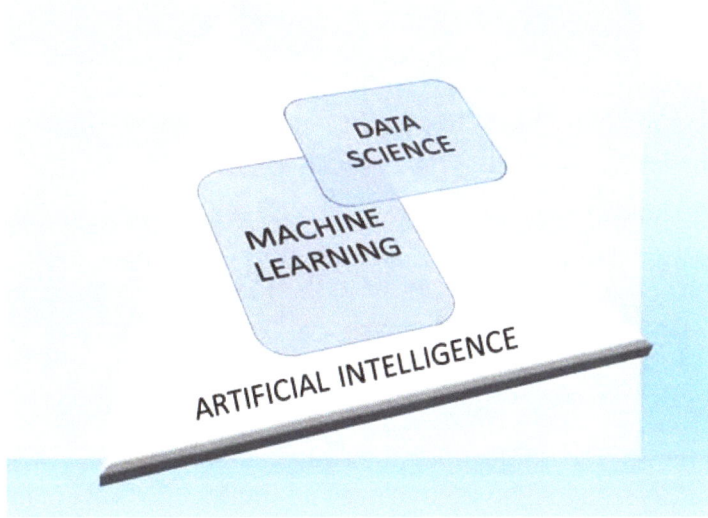

**Fig. (2).** Schematic representation of the relationship between artificial intelligence, machine learning, and data science [45].

The use of artificial intelligence neural networks (*e.g.,* deep learning) forms the basis for rapid and effective machine learning in solving a number of scientific problems in plant science and many physiological problems. For example, deep learning technologies have recently been useful in a variety of predictive tasks, such as species identification, determination of natural ranges, modeling of plant species distributions, diagnosis of diseases, and identification of disease agents. In addition, studies on comparative genomics and gene expression have been highly effective in understanding metabolic pathways and physiological activities in the cell. Moreover, new approaches to combining high-resolution imaging technologies with artificial intelligence and machine learning and digital repositories are now poised to revolutionize the study of plant phenology and functional traits. Artificial intelligence technologies are advancing rapidly due to their advantages in plant physiology and cellular metabolic processes, combining the development of alternative approaches best suited to specific questions, data sources and analytical techniques [46, 47].

## CONCLUSION

Biological sciences are developing very rapidly, with the help of technological developments such as genomics, molecular genetics, high-throughput genomes, and image analysis in the post-genomic era we are in. As a result, unknowns regarding the structure, function, and development of plants are rapidly being clarified. Especially since the early 2000s, our knowledge about the relationships between signal pathways and gene networks that complement each other and work in these characteristics of plants has been increasing day by day. Considering that the metabolism, growth, and development of the plant are a whole and continuous interaction, many unexplained mechanisms, concepts, and metabolic pathways in plant physiology will be elucidated in the future, thanks to the methods updated with the developing technology.

The ever-increasing world population confronts agriculture with new challenges, which necessitates greatly increasing the productivity of future agriculture. To continue to increase crop yields sufficiently in the future, a new green revolution seems inevitable. However, as the improved yields achieved gradually increase, conventional breeding is increasingly reaching its limits but is unlikely to keep pace with the rate of population growth in the long term.

Advanced technologies and molecular plant physiology approaches have now begun to open new horizons in the fight against an ineffective process in the plant that competes with photosynthesis. This refers to the plant's respiratory processes (photorespiration) that cause energy loss and therefore limit growth, and researchers want to modify metabolic pathways or develop new ways to bypass or improve photorespiration. This suggests that this would be possible, for example, by changing enzyme properties or adding new proteins.

# REFERENCES

[1]   Hopkins WG, Hüner NPA. Introduction to Plant Physiology. 4th ed., Hoboken, USA: John Wiley and Sons 2009.

[2]   Salisbury FB, Ross CW. Plant Physiology, Hormones and Plant Regulators: Auxins and Gibberellins. 4th ed., Belmont, California: Wadsworth Publishing Company 1992.

[3]   Taiz L, Zeiger E. 2010.

[4]   DeJong TM, Da Silva D, Vos J, Escobar-Gutiérrez AJ. Using functional–structural plant models to study, understand and integrate plant development and ecophysiology. Ann Bot (Lond) 2011; 108(6): 987-9.
      [http://dx.doi.org/10.1093/aob/mcr257] [PMID: 22084818]

[5]   Olas JJ, Fichtner F, Apelt F. All roads lead to growth: imaging-based and biochemical methods to measure plant growth. J Exp Bot 2020; 71(1): 11-21.
      [http://dx.doi.org/10.1093/jxb/erz406] [PMID: 31613967]

[6]   Feng Y, Zheng K, Lin X, Huang J. Plant growth, physiological variation and homological relationship of *Cyclocarya* species in *ex situ* conservation. Conserv Physiol 2022; 10(1): coac016.
      [http://dx.doi.org/10.1093/conphys/coac016] [PMID: 35539008]

[7]   Antonkiewicz J, Łabętowicz J. Chemical innovation in plant nutrition in a historical continuum from ancient greece and rome until modern times. Chemistry-Didactics-Ecology-Metrology 2016; 21(1-2): 29-43.
      [http://dx.doi.org/10.1515/cdem-2016-0002]

[8]   Stokes MA. Antoine Lavoisier and the study of respiration: 200 years old. Aust N Z J Surg 1991; 61(3): 229-32.
      [http://dx.doi.org/10.1111/j.1445-2197.1991.tb07597.x] [PMID: 2003841]

[9]     Martin D, Thompson A, Stewart I, *et al.* A paradigm of fragile Earth in Priestley's bell jar. Extrem
        Physiol Med 2012; 1(1): 4.
        [http://dx.doi.org/10.1186/2046-7648-1-4] [PMID: 23849304]

[10]    Pennazio S. The dawn of photosynthesis. Theor Biol Forum 2011; 104(2): 47-63.
        [PMID: 25095597]

[11]    Hart H. Nicolas theodore de saussure. Plant Physiol 1930; 5(3): 424-9.
        [http://dx.doi.org/10.1104/pp.5.3.424] [PMID: 16652672]

[12]    Pennazio S. Photosynthesis: from de saussure to liebig. Theor Biol Forum 2017; 110(1-2): 95-113.
        [http://dx.doi.org/10.19272/201711402006] [PMID: 29687833]

[13]    Tang KH, Tang YJ, Blankenship RE. Carbon metabolic pathways in phototrophic bacteria and their
        broader evolutionary implications. Front Microbiol 2011; 2: 165.
        [http://dx.doi.org/10.3389/fmicb.2011.00165] [PMID: 21866228]

[14]    Niklas KJ. Historical roots and current status of plant physiology. Plant Signal Behav 2019; 14(1):
        1552058.
        [http://dx.doi.org/10.1080/15592324.2018.1552058]

[15]    Gorban AN, Pokidysheva LI, Smirnova EV, Tyukina TA. Law of the Minimum paradoxes. Bull Math
        Biol 2011; 73(9): 2013-44.
        [http://dx.doi.org/10.1007/s11538-010-9597-1] [PMID: 21088995]

[16]    Van Lijsebettens M, Van Montagu M. Historical perspectives on plant developmental biology. Int J
        Dev Biol 2005; 49(5-6): 453-65.
        [http://dx.doi.org/10.1387/ijdb.041927ml] [PMID: 16096956]

[17]    Chastain Ben B. Jan Baptista van Helmont. Encyclopedia Britannica, 2024, Available from:
        https://www.britannica.com/biography/Jan-Baptista-van-Helmont

[18]    Britannica, The Editors of Encyclopaedia. Stephen Hales. Encyclopedia Britannica, 2024, Available
        from: https://www.britannica.com/biography/Stephen-Hales

[19]    Rockwell F, Sage RF. Plants and water: the search for a comprehensive understanding. Ann Bot
        (Lond) 2022; 130(3): i-viii.
        [http://dx.doi.org/10.1093/aob/mcac107] [PMID: 35997781]

[20]    Boursiac Y, Protto V, Rishmawi L, Maurel C. Experimental and conceptual approaches to root water
        transport. Plant Soil 2022; 478(1-2): 349-70.
        [http://dx.doi.org/10.1007/s11104-022-05427-z] [PMID: 36277078]

[21]    Volkov V, Schwenke H. Quest for mechanisms of plant root exudation brings new results and models,
        300 years after Hales. Plants 2020; 10(1): 38.
        [http://dx.doi.org/10.3390/plants10010038] [PMID: 33375713]

[22]    Jensen KH, Berg-Sørensen K, Friis SMM, Bohr T. Analytic solutions and universal properties of sugar
        loading models in Münch phloem flow. J Theor Biol 2012; 304: 286-96.
        [http://dx.doi.org/10.1016/j.jtbi.2012.03.012] [PMID: 22774225]

[23]    Kutschera U, Niklas KJ. Julius Sachs (1868): The father of plant physiology. Am J Bot, 2018; 105(4):
        656-666.
        [http://dx.doi.org/10.1002/ajb2.1078]

[24]    Raikhel NV. Plant physiology: past, present, and future. Plant Physiol 2001; 125(1): 1-3.
        [http://dx.doi.org/10.1104/pp.125.1.1] [PMID: 10806219]

[25]    Fatma M, Asgher M, Iqbal N, *et al.* Ethylene signaling under stressful environments: analyzing
        collaborative knowledge. Plants 2022; 11(17): 2211.
        [http://dx.doi.org/10.3390/plants11172211] [PMID: 36079592]

[26]    Lymperopoulos P, Msanne J, Rabara R. Phytochrome and phytohormones: working in tandem for

plant growth and development. Front Plant Sci 2018; 9: 1037.
[http://dx.doi.org/10.3389/fpls.2018.01037] [PMID: 30100912]

[27] Mandal S, Ghorai M, Anand U, *et al.* RETRACTED: Cytokinin and abiotic stress tolerance -What has been accomplished and the way forward? Front Genet 2022; 13: 943025.
[http://dx.doi.org/10.3389/fgene.2022.943025] [PMID: 36017502]

[28] Sussex IM. The scientific roots of modern plant biotechnology. Plant Cell 2008; 20(5): 1189-98.
[http://dx.doi.org/10.1105/tpc.108.058735] [PMID: 18515500]

[29] Abel S, Theologis A. Odyssey of Auxin. Cold Spring Harb Perspect Biol 2010; 2(10): a004572.
[http://dx.doi.org/10.1101/cshperspect.a004572] [PMID: 20739413]

[30] Dalton SJ. A reformulation of Murashige and Skoog medium (WPBS medium) improves embryogenesis, morphogenesis and transformation efficiency in temperate and tropical grasses and cereals. Plant Cell Tissue Organ Cult 2020; 141(2): 257-73.
[http://dx.doi.org/10.1007/s11240-020-01784-8] [PMID: 32308245]

[31] Hayama R, Coupland G. The molecular basis of diversity in the photoperiodic flowering responses of Arabidopsis and rice. Plant Physiol 2004; 135(2): 677-84.
[http://dx.doi.org/10.1104/pp.104.042614] [PMID: 15208414]

[32] Golembeski GS, Kinmonth-Schultz HA, Song YH, Imaizumi T. Photoperiodic flowering regulation in *Arabidopsis thaliana.* Adv Bot Res 2014; 72: 1-28.
[http://dx.doi.org/10.1016/B978-0-12-417162-6.00001-8] [PMID: 25684830]

[33] Alberts B, Johnson A, Lewis J, *et al.* Molecular Biology of the Cell. 4th ed., New York: Garland Science 2002.

[34] Stirbet A, Lazár D, Guo Y, Govindjee G. Photosynthesis: basics, history and modelling. Ann Bot (Lond) 2020; 126(4): 511-37.
[http://dx.doi.org/10.1093/aob/mcz171] [PMID: 31641747]

[35] Dismukes GC, Klimov VV, Baranov SV, Kozlov YN, DasGupta J, Tyryshkin A. The origin of atmospheric oxygen on Earth: The innovation of oxygenic photosynthesis. Proc Natl Acad Sci USA 2001; 98(5): 2170-5.
[http://dx.doi.org/10.1073/pnas.061514798] [PMID: 11226211]

[36] Ebenhöh O, Spelberg S. The importance of the photosynthetic Gibbs effect in the elucidation of the Calvin–Benson–Bassham cycle. Biochem Soc Trans 2018; 46(1): 131-40.
[http://dx.doi.org/10.1042/BST20170245] [PMID: 29305411]

[37] Clemente-Suárez VJ, Mielgo-Ayuso J, Martín-Rodríguez A, Ramos-Campo DJ, Redondo-Flórez L, Tornero-Aguilera JF. The burden of carbohydrates in health and disease. Nutrients 2022; 14(18): 3809.
[http://dx.doi.org/10.3390/nu14183809] [PMID: 36145184]

[38] Burkholder PR. The rôle of light in the life of plants. Bot Rev 1936; 2(1): 1-52.http://www.jstor.org/stable/4353120
[http://dx.doi.org/10.1007/BF02869924]

[39] Somssich M. The dawn of plant molecular biology: How three key methodologies paved the way. Curr Protoc 2022; 2(4): e417.
[http://dx.doi.org/10.1002/cpz1.417] [PMID: 35441802]

[40] Koornneef M, Meinke D. The development of Arabidopsis as a model plant. Plant J 2010; 61(6): 909-21.
[http://dx.doi.org/10.1111/j.1365-313X.2009.04086.x] [PMID: 20409266]

[41] Obeng E. Apoptosis (programmed cell death) and its signals - A review. Braz J Biol 2021; 81(4): 1133-43.
[http://dx.doi.org/10.1590/1519-6984.228437] [PMID: 33111928]

[42] John Bright J, Khar A. Apoptosis: Programmed cell death in health and disease. Biosci Rep 1994;

14(2): 67-81.
[http://dx.doi.org/10.1007/BF01210302] [PMID: 7948772]

[43]    Nabwire S, Suh HK, Kim MS, Baek I, Cho BK. Application of artificial intelligence in phenomics. Sensors (Basel) 2021; 21(13): 4363.
[http://dx.doi.org/10.3390/s21134363] [PMID: 34202291]

[44]    Freeman D, Gupta S, Smith DH, *et al.* Watson on the farm: Using cloud-based artificial intelligence to identify early indicators of water stress. Remote Sens (Basel) 2019; 11(22): 2645.
[http://dx.doi.org/10.3390/rs11222645]

[45]    Galatali S, Balci MA, Akguller O *et al.* Production of Disease-Free Olive Seedlings with Artificial Intelligence and Biotechnological Methods. European j biol biotechnol 2021; 2(3): 79–84.
[http://dx.doi.org/10.24018/ejbio.2021.2.3.172]

[46]    Rico-Chávez AK, Franco JA, Fernandez-Jaramillo AA, Contreras-Medina LM, Guevara-González RG, Hernandez-Escobedo Q. Machine learning for plant stress modeling: A Perspective towards hormesis management. Plants 2022; 11(7): 970.
[http://dx.doi.org/10.3390/plants11070970] [PMID: 35406950]

[47]    Niazian M, Niedbała G. Machine learning for plant breeding and biotechnology. Agriculture 2020; 10(10): 436.
[http://dx.doi.org/10.3390/agriculture10100436]

CHAPTER 2

# Stress Physiology and Current Approaches

## Ergun Kaya[1,*] and Selin Galatalı[1]

[1] *Muğla Sıtkı Koçman University, Faculty of Science, Molecular Biology and Genetics Department, 48000, Menteşe, Muğla, Türkiye*

**Abstract:** Plants often encounter environmental stressors in both wild and cultivated environments. Certain environmental stressors, like air temperature, only last a few minutes, but others, like soil water content, might persist for several days. Stress might last for months if there is a mineral shortage in the soil. This chapter gives an overview of the ways that soil, climate, and stress affect the spread of different plant species. Thus, it is crucial for agriculture and the environment to comprehend the physiological mechanisms that underlie plants' methods of adaptation and acclimatization to environmental challenges. A common definition of stress is an outside influence that negatively impacts plants. Stress tolerance and the concept of stress are closely related. The capacity of a plant to withstand adverse environmental conditions is known as stress tolerance. One plant may not find stress in the same environment as another. Based on the fundamental ideas of stress physiology in plants, this chapter seeks to provide a modern and fundamental explanation of the metabolic processes that occur in cells.

**Keywords:** Antioxidant mechanism, Heat stress, Salt stress, Toxicity, Water stress.

## INTRODUCTION

Living things, by their nature, are in constant contact with the external environment. If inappropriate conditions occur in their environment, they are exposed to stress conditions due to lack of adaptation. When environmental conditions change so much that they negatively affect the normal growth and development of a plant, the situation that occurs in the plant is called stress. In other words, it is defined as external factors that have negative effects on the plant. In many cases, stress is a concept that needs to be explained by relating it to the survival of the plant, its ability to produce products, biomass accumulation and assimilation [1, 2].

---

\* **Corresponding author Ergun Kaya:** Muğla Sıtkı Koçman University, Faculty of Science, Molecular Biology and Genetics Department, 48000, Menteşe, Muğla, Türkiye; E-mail: ergunkaya@mu.edu.tr

As a matter of fact, abiotic stresses such as drought, salinity, extreme temperatures, chemical toxicity, and oxidative stress are serious threats that disrupt agricultural activities and deteriorate the environment. For example, abiotic stress is the primary cause of crop yield loss worldwide, greatly reducing average crop yields in the most productive crop plants. Accumulated data on how plants respond to biotic and abiotic stress and how stress affects the developmental processes in the plant life cycle have enabled the development of new approaches on the subject [3, 4].

Plants encounter many stress factors during their lives. According to Levitt, stress factors are divided into two: biotic and physicochemical [5]. Biotic factors include stress factors caused by the infection of microorganisms (fungi, bacteria, and viruses) and attacks by harmful animals. Abiotic factors are environmental factors such as water, temperature, radiation, chemicals, magnetic and electrical fields [6]. Plants, which do not have the option of avoiding the stressor by moving away from it due to their sessile nature, are directly exposed to stress, unlike animals. This direct effect negatively affects growth and development and causes plant organs to lose their vitality [7, 8].

Damage caused by stress factors varies depending on the plant type, tolerance, and adaptation ability [9, 10]. Considering that plants encounter many stress factors in nature throughout their lives, it is very important to elucidate stress-related mechanisms and develop tolerant species and varieties. For this purpose, in this chapter, the molecular and biochemical events that occur in plants under stress conditions will be discussed and the responses to stress will be tried to be explained [7, 11].

## RESISTANCE TO WATER SHORTAGE AND DROUGHT

Drought resistance mechanisms are divided into several types. First, postponement of drying (the ability to retain water in the tissue) and tolerance to drying. These two are sometimes referred to as drought tolerance at high and low water potentials, respectively. A third mechanism is escape from drought. This mechanism involves completing the life cycle during the rainy season, before drought occurs. There are two categories of people who delay drying: those who do not waste water and those who do. Individuals who do not waste water use it in moderation. In order to use it later in life, these plants store part of the water in the soil. Individuals that wastewater consume a lot of water. The mesquite tree (*Prosopis glandulosa* Torr.) is an example of a water waster. This plant, whose roots can reach very deep, has destroyed the semi-arid grasslands in the southwestern United States [12]. Since it uses excessive water, it prevents grasses of agricultural value from settling in that area. A plant with a high ability to gain and use water will be more resistant to drought [12, 13].

Some plants have adaptations such as the C4 and CAM photosynthesis pathways (Fig. 1). These adaptations allow plants to use more water. Additionally, plants have acclimation mechanisms that are activated to respond to water stress. Water scarcity is defined as any water content of a tissue or cell below the highest water content in the plant. Water stress has some effects on growth: it is the limitation of leaf expansion (*i*), water shortage stimulates leaf abscission (*ii*), during water scarcity, roots grow towards moist regions deep in the soil (*iii*), stomata close in response to abscisic acid during water scarcity (*iv*), water shortage limits photosynthesis (*v*), osmotic stress promotes *Crassulacean Acid Metabolism* (CAM) in some plants (*vi*), and leads to changes in gene expression (*vii*) [14, 15].

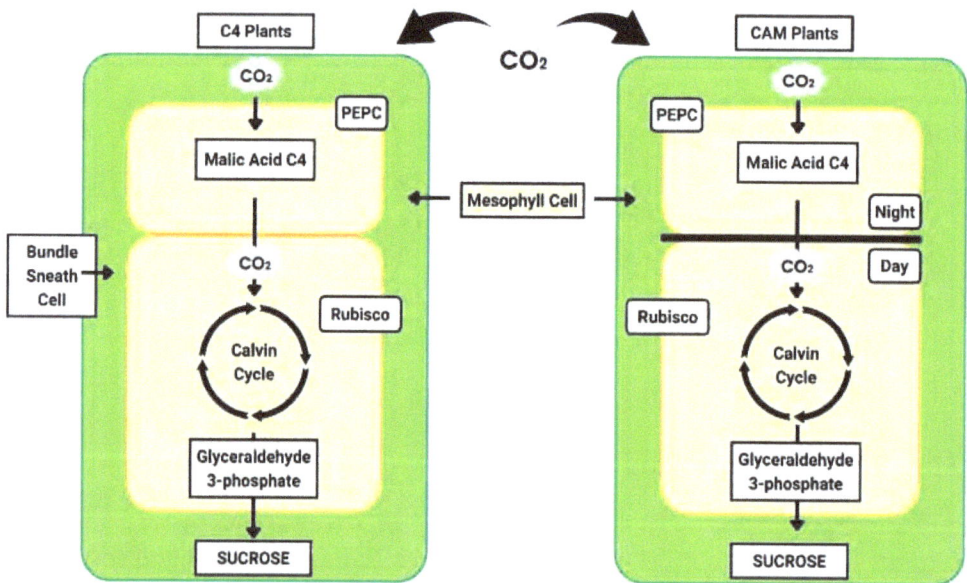

**Fig. (1).** Types of photosynthesis in C4 and CAM [18].

## Water Scarcity Limits the Leaf Expansion

As the water content of the plant decreases, the cells shrink and the cell walls become looser. This decrease in cell volume causes the turgor pressure in the cells and subsequently the solute concentration to decrease. As the area it covers decreases, the plasma membrane thickens and the pressure on it increases. Decreased turgor is the first and most important biophysical effect of water stress. Inhibition of cell expansion in the early stages of water scarcity slows leaf expansion. As the leaf area decreases, water loss through transpiration decreases.

Thus, the limited amount of water in the soil is effectively preserved for a long time. Therefore, the reduction in leaf area can be considered the first line of defense against drought [16, 17].

## Water Shortage Stimulates Leaf Abscission

Water has a significant impact on the physiological and morphological development of plants. Water stress can also affect plants. Plants may organalise in response to water stress in order to preserve the dynamic water balance. The total leaf area of a plant (number of leaves × surface area of each leaf) does not remain constant after all the leaves have matured. If plants are exposed to stress after the leaf area has been formed, the leaves will turn yellow and eventually fall off [19, 20].

## Root Growth Response to Moisture During Water Scarcity

One significant environmental element limiting crop yields and plant development is limited water availability. The ability of plants to maintain root growth in order to have continuous access to soil water is one of their most notable responses to water shortages. The innate complexity of root systems and their interactions with the soil environment has hindered understanding advances even though the adaptive significance of maintaining root growth under water deficiencies was recognized early on. Moderate water scarcity also affects the development of the root system. The ratio of the mass of the root to that of the trunk is determined by the functional balance between water uptake from the roots and photosynthesis by the above-ground parts [21, 22].

## Abscisic Acid-Induced Stomatal Closure During Water Scarcity

Closing of stomata constitutes another line of defense against drought. The uptake and loss of water into the guard cells changes the turgor of these cells, allowing the stomata to open and close. Since guard cells are located in the epidermis, they lose their turgor by losing water directly to the atmosphere through evaporation. The decrease in turgor causes the stomata to close hydropassively. A second mechanism, called hydroactive closure, allows the stomata to close when all leaves and roots lose water. Hydroactive closure depends on metabolic processes in guard cells. The decrease in the amount of dissolved substances in the guard cells causes water loss and turgor decrease and causes the stomata to close. Therefore, the hydraulic mechanism of hydroactive closure is the opposite of the stomatal opening mechanism. However, hydroactive closure is controlled differently than stoma opening. Decreasing the water content of the leaf initiates the loss of dissolved substances from the guard cells. Abscisic acid (ABA) plays a major role in this process. ABA (Fig. **2**)., which is synthesized at a low rate and continuously in mesophyll cells, tends to accumulate in chloroplasts. When mesophyll cells lose moderate amounts of water, two events occur: (a) Some of the ABA accumulated in chloroplasts is released into the apoplast (cell wall space) of the mesophyll cell. The unequal distribution of ABA depends on pH gradients within the leaf, the weakly acidic nature of the ABA molecule, and the permeability of cell membranes. The unequal distribution of ABA ensures that a portion of ABA is transported to guard cells by transpiration flow. (b) ABA is synthesized at a higher rate and more ABA accumulates in the leaf apoplast. The increase in ABA concentration as a result of rapid synthesis increases or prolongs the previously formed sealing effect of accumulated ABA [23, 24].

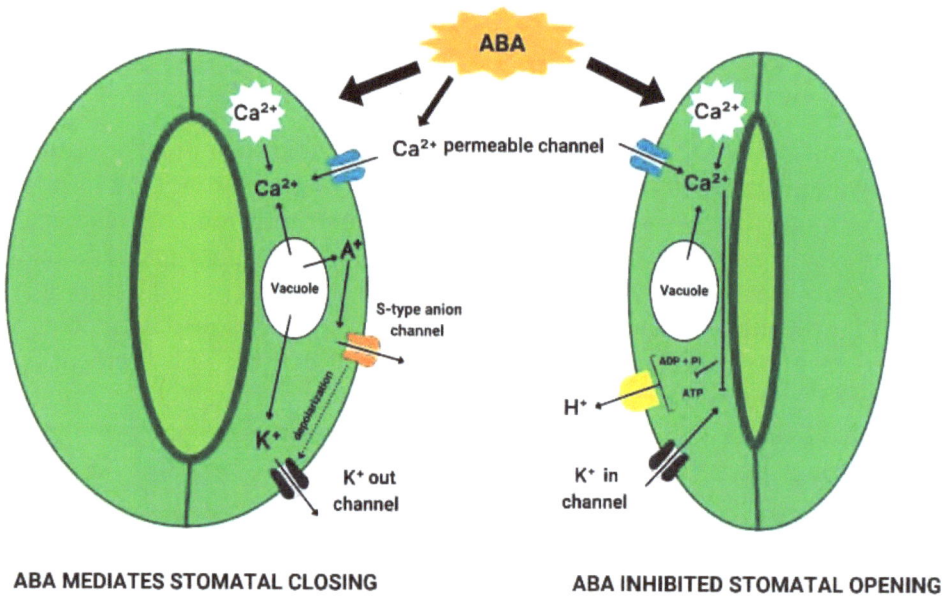

**Fig. (2).** Function of ABA in stomatal defense against water shortage [25].

## Water Shortage Limits Photosynthesis

Photosynthesis is much less sensitive to turgor change than leaf expansion. Therefore, the rate of photosynthesis in the leaf (expressed per unit leaf area) does not respond as well to leaf expansion to moderate water stress. However, moderate water stress generally affects both photosynthesis and stomatal conductance. In the early stages of water stress, water use efficiency may increase as stomata close (*i.e.,* more $CO_2$ can be taken in per unit amount of water lost by transpiration); because stomatal closure inhibits transpiration more than intercellular $CO_2$ concentration [26, 27].

## Osmotic Stress Promotes Crassulacean Acid Metabolism

CAM can be induced by salt or water stress, which turns on gene expression for the manufacture of the component enzymes. CAM is also a plant adaptation in which stomata are kept open at night and closed during the day. The difference between the vapor pressure of the leaf and the air, which enables transpiration to occur, decreases significantly when both the leaf and the air cool. Therefore, the water useability of CAM plants is the highest of all measured. CAM is very common in succulent plants such as cacti. Some succulent species show facultative CAM properties. That is, they turn to GLASS when exposed to water shortage or saline conditions [28, 29].

## Osmotic Stress Alters Gene Expression

As previously noted, the accumulation of compatible solutes to respond to osmotic stress requires the activation of metabolic pathways that enable the synthesis of such substances. Some genes encoding enzymes responsible for osmotic regulation are switched on under the influence of osmotic stress and/or salinity and cold stress. These genes encode the following enzymes: *Δ'1-Pyrroline-5-carboxylate synthase* (*i*), a key enzyme in the proline biosynthesis pathway; *Betaine aldehyde dehydrogenase* (*ii*), which is involved in the accumulation of glycine betaine; *Myo-Inositol 6-O-methyltransferase* (*iii*), a rate-limiting enzyme in the accumulation of pinitol; *Glyceraldehyde-3-phosphate dehydrogenase* (*iv*), which is expressed more during osmotic stress [30, 31].

Other genes regulated by osmotic stress encode proteins involved in membrane transport. ATPases and channel proteins that allow water to pass through, aquaporins, are among these. Additionally, stress stimulates some protease enzymes. These enzymes can break down (remove and recycle) other proteins whose structure is disrupted during stress. The protein called ubiquitin labels target proteins for proteolytic degradation. In Arabidopsis [31], mRNA synthesis responsible for ubiquitin increases after drought stress. Additionally, some osmotically stimulated heat shock proteins protect or renaturate proteins inactivated by drying [32, 33].

## HEAT STRESS AND HEAT SHOCK

Most of the tissues of higher plants cannot survive prolonged temperatures above 45°C. Growth-arrested cells and dehydrated tissues (*e.g.,* seeds and pollen) can survive at much higher temperatures than growing vegetative cells that contain water. While actively growing tissues do not survive temperatures above 45°C, dry seeds can withstand 120°C and pollen of some species can withstand 70°C. In general, only single-celled eukaryotes can complete their life cycles at temperatures above 50°C, and only prokaryotes can divide and grow at temperatures above 60°C. Periodic application of short-term sublethal heat stress often stimulates tolerance to lethal temperatures. We refer to this phenomena as induced thermotolerance [34, 35].

## High Leaf Temperature and Water Shortage Lead to Heat Stress

Many CAM and succulent higher plants, such as Opuntia and Sempervivum, are adapted to high temperatures. These plants can withstand tissue temperatures of 60-65°C under intense sunlight in summer. Since CAM plants close their stomata during the day, they cannot cool down through transpiration. Instead, they radiate heat from sunlight by reflecting back long wavelength (infrared) rays and also lose heat by conduction and convection [36, 37].

The increase in leaf temperature during the day is striking in plants of arid and semi-arid climate regions where drought and sunlight are intense. Heat stress is also a potential hazard in greenhouses. Because air movement in greenhouses is slow and humidity is high, leaf cooling decreases. Moderate heat stress slows the growth of the entire plant [38, 39].

## At High Temperatures, Photosynthesis is Inhibited Before Respiration

The unique change in the redox state linked to thylakoid membranes is indicative of a cellular energy imbalance induced by environmental stress, which is sensed globally by photosynthesis. Because photosynthesis is extremely vulnerable to stress from high temperatures, it frequently stops before other cell activities are compromised. The ability of maximum plants to precisely adjust their photosynthetic traits to their growing temperatures is demonstrated in great detail. The most notable observable fact is that plants can increase their photosynthetic efficiency at their new growth temperature by shifting the optimal temperature of photosynthesis in response to changes in growth temperature or seasonal temperature swings. Both photosynthesis and respiration are inhibited at high temperatures; however, as temperature increases, photosynthesis rates decrease before respiration rates. The temperature at which the amount of $CO_2$ fixed by photosynthesis is equal to the amount of $CO_2$ released in respiration within a certain time interval is called the temperature compensation point [40, 41].

## Temperature Decreases the Stability of the Membrane Structure

Maintaining the stability of the structure of various cellular membranes is important during cold and freezing, as well as during high temperatures. Hyperfluidity of membrane lipids at high temperatures indicates loss of physiological function of the membrane. The ability of oleander (*Nerium oleander*) to survive at high temperatures is a result of the high amount of saturated fatty acids in its membrane lipids. Saturated fatty acids reduce fluidity in membranes. At high temperatures, hydrogen bonds and electrostatic interactions between polar groups of proteins in the fluid phase of the membrane decrease. Therefore, high temperatures change the composition and structure of the membrane and can cause ions to leak. Photosynthesis is particularly sensitive to high temperature; the temperatures at which enzymes denature and lose their activity are significantly higher than the temperatures at which photosynthesis begins to decline. These results indicate that the initial stages of heat damage in photosynthesis are associated with changes in the structure of the membrane and the coupling of energy transfer mechanisms in chloroplasts rather than with a general deterioration in proteins [42, 43].

## Plants Produce Heat Shock Proteins at High temperatures

In response to sudden increases in temperature of 5-10°C, plants produce a series of unique proteins called heat shock (HSPs). Most HSPs act as molecular chaperones. Heat stress causes misfolding or prevents the folding of many proteins that function as enzymes or structural components in the cell. As a result, the structure of the enzymes is disrupted and their activities are lost [34, 44].

Such misfolded proteins often aggregate and precipitate, creating significant problems in the cell. HSPs function as molecular chaperones. It also ensures proper folding of misfolded and aggregated proteins and prevents them from misfolding. Therefore, it enables cells to maintain their functions in stressful environments [45, 46].

Although heat shock proteins in plants have been determined under sudden temperature changes of 25 to 45°C, which are rarely seen in nature, they are also synthesized with a relative increase in temperature, as in the natural environment. They also occur in plants under field conditions. Some HSPs also occur in cells in a stress-free environment. Although some important proteins in the cell are homologous to HSPs, they are not formed in response to heat stress. Low Temperature Stress are not sufficient for normal growth, nor are they sufficient to form ice in tissues. Tropical and subtropical species can be very sensitive to low temperatures. Among the cultivated plants, corn, beans, rice, tomatoes, cucumbers, sweet potatoes and cotton, and among the ornamental plants, Passiflora, Coleus and Gloxinia are sensitive to low temperatures. Chilling damage occurs to plants growing at relatively high temperatures (25 to 35°C) when they are brought to 10 to 15°C: Growth slows down, discoloration or rot occurs in the leaves. The plant may wilt when the roots are exposed to low temperatures [34, 44].

Plants sensitive to chilly temperatures vary in their response to low temperatures. Freezing damage, on the other hand, occurs at temperatures below the freezing point of water. As with freezing temperatures, a period of acclimatization to low temperatures is necessary to induce freezing tolerance. The proportion of unsaturated fatty acids in the membrane lipids of plants resistant to chilling temperatures is higher than that of plants sensitive to chilling temperatures. During acclimation to low temperatures, the rate of desaturating enzymes and unsaturated lipids increases. This change lowers the temperature at which membrane lipids change from a semifluid state to a semicrystalline state. This ensures that the membranes remain fluid at low temperatures. Therefore, the desaturation of fatty acids provides some protection against low-temperature damage [47, 48].

## FREEZING STRESS

Low temperature causes effects such as necrosis, wilting, tissue destruction, browning, a decrease in growth, and a decrease in germination in plants. The proportion of unsaturated fatty acids in the membrane lipids of plants resistant to chilling temperatures is higher than that of plants sensitive to chilling temperatures. During acclimation to low temperatures, the rate of desaturating enzymes and unsaturated lipids increases. This change lowers the temperature at which membrane lipids change from a semifluid state to a semicrystalline state. This ensures that the membranes remain fluid at low temperatures. Therefore, the desaturation of fatty acids provides some protection against low-temperature damage [49, 50].

### The Formation of Ice Crystals and Water Loss Causes Cell Death

The ability to tolerate freezing temperatures under natural conditions varies significantly among plant species and tissues. Seeds, other dehydrated tissues and fungal pores can be preserved indefinitely at temperatures near absolute zero. This shows that very low temperatures are not actually harmful. Vegetative cells that have not lost any water can maintain their viability when cooled quickly enough to prevent the formation of large and slow-growing ice crystals that can break down intracellular organelles. Ice crystals formed during rapid freezing are too small to cause mechanical damage. In contrast, frozen tissue must be heated rapidly to prevent small ice crystals from growing to damaging sizes or loss of water vapor by sublimation. Both of these occur at moderate temperatures (-10 to 100°C) [51, 52].

During rapid freezing, the protoplast, including the vacuole, becomes supercooled; that is, the water inside the cell remains liquid at temperatures a few degrees below the theoretical freezing point. It takes several hundred molecules for an ice crystal to form. This process, initiated by hundreds of water molecules to form a permanent ice crystal, is called ice nucleation. Ice nucleation is very dependent on the properties of the surfaces involved in this process. Some large polysaccharides and proteins that facilitate icing are ice nuclei formers. Some ice nucleators produced by bacteria facilitate ice nucleation by aggregating water molecules along amino acid sequences within the protein. In plant cells, ice crystals begin to develop from internal ice nuclei formers. Then, relatively large ice crystals inside the cell cause great damage to the cell. The big crystals of ice can be fatal [53, 54].

## Limiting Ice Formation Increases Freezing Tolerance

Some specialized plant proteins prevent the growth of ice crystals by inhibiting their formation. This effect is independent of lowering the freezing point of water and occurs due to the presence of dissolved substances. These antifreeze proteins, formed under low temperatures, bind to the surfaces of ice crystals and prevent their growth. In rye leaves, antifreeze proteins are found in the cells surrounding the intercellular spaces. Therefore, they prevent ice formation in the epidermis cells and outside the cells. The formation of ice crystals in plants and animals may be limited by similar mechanisms. A gene induced by low temperature in *Arabidopsis* has a DNA equivalent to a gene encoding an antifreeze protein in winter sole [55, 56].

Sugars and some proteins stimulated by low temperatures are thought to have protective effects against cold (cryoprotective, cryo = 'cold'). These proteins [such as cold-responsive (COR) proteins, and diacylglycerol (DAG) kinase] increase the stability of proteins and membranes during water loss induced by low temperature. The increase in sucrose concentration in winter wheat increases freezing tolerance. Sucrose is the main sugar involved in the formation of freezing tolerance. Sucrose acts as a colligative; however, in some species raffinose, fructans, sorbitol or mannitol function similarly. As winter grains acclimate to the cold, soluble sugars accumulate in their cell walls, limiting ice growth. A glycoprotein that protects against cold has been isolated from cold-acclimated cabbage (*Brassica oleraceae*) leaves. This protein-protected naive spinach (*Spinacia oleracea*) leaves against freezing damage *in vitro* [57 - 59].

## ABA and Protein Synthesis are Involved in Freezing Acclimation

External ABA application without prior low-temperature application or with low-temperature application greatly increased the tolerance of common alfalfa (*Medicago sativa* L.) to freezing at -10°C. These applications cause changes in the properties of newly synthesized proteins that can be resolved in two-dimensional gels. Some of these changes are specific to a particular treatment (cold or ABA), but some of the newly synthesized proteins stimulated by cold appear to be the same as those stimulated by ABA or moderate water scarcity [60, 61].

Protein synthesis is necessary for the establishment of tolerance to freezing. For tolerance to occur, certain proteins accumulate as a result of changes in gene expression. By isolating the genes related to these proteins, it was revealed that some of the proteins stimulated by low temperature were similar to the RAB/LEA/DHN (respectively, ABA-responsive, abundant in the last period of the embryo, dehydrin) protein family. As explained previously in the section on gene regulation by osmotic stress, these proteins accumulate in tissues exposed to different stresses such as osmotic stress. The functions of these proteins are being investigated [62, 63].

ABA is thought to stimulate tolerance to freezing. When plants such as winter wheat, rye, spinach and Arabidopsis thaliana, which are all cold-tolerant, become resistant to water shortage, their tolerance to freezing also increases. Freezing tolerance increases at low temperatures or at temperatures where acclimatization is not possible under moderate water scarcity. Both of these conditions increase ABA concentration in leaves. When ABA is applied to plants, tolerance to freezing in unusual temperatures develops. Many genes and proteins expressed at low temperatures or under water scarcity are also stimulated by ABA at unaccustomed temperatures. All this supports the role of ABA in freezing tolerance [59, 64].

**Many Genes are Expressed During Cold Acclimation**

Although the expression of certain genes and the synthesis of specific proteins are common to both heat stress and cold stress, the expression of genes in cold is different from the effect of heat stress in some respects. While the expression of genes related to the synthesis of 'housekeeping' proteins (proteins made in the absence of stress) does not decrease during cold periods, housekeeping protein synthesis stops significantly during heat stress [65, 66].

On the other hand, the expression of genes related to some heat shock proteins that function as molecular chaperones increases under both heat stress and low temperature stress. This shows that the structure of proteins is disrupted during both heat and cold stress, and that the mechanisms that ensure the stability of proteins in both hot and cold periods are important for the continuation of life [67, 68].

Another important class of proteins with increased expression due to cold stress are antifreeze proteins. Antifreeze proteins were first discovered in fish living under polar ice caps. These proteins prevent ice formation by preventing crystals from coming together. Thus, it prevents freezing damage in moderate freezing temperatures. Antifreeze proteins are sometimes called thermal hysteresis proteins (THP) because they provide thermal hysteresis to the aqueous solution (the transition from liquid to solid increases at a lower temperature than the transition from solid to liquid) [69, 70].

Several types of low-temperature-induced antifreeze proteins have been discovered in some cold-acclimated, winter-resistant monocots. After specific genes encoding these proteins were cloned and sequence analyzed, it was found that all antifreeze proteins belong to the class of proteins such as endochitinase and endonuclease. Endochitinase and endonucleases are stimulated by the infection of various pathogens. These proteins, called pathogen-related (PR) proteins, are thought to protect plants against pathogens. Therefore, the finding of two different roles of these proteins, antifreeze- and pathogen-related, suggests that they may protect plant cells against both low-temperature stress and pathogen attack, at least in monocots [71, 72].

Another group of proteins associated with osmotic stress also increases during cold stress. This group includes proteins that are necessary to maintain the stability of the membrane and are involved in the synthesis of osmolytes and late embryogenesis abundant proteins (LEA) proteins. Since ice crystals in the intercellular spaces create a significant osmotic stress for the part of the cell, osmotic stress must also be dealt with in order to cope with freezing stress [51, 73].

## SALINITY STRESS

Under natural conditions, terrestrial higher plants encounter high salt concentrations on seashores and deltas, where seawater and freshwater mix or are displaced by tides. In more inland areas, natural salt seeps from geological sea accumulations spread into the environment, making those areas unsuitable for agriculture. However, the accumulation of salts in irrigation water is a more common problem in agriculture [74, 75].

Evaporation and transpiration remove pure water from the soil (as steam), and this loss increases the concentration of dissolved substances in the soil. If the concentration of dissolved substances in irrigation water is high and accumulated salts are not washed away by a drainage system, salinity can quickly reach levels that harm salt-sensitive species. It has been calculated that approximately one-third of the irrigated land in the world is affected by salt [76, 77].

## Salinity Suppresses Growth and Photosynthesis in Sensitive Species

Plants are divided into two large groups in terms of their response to high salt concentrations. Halophytes, which are specific to salty soils, complete their entire life cycle in that environment. Plants other than glycophytes (sweet plants) or halophytes are not as resistant to salts as halophytes. Generally, there are salt-threshold concentrations at which growth in glycophytes is inhibited, discoloration of leaves and a decrease in dry weight begin to occur. Among crop plants, corn, onions, lemons, lettuce, and beans are very sensitive to salt; cotton and corn are moderately tolerant; Sugar beets and palm trees are very tolerant [78, 79].

## Salt Damage Occurs Due to the Osmotic and Specific Ion Effect

The major difference between low water potential environments caused by salinity and soil drying is the total amount of available water. During water loss from the soil, the plant can take very little water from the soil. Thus, the water potential decreases even more. Although most saline environments have a large (in fact unlimited) amount of water, the water potential in these environments is low [75, 80].

In addition to low water potential, the problem of specification toxicity arises when ions, especially $Na^+$, $Cl^-$ or $SO_4^{2-}$, accumulate in harmful concentrations in the cell. In nonsaline conditions, the cytosol of higher plants contains 100 to 200 mM $K^+$ and 1 to 10 mM $Na^+$. Many enzymes (such as catalases, peroxidases, *etc.*) function optimally in such an ionic environment. An abnormally high ratio of $Na^+$ to $K^+$ inactivates enzymes and prevents protein synthesis [80, 81].

## Plants Use Different Strategies to Fight Salt Damage

Plants suffer from salt damage when salt is sprayed on stems and buds of deciduous woody plants and on leaves, needles, buds, and stems of evergreen plants by passing cars. On buds, leaves, and small twigs, salt spray can result in salt burn. Desiccating the bud scales and exposing the fragile tissues of the budding leaves and flowers is another way that salt spray can harm plants. The chilly winter winds frequently dry out and destroy the unprotected budding leaves and flower buds. The damage is frequently not noticeable until late winter or early spring. Plants minimize the damage of salt by expelling salt, especially from shoots, meristems and actively growing and photosynthesizing leaves. In salt-sensitive plants, resistance to moderate levels of salinity in the soil is achieved partly by preventing the transfer of harmful ions from the roots to the shoots [75, 82].

## Excretion of Ions is a Critical Factor for Acclimation and Adaptation to Salinity Stress

In relation to the response of plants to salinity stress, these pumps must be activated for secondary active transport of excess ions. The activity of these $H^+$ pumps has been shown to be increased by salinity. Additionally, induction of relevant genes may be responsible for part of this increase [83, 84].

### Sodium is Transported Across the Plasma Membrane and Tonoplast

The energy-consuming excretion of $Na^+$ from the cytosol of plant cells through the plasma membrane is mediated by a gene product of the SOS1 (salt hypersensitive) gene. This gene product functions as a $Na^+$-$H^+$ antiporter. The SOS1 antiporter is regulated by two other gene products referred to as SOS2 and SOS3. SOS2 is a serine/threonine kinase that is activated by calcium through the recruitment of SOS3. SOS3 is a calcium-regulated protein phosphatase [85, 86]. The intracellular membrane $Na^+/H^+$ antiporter, also shown in Fig. (3), proteins involved in ion balance; however, its molecular identity is unknown or has not yet been identified in plants. These include plasma membrane and tonoplast calcium channel proteins and vacuolar proton pumping ATPases and pyrophosphatases [87, 88].

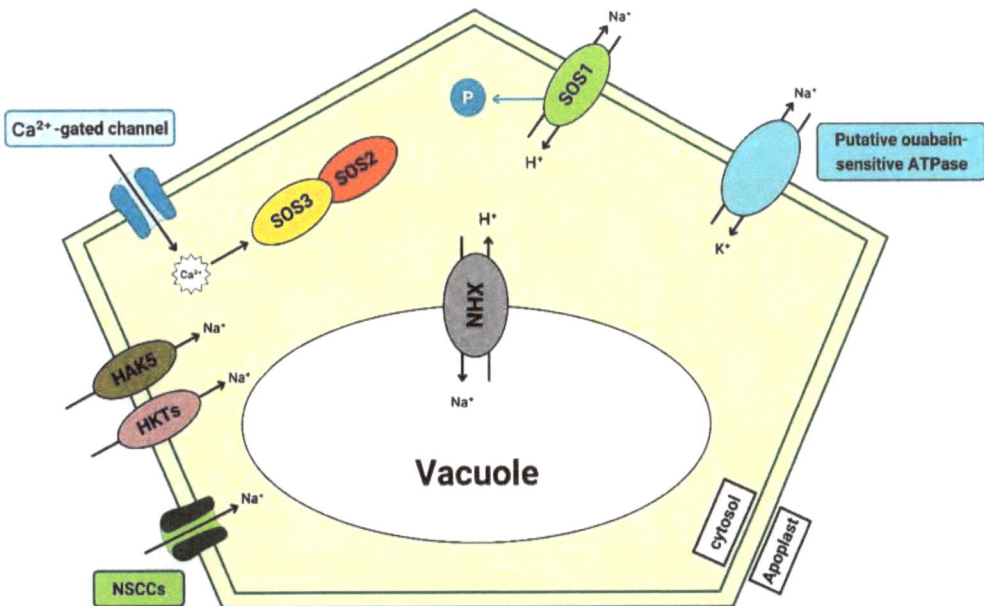

**Fig. (3).** Transport proteins in the membrane that direct sodium, potassium and calcium transport during salt stress in plants [89].

Understanding the mechanisms of sodium ($Na^+$) uptake, manipulation, and excretion in plant cells is crucial for controlling Na+ accumulation and maintaining decreasing $Na^+$ concentrations in the cytosol, thereby strengthening the plant's tolerance to salt stress. $Na^+$ transfer across the plasma membrane of the lowest roots occurs mainly through non-selective cation channels (NSCCs). $Na^+$ is delivered to vacuoles *via* $Na^+/H^+$ exchangers (NHXs). $Na^+$ transfer from plant roots is mediated by the activity of $Na^+/H^+$ antiporters catalyzed by the salt hypersensitive 1 (SOS1) protein. In animal cells from other living groups, ouabain (OU)-sensitive $Na^+$, $K^+$-ATPase (a P-type ATPase) mediates sodium transfer. The evolution of P-type ATPases in cells of higher plants does not exclude the possibility of sodium release mechanisms similar to the $Na^+$, $K^+$-ATPase-dependent mechanisms characteristic of animals. With the advancement of technology and using new fluorescence imaging and spectrofluorometric methodologies, an OU-sensitive sodium transfer system has recently been reported to be physiologically active in roots [89].

## HEAVY METAL TOXICITY

Pollution of the biosphere with toxic metals as a result of human activities poses a major problem for both human health and the ecosystem. Toxic metals can be found naturally in our environment, but they reach dangerous levels as a result of activities such as the use of fossil fuels, mining, and the use of pesticides [90]. As a result of human activities, on average, 7.6 tons of Cd, 35 tons of Cu, 38 tons of Mn, 332 tons of Pb, and 19 tons of as are released into the world every year. It spreads to organisms through the food chain and has negative effects [91, 92].

Heavy metals also have toxic effects on plants. They cause toxicity symptoms such as growth arrest in plants, inhibition of seed germination, leaf curling, chlorosis, inhibition of mineral uptake, and deterioration of the structure of enzymes. The effects of these metals on plants vary according to species; Some species are tolerant and may accumulate heavy metals. Some of the heavy metals are micronutrient elements and are absolutely necessary for plant development [93]. Toxic effects occur on plants in case of micronutrient deficiency or exposure to high concentrations of these elements. Cu, Fe, B, Cl, Ni, Zn, Mn, and Mo are micronutrient elements. Some morphological changes can be seen in micronutrient deficiency. For example, in Fe and Mn deficiency, changes occur especially in the epidermal cells of the root. In zinc deficiency, not enough auxin can be synthesized for plant growth and internode elongation stops. Zinc is a cofactor of some enzymes such as dehydrogenase, peroxidase, and oxidase in plants. It also functions in the regulation of nitrogen metabolism, photosynthesis and auxin synthesis in plants [94, 95].

Power plants, heating systems, city traffic, mining operations, and phosphate fertilizers cause the release of Cd all over the earth. In addition, it may be naturally released into the environment as a result of the mineralization of rocks, but this rate is quite low compared to human-induced Cd release. This element, whose density is 8.6 gcm$^{-3}$, is toxic to almost all organisms. Cadmium causes chlorosis in plants, disruption of water balance, inhibition of the uptake of some minerals, and the activities of some enzymes, resulting in iron deficiency. Additionally, Cd prevents the opening of stomata and causes a decrease in the H$^+$/K$^+$ exchange rate in the plasma membrane [96, 97].

Moderate Cd pollution is observed in soils in many parts of the world. Some edible plants, such as rice, show Cd accumulation in various organs. These plants take up Cd from the soil through their roots. Cd chelated in the roots is loaded into the xylem and transported to the stem, then transported to the seeds in the plant *via* the phloem, or redistributed in the leaves and stored in these organs. Plants do not need Cd for their reproduction and development, but the bioaccumulation index of Cd is as high as trace elements. Even if cadmium does not cause phytotoxicity, its concentrations in plant tissues pose a danger to humans and animals. Therefore, considering the food chain, Cd is one of the most dangerous metals [96, 98].

## Heavy Metals

There are many definitions for the term heavy metal, generally made by taking into account various chemical properties of the elements such as density, atomic weight, and atomic number. According to these definitions, 53 of the 90 naturally occurring elements are heavy metals. But not all of them are biologically important. Considering their solubility under physiological conditions, 17 heavy metals are important for living cells, organisms, and ecosystems. Among these metals; Fe, Mo, and Mn are important micronutrient elements. Zn, Ni, Cu, V, and Co are trace elements of high or low importance. The functions of As, Hg, Ag, Sb, Cd, Pb and U as nutritional elements are unknown [99, 100].

In terrestrial ecosystems, there are two main sources of heavy metals: parent material and atmosphere. Its natural resources are volcanoes and land masses. Anthropogenic activities such as mining, consumption of fossil fuels, metal industry, and use of phosphate fertilizers increase the spread of heavy metals and the accumulation of these compounds in the ecosystem [99].

It is known that the vegetative organs of plants exposed to heavy metals at high concentrations or low concentrations for a long time are affected morphologically and physiologically [101]. Some heavy metals such as copper and zinc are components of many enzymes and proteins and are necessary for plant growth and development (Fig. **4**). However, these elements inhibit growth in many plant species at high concentrations and cause toxicity symptoms [102]. Heavy metal toxicity is based on three main causes: Metals interact directly with proteins due to their tendency to bind to -thiol, -histidine and -carboxyl groups, causing the structure, catalytic and transport region of the cell to change (*i*), increases the formation of free radicals and reactive oxygen species that can lead to oxidative stress (*ii*), and it replaces the main cations in biomolecules and causes their functions to deteriorate (*iii*). Due to the different chemical properties of metals and their different behaviors in biological systems, these three mechanisms may not be sufficient to explain the causes of toxicity [103 - 105].

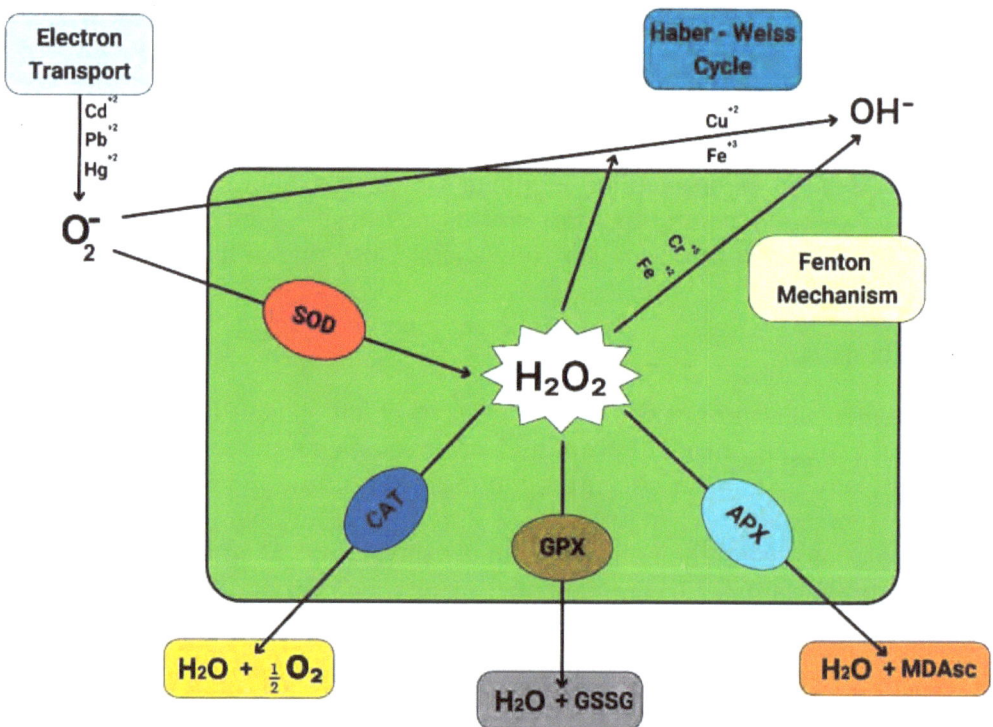

**Fig. (4).** Heavy metal stress induces cellular generation of ROS. GSSG; The resulting oxidized form of the antioxidant glutathione consists of two molecules disulfide-bonded together, MDAsc; Monodehydroascorbate [104].

## Heavy Metal Retrieval and Transport

Metals are found in the soil either attached to colloids or bound to organic matter. Plants can only take up ionized metals in soil solution. Heavy metal uptake varies depending on the plant species. Characteristics such as root cation exchange capacity, root surface area, soil pH, temperature, and metal concentration affect heavy metal uptake. In addition, plants increase heavy metal absorption by changing the pH of the rhizosphere, that is, by secreting substances such as malate, citrate, and mucilage (phytosiderephore) into the rhizosphere. At low pH, hydrogen ions compete with metals to adhere to colloids. Thus, hydrogen ions cling to the colloid, and metals remain in the soil water and can be taken up by plants [107, 108].

The mutual interactions of ions in the environment during metal uptake by plants are known. Excessive exposure of plants to Mn reduces Mg uptake by 50%. Manganese and Mg compete for binding sites in the stem cell membrane during absorption and affect each other's uptake. Likewise, Zn and Cu ions in the environment reduce Mn uptake. Additionally, Se reduces Fe and Mg uptake in plants. It has been determined that Cd competes with nutritional elements such as Fe, Mn, Cu, Zn, and Ni for binding sites. Some synthetic chelators such as EDTA and HEDTA increase the solubility of metals in soil and their uptake by plants [109, 110].

Metals first bind to the cell walls in the root. Metal ions are taken into the cell by secondary transporters such as carrier proteins and/or channel proteins. Some cation transporters have been identified in Saccharomyces cerevisiae using molecular techniques. They belong to the ZIP (ZRT, IRT-like protein) and Nramp (natural resistance macrophage protein) families, which are responsible for micronutrient uptake. Metal ions taken into the root cells are chelated here and loaded into the xylem. The walls of the endodermal cell layer form a barrier during the apoplastic movement of metal ions into the xylem. Metal ions must use the symplastic pathway to overcome this obstacle. Cations are loaded into the xylem from root cells *via* carrier proteins [111, 112].

Transport of heavy metals in the xylem varies depending on plant species and metal type. For example, while Ni is transported as a Ni-peptide complex in the xylem of some plants, it is transported by forming a complex with the amino acid histidine in metal-accumulating plants. Although the distribution of heavy metals within the plant occurs primarily through the xylem, the distribution of these elements within the leaf and other parts of the plant occurs through the phloem [109, 113].

## OXIDATIVE STRESS AND REACTIVE OXYGEN SPECIES

Oxidative stress can be defined as the shift in the balance between prooxidative and antioxidative reactions as a result of the action of various agents in living organisms. Some environmental and biological factors such as temperature, heavy metals, pathogen infection, pollution, light, excessive salinity, drought, and cold can cause oxidative stress [114].

In plants, ROS are constantly produced in chloroplasts, mitochondria, and peroxisomes as secondary products of aerobic metabolism. However, heavy metals increase the formation of ROS [115]. The balance between ROS production and destruction must be kept strictly under control. Abiotic stress factors disrupt the balance between ROS scavengers and production. It affects many cellular functions by causing lipid peroxidation, protein oxidation, and damaging nucleic acids. In addition to damaging cells, ROS can also cause some responses, such as the expression of new genes [116, 117].

## Damages Caused by ROS

Atmospheric $O_2$, which enables respiration and is used as the final electron catcher in energy production systems, causes ROS to form in cells. Although atmospheric $O_2$ is relatively unreactive, it leads to the formation of ROS such as $O_2^{\cdot}$, $H_2O_2$, $OH^{\cdot}$, $^1O_2$ (Fig. **5**). In the presence of metals such as copper and zinc, $OH^{\cdot}$ radical is formed by the Haber-Weiss mechanism or Fenton reactions. $OH^{\cdot}$ is the most reactive and dangerous chemical species in the biological world [118].

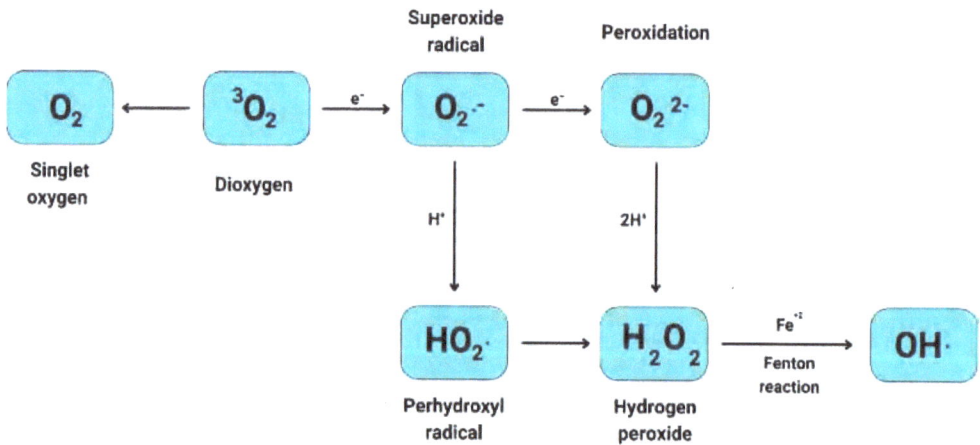

**Fig. (5).** Generation of ROS by energy transfer [116].

## *Singlet Oxygen ($^1O_2$)*

Singlet oxygen is the form of molecular oxygen formed by the rearrangement of its electrons, that is, by energy transfer. Since the spin restriction in singlet oxygen is removed, it is more reactive than molecular oxygen and can directly oxidize proteins, DNA, and lipids. There are two types of singlet oxygen: sigma and delta form. Delta singlet oxygen has two unpaired electrons and is its long-lived primary form. Sigma singlet oxygen has no unpaired electrons, has high energy, is short-lived, and is not a free radical [119].

## Hydrogen Peroxide ($H_2O_2$)

Hydrogen peroxide is formed as a result of the enzymatic reduction of oxygen or the dismutation of superoxides. In the presence of transition metals such as iron and copper, hydrogen peroxide leads to the formation of hydroxyl radicals *via* the Haber-Weiss Reaction. Hydrogen peroxide, which is itself unreactive because it does not contain unshared electrons, is considered an oxidizer due to this feature [120].

Haber-Weiss reaction. These are reactions in which highly reactive hydroxyl radicals are formed by the interaction of $O_2^{\cdot}$ and $H_2O_2$. As a result of this reaction, more toxic radicals are formed enzymatically [121].

## Super Oxide ($O_2^{\cdot-}$)

Superoxide is formed by partial reduction of $O_2$ or electron transfer in chloroplasts during photosynthesis. The main production site is the electron capturers of Photosystem I (PSI) attached to the thylakoid membrane. The half-life of the moderately reactive superoxide is 2-4 µs. Superoxide formation triggers the formation of hydroxyl radical, a more reactive ROS. The superoxide reacts with Fe (II) and initiates Haber-Weiss reactions, which will ultimately lead to the formation of the hydroxyl radical [122].

## Hydroxyl Radical (OH)

Hydroxyl radical is the most active of the reactive oxygen species in biological systems and is generally formed as a result of Fenton Reactions catalyzed by transition metals, especially Fe(II) and Fe(III). Hydroxyl radicals, which have a short half-life and are extremely reactive, react with the first molecule they encounter. There is no specific scavenger for this radical and it can react with all biological molecules including DNA, RNA, and proteins [123].

## Production of Reactive Oxygen Species

Photosynthetic plants are at greater risk of oxidative damage than other organisms. Chloroplasts, peroxisomes and mitochondria are the main production sites of ROS [124].

### Production of Reactive Oxygen Species in Mitochondria

In addition to being a target of ROS, plant mitochondria are also one of the main ROS production sites. Plant mitochondria differ significantly from animal mitochondria with their specific components of the electron transport system (ETS) and their functions in processes such as photorespiration. The environment of plant mitochondria is rich in $O_2$ and carbohydrates, which are the products of photosynthesis. Under normal respiratory conditions, ROS production occurs in mitochondria, but this production rate increases under various biotic and abiotic stress conditions [125].

During respiration, $O_2$ is used as an electron scavenger in the mitochondrial ETS, especially complex I and complex II. Thus, super oxide ions are formed. Mn-SOD converts superoxide into hydrogen peroxide. The resulting $H_2O_2$ reacts with Fe and Cu and forms OH as a result of Fenton reactions [126].

## Production of Reactive Oxygen Species in Chloroplast

$O_2$ produced in chloroplasts by photosynthesis can randomly retain electrons in the photosystem; thus $O_2^{\cdot}$ is formed. Under various stress conditions, ROS production increases in chloroplasts. Normally, electron flow is from the photosystem center to $NADP^+$. $NADP^+$ is reduced to NADPH by gaining electrons and enters the Calvin cycle, where the last electron acceptor reduces $CO_2$. In case of overload of the ETS, some of the electrons deviate from ferredoxin to $O_2$, and $O_2^{\cdot}$ is formed [127].

## Production of Reactive Oxygen Species in the Peroxisome

Peroxisomes are spherical microbodies surrounded by a double-layered lipid membrane. Peroxisomes are one of the main areas in the cell where ROS are produced. $O_2^{\cdot}$ radical is also produced as a product of normal metabolism in peroxisomes such as chloroplasts and mitochondria. In peroxisomes, $O_2^{\cdot}$ is produced in two regions. One of these is the organelle matrix containing xanthine oxidase, which catalyzes the oxidation of xanthine and hypoxanthine to uric acid. The other is the peroxisome membrane, which contains a small ETS consisting of a flavoprotein NADH and cytochrome b [128]. Increased $H_2O_2$ and $O_2^{\cdot}$ production in peroxisomes causes oxidative damage that can lead to cell death. But it also plays a role as a signaling molecule for pathogen-induced programmed cell death in plants at low levels of $H_2O_2$ and $O_2^{\cdot}$ produced in peroxisomes [129].

## Lipid Peroxidation

Lipid peroxidation (LP) is the most damaging process for all living things. Under various stress factors, membrane damage is used as a parameter to determine lipid degradation. During lipid peroxidation, ketones, malondialdehyde (MDA), and small hydrocarbon fragments are formed. Some of these components react with thiobarbituric acid to form colored products called thiobarbituric acid reactive substances (TBARS). When reactive oxygen species production exceeds a certain level, LP occurs in cell and organelle membranes. Lipid peroxidation not only disrupts the normal functions of the cell during oxidative stress but also increases the production of lipid-derived radicals. Lipid peroxidation reduces membrane fluidity and increases leakage. They also cause damage to membrane proteins and inactivation of ion channels, receptors and enzymes. As a result of lipid peroxidation, aldehydes such as 4-hydroxy-2-nonenal, MDA, as well as keto fatty acids are formed. When plants are exposed to abiotic stress, the accumulation of LP and lipid peroxidase under Cd stress in different plants increases significantly due to the increase in ROS production [130, 131].

## Protein Oxidation

Protein oxidation has been defined as the byproduct of oxidative stress or covalent modification of protein induced by ROS. Many types of protein oxidation are essentially irreversible. However, the oxidation of a few containing sulfur amino acids is reversible. Protein carbonylation is widely used as a marker of protein oxidation. Protein carbonylation can occur due to the direct oxidation of bonds between amino acids. Regardless of the site of ROS synthesis and action, proteins containing thiol groups and sulfur amino acids are possible ROS targets. Cys and Met are particularly reactive with $^1O_2$ and $OH^-$ [132].

## DNA Oxidation

Plant DNA is very robust but can be damaged as a result of abiotic or biotic stress. High ROS levels can damage cellular structures, nucleic acids, lipids, and proteins. $OH^-$ can cause damage to all molecules in the structure of DNA, such as purine, pyrimidine bases, and the deoxyribose backbone of DNA. $^1O_2$ reacts primarily with guanine; $H_2O_2$ and $O_2^-$ do not react with any DNA components. Reactive oxygen species can cause base modifications such as base deletions, oxidation, and alkylation in DNA. Damage to DNA causes physiological effects such as damage to photosynthetic proteins that affect the growth and development of the organism, destruction of the cell membrane, and decreased protein synthesis [133, 134].

## Antioxidant Defense System

Exposure of plants to unfavorable environmental conditions such as high temperatures, heavy metals, drought, excess water, air pollution, nutrient deficiency or salt stress can increase the production of reactive oxygen species such as $^1O_2$, $O_2^-$, $H_2O_2$ and $OH^-$. Plants use antioxidant defense systems in their cells and organelles such as chloroplasts, mitochondria and peroxisomes to protect themselves against these toxic oxygen intermediates. Many studies have proven that stimulation of the cellular antioxidant mechanism is important for protection against various stresses. The components of the antioxidant defense system are enzymatic or non-enzymatic antioxidants [135, 136].

## Superoxide Dismutase (SOD)

SOD is found in all intracellular regions where oxidative stress can be seen and is the most effective enzymatic antioxidant. Various environmental stresses cause an increase in ROS production. SOD constitutes the first line of defense against increasing ROS levels. Therefore, it is very important in plant stress tolerance. Superoxide dismutase catalyzes $O_2^{\cdot}$ dismutation, enabling one $O_2^{\cdot}$ to be reduced to $^{\cdot}H_2O_2$ and the other to be oxidized to $O_2$. It prevents the formation of OH by Haber-Weiss reactions by eliminating $O2^{\cdot}$. This reaction is ten thousand times faster than natural dismutation. Superoxide dismutase is grouped into three classes according to its metal cofactors, these types are Cu/Zn-SOD, Mn-SOD, and Fe-SOD, which are found in different cellular regions [118]. *Arabidopsis thaliana* has three Fe-SOD genes (FSD1), three Cu/Zn-SOD genes (CSD1, CSD2, CSD3), and one Mn-SOD gene (MSD1) [137].

Some SOD isoenzymes are sensitive to KCN and $H_2O_2$, and the activity of SOD isoenzymes can be determined by the reverse staining method based on this. While Mn-SOD is resistant to both of these inhibitors, Fe-SOD is resistant to KCN and sensitive to $H_2O_2$. Cu/Zn-SOD is sensitive to both inhibitors. Mn-SOD is found in the mitochondria and peroxisomes of eukaryotic cells, while Cu/Zn-SOD is found in the cytosol and chloroplasts of higher plants. Fe-SOD is rarely found in plants. In plants containing Fe-SOD, this enzyme is found in chloroplasts [138, 139].

## Ascorbate Peroxidase (APX)

Ascorbate peroxidase plays an important role in clearing ROS and protecting cells in higher plants, algae and other organisms. Ascorbate peroxidase breaks $H_2O_2$ into two waters and takes part in the ascorbate-glutathione cycle. Ascorbate peroxidase has 5 different isoenzymes. These are thylakoid (tAPX) and glyoxysome membrane that forms (gmAPX), as well as chloroplast stromal soluble form (sAPX), and cytosolic form (cAPX). Ascorbate peroxidase has a higher affinity for $H_2O_2$ than CAT, SOD. It has a critical role in clearing ROS during stress. APX expression increases under different stress conditions [140, 141].

Ascorbate peroxidase reduces $H_2O_2$ to water, using ascorbate (AA) as an electron donor, and then AA is oxidized to monodehydroascorbate (MDHA). Monodehydroascorbate can be spontaneously reduced or oxidized to dehydroascorbate (DHA) or AA (Fig. **6**). Additionally, MDHA can be directly reduced to AA by MDHA reductase. During this process, MDHA uses NADPH as an electron donor. DHA reductase uses glutathione (GSH) while reducing DHA and causes its oxidation. While oxidized GSH is reduced by GSH reductase, it is also reduced by NADPH. Thus, the accumulation of $H_2O_2$ at toxic levels is prevented by the ascorbate-glutathione cycle [142].

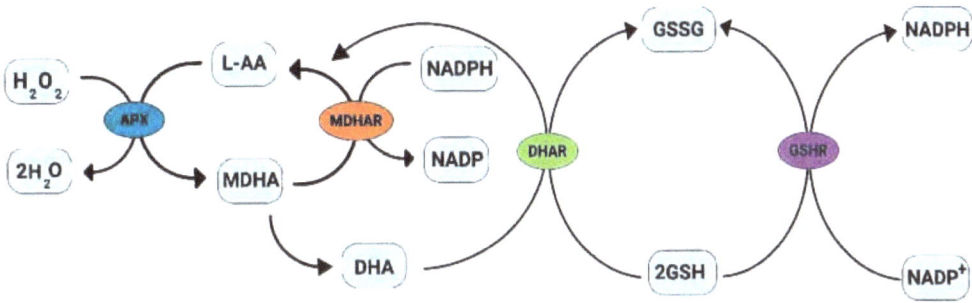

**Fig. (6).** The ascorbate-glutathione cycle. APX; Ascorbate peroxidase, MDHAR; Monodehydroascorbate reductase, DHAR; Dehydroascorbate, GSHR; glutathione reductase [149].

## Catalase (CAT)

CAT can directly break down $H_2O_2$ into $H_2O$ and $O_2$. Catalase is absolutely necessary for ROS detoxification under stress conditions. In addition, CAT has the highest turnover rate among all enzymes; one CAT molecule can convert six million $H_2O_2$ molecules into $H_2O$ and $O_2$ in one minute. Catalase is important for the retention of ROS occurring in peroxisomes. A great deal of information is available about CAT isoenzymes in higher plants; for example, there are two isoforms in H. vulgare and four isoforms in *Helianthus annuus* cotyledons [143, 144].

## Guaiacol Peroxidase (GPOX)

Guaiacol peroxidase breaks down to indole-3-acetic acid (IAA) and plays a role in lignin biosynthesis. It also serves to protect against biotic stress by breaking down $H_2O_2$. Guaiacol peroxidase uses aromatic compounds such as guaiacol as electron donors. Guaiacol peroxidase activity varies depending on stress conditions and plant species [145].

## Proline

Proline is a powerful antioxidant and inhibitor of programmed cell death. Proline reduces the negative effects of ROS in microorganisms, plants, and animals. Proline is an important redox signaling molecule in addition to scavenging ROS in plants and algae under salt, metal, and drought stress. It is also an important osmotic regulator in some plants such as potatoes. In other plants such as tomatoes, a very small portion of the total osmotic regulators is proline [146, 147].

The first step in proline synthesis is the phosphorylation of glutamate. In plants, pyrroline-5-carboxylate synthetase catalyzes the conversion of glutamate to glutamate-γ-semialdehyde. GSA spontaneously converts to pyrroline-5-carboxylate (P5C). P5C is reduced to proline by P5C reductase. In plants, proline is not only synthesized from glutamate, it can also be synthesized *via* arginine and ornithine [148, 149].

### *Ascorbic Acid (Vitamin C)*

Ascorbic acid (AA) occurs in almost all plant tissues except dry seeds. Ascorbic acid concentration is higher in leaves and meristem than in roots. High AA concentrations are seen in the fruits of some plants, especially citrus fruits. However, in every plant, AA concentration is not higher in fruits than in leaves. Ascorbic acid occurs in all intracellular regions, including the cell wall, but is found in low concentrations in vacuoles. Ascorbic acid is reduced or oxidized while performing various activities in the cell [150, 151].

Ascorbic acid is an abundant, water-soluble and powerful antioxidant that plays a role in reducing or preventing damage caused by ROS. In addition to playing a role in enzyme cofactor, antioxidant, oxalate, and tartrate synthesis, it is also involved in photosynthesis. Ascorbic acid takes part as an enzyme cofactor in many important enzymatic reactions. These enzymes are mono- or dioxygenases with iron or copper in their active sites. These enzymes need AA to show maximum activity. Ascorbic acid can quickly react with ROS such as singlet oxygen, hydrogen peroxide, superoxide and hydroxyl radical and is responsible for their detoxification. Ascorbic acid is an electron acceptor/donor in electron transport in the plasma membrane and chloroplasts. Ascorbic acid serves as a substrate for APXs in chloroplasts. It ensures the removal of peroxides occurring in the thylakoids [152, 153].

## BIOTIC STRESSES AND RESPONSES IN PLANTS

A multitude of biotic stress conditions can affect plants. A complex web of morphological, physiological, and biochemical systems is activated by the plant's perception of stress, which sets off a series of molecular and cellular processes. Complex sensory mechanisms have been established by plants to recognize biotic invasion and counteract its detrimental effects on growth, productivity, and survival. Plants have therefore developed an abundance of defense mechanisms to fend off invasions by a wide range of pathogens and pests, such as bacteria, fungi, viruses, nematodes, and herbivorous insects. Therefore, in order to counteract the detrimental impact on their survival, plants typically find a balance between their reaction and biotic stress. The molecular processes behind plant defense responses have been thoroughly explained. However, it is still unclear how and why many signaling pathways come together to produce biotic stress responses [154 - 158].

Plant infections and biotic stress are caused by a variety of pests, parasites, and diseases. Fungal parasites can be classified as either biotrophic (feeding on living host cells) or necrotrophic (killing host cell by toxin production). They can cause cankers, leaf spots, and vascular wilts in plants. Nematodes are mostly responsible for soil-borne illnesses that result in nutrient deficiencies, stunted growth, and wilting. Nematodes feed on plant components. Likewise, viruses can cause harm both locally and systemically, which can lead to stunting and chlorosis. Conversely, insects and mites damage plants by either laying eggs on them or feeding on them (piercing and sucking). Additionally, the insects may serve as carriers of several germs and viruses. To fend against these kinds of stressors, plants have evolved a complex immune system. In order to ward off infections and insects, plants have a passive first line of defense that consists of physical barriers like trichomes, wax, and cuticles. Additionally, plants have the ability to produce chemical defenses against diseases. Furthermore, plants use two layers of pathogen detection to initiate defense against biotic invaders [154, 159 - 162].

## CONCLUSION

Plants try to survive by developing many reversible or irreversible responses to all biotic and abiotic environmental factors that they perceive as stressors. The responses developed vary depending on the size of the stress factor and the genetic and ontogenic characteristics of the plant. Thus, plants can adapt to the factors of the environment they live in in order to survive, thanks to the stress responses they have developed. In this context, in the light of current technological developments, understanding the defense mechanisms, especially in plant species resistant to stress factors, will be a very important step in minimizing crop losses. In this context, in the studies carried out so far in which the molecular responses of plants to stress were evaluated, genes that may be related to stress were tried to be determined, and the expression levels of genes thought to be related were examined under different plants and stress conditions. Following the identification of stress-related target genes and their behavior against stress, studies are continuing to develop stress-resistant biotechnological products through special molecular methods such as gene transfer or gene silencing.

# REFERENCES

[1]     Chu EW, Karr JR. Environmental impact: concept, consequences, measurement. Reference Module in Life Sci. 2017; B978-0-12-809633-8.02380-3.
[http://dx.doi.org/10.1016/B978-0-12-809633-8.02380-3]

[2]     Schneiderman N, Ironson G, Siegel SD. Stress and health: psychological, behavioral, and biological determinants. Annu Rev Clin Psychol 2005; 1(1): 607-28.
[http://dx.doi.org/10.1146/annurev.clinpsy.1.102803.144141]

[3]     Wang W, Vinocur B, Altman A. Plant responses to drought, salinity and extreme temperatures: towards genetic engineering for stress tolerance. Planta 2003; 218(1): 1-14.
[http://dx.doi.org/10.1007/s00425-003-1105-5]

[4]     Kopecká R, Kameniarová M, Černý M, Brzobohatý B, Novák J. Abiotic stress in crop production. Int J Mol Sci 2023; 24(7): 6603.
[http://dx.doi.org/10.3390/ijms24076603]

[5]     Levitt J. Responses of Plants to Environmental Stresses. New York, London: Academic Press 1972; p. 697.

[6]     Lichtenthaler HK. Vegetation stress: an introduction to the stress concept in plants. J Plant Physiol 1996; 148(1-2): 4-14.
[http://dx.doi.org/10.1016/S0176-1617(96)80287-2]

[7]     Seleiman MF, Al-Suhaibani N, Ali N, et al. Drought stress impacts on plants and different approaches to alleviate its adverse effects. Plants 2021; 10(2): 259.
[http://dx.doi.org/10.3390/plants10020259]

[8]     Mareri L, Parrotta L, Cai G. Environmental stress and plants. Int J Mol Sci 2022; 23(10): 5416.
[http://dx.doi.org/10.3390/ijms23105416]

[9]     Madhova Rao KV, Raghavendra AS, Janardhan Reddy K. Physiology and Molecular Biology of Stress Tolerance in Plants. Netherlands: Springer 2005; p. 345.

[10]    Dubey RS. Handbook of Plant and Crop Stress. New York: Marcel Dekker 1994; p. 227.

[11]    Paes de Melo B, Carpinetti PA, Fraga OT, et al. Abiotic stresses in plants and their markers: a practice view of plant stress responses and programmed cell death mechanisms. Plants 2022; 11(9): 1100.
[http://dx.doi.org/10.3390/plants11091100]

[12]    Oladosu Y, Rafii MY, Samuel C, et al. Drought resistance in rice from conventional to molecular breeding: a review. Int J Mol Sci 2019; 20(14): 3519.
[http://dx.doi.org/10.3390/ijms20143519]

[13]    Mahmood T, Khalid S, Abdullah M, et al. Insights into drought stress signaling in plants and the molecular genetic basis of cotton drought tolerance. Cells 2019; 9(1): 105.
[http://dx.doi.org/10.3390/cells9010105]

[14]    Yang X, Lu M, Wang Y, Wang Y, Liu Z, Chen S. Response mechanism of plants to drought stress. Horticulturae 2021; 7(3): 50.
[http://dx.doi.org/10.3390/horticulturae7030050]

[15]    Lüttge U. Ecophysiology of crassulacean acid metabolism (CAM). Ann Bot (Lond) 2004; 93(6): 629-52.
[http://dx.doi.org/10.1093/aob/mch087]

[16]    Munns R, Passioura JB, Guo J, Chazen O, Cramer GR. Water relations and leaf expansion: importance of time scale. J Exp Bot 2000; 51(350): 1495-504.
[http://dx.doi.org/10.1093/jexbot/51.350.1495]

[17]    Chen JJ, Sun Y, Kopp K, Oki L, Jones SB, Hipps L. Effects of water availability on leaf trichome density and plant growth and development of Shepherdia ×utahensis. Front Plant Sci 2022; 13: 855858.

[http://dx.doi.org/10.3389/fpls.2022.855858]

[18]   Sage RF. Are crassulacean acid metabolism and C4 photosynthesis incompatible? Funct Plant Biol 2002; 29(6): 775-85.
[http://dx.doi.org/10.1071/PP01217]

[19]   Gomez-Cadenas A, Tadeo FR, Talon M, Primo-Millo E. Leaf abscission induced by ethylene in water-stressed intact seedlings of cleopatra mandarin requires previous abscisic acid accumulation in roots. Plant Physiol 1996; 112(1): 401-8.
[http://dx.doi.org/10.1104/pp.112.1.401]

[20]   Guinn G. Water deficit and ethylene evolution by young cotton bolls. Plant Physiol 1976; 57(3): 403-5.
[http://dx.doi.org/10.1104/pp.57.3.403]

[21]   Fromm H. Root plasticity in the pursuit of water. Plants 2019; 8(7): 236.
[http://dx.doi.org/10.3390/plants8070236]

[22]   Kou X, Han W, Kang J. Responses of root system architecture to water stress at multiple levels: A meta-analysis of trials under controlled conditions. Front Plant Sci 2022; 13: 1085409.
[http://dx.doi.org/10.3389/fpls.2022.1085409]

[23]   Radin JW. Stomatal responses to water stress and to abscisic acid in phosphorus-deficient cotton plants. Plant Physiol 1984; 76(2): 392-4.
[http://dx.doi.org/10.1104/pp.76.2.392]

[24]   Agurla S, Gahir S, Munemasa S, Murata Y, Raghavendra AS. Mechanism of stomatal closure in plants exposed to drought and cold stress. Adv Exp Med Biol 2018; 1081: 215-32.
[http://dx.doi.org/10.1007/978-981-13-1244-1_12]

[25]   Christmann A, Hoffmann T, Teplova I, Grill E, Müller A. Generation of active pools of abscisic acid revealed by *in vivo* imaging of water-stressed Arabidopsis. Plant Physiol 2005; 137(1): 209-19.
[http://dx.doi.org/10.1104/pp.104.053082]

[26]   Flexas J, Bota J, Loreto F, Cornic G, Sharkey TD. Diffusive and metabolic limitations to photosynthesis under drought and salinity in C(3) plants. Plant Biol 2004; 6(3): 269-79.
[http://dx.doi.org/10.1055/s-2004-820867]

[27]   Flexas J, Barón M, Bota J, *et al.* Photosynthesis limitations during water stress acclimation and recovery in the drought-adapted Vitis hybrid Richter-110 (V. berlandieri×V. rupestris). J Exp Bot 2009; 60(8): 2361-77.
[http://dx.doi.org/10.1093/jxb/erp069]

[28]   Ruess BR, Eller BM. The correlation between crassulacean acid metabolism and water uptake in Senecio medley-woodii. Planta 1985; 166(1): 57-66.
[http://dx.doi.org/10.1007/BF00397386]

[29]   Taybi T, Cushman JC. Signaling events leading to crassulacean acid metabolism induction in the common ice plant. Plant Physiol 1999; 121(2): 545-56.
[http://dx.doi.org/10.1104/pp.121.2.545]

[30]   Finan JD, Guilak F. The effects of osmotic stress on the structure and function of the cell nucleus. J Cell Biochem 2010; 109(3): 460-7.
[http://dx.doi.org/10.1002/jcb.22437]

[31]   Xiong L, Ishitani M, Lee H, Zhu JK. HOS5–A negative regulator of osmotic stress-induced gene expression in Arabidopsis thaliana. Plant J 1999; 19(5): 569-78.
[http://dx.doi.org/10.1046/j.1365-313X.1999.00558.x]

[32]   Cohen BE. Functional linkage between genes that regulate osmotic stress responses and multidrug resistance transporters: challenges and opportunities for antibiotic discovery. Antimicrob Agents Chemother 2014; 58(2): 640-6.
[http://dx.doi.org/10.1128/AAC.02095-13]

[33]   Gill RA, Ahmar S, Ali B, *et al.* The role of membrane transporters in plant growth and development, and abiotic stress tolerance. Int J Mol Sci 2021; 22(23): 12792.
[http://dx.doi.org/10.3390/ijms222312792]

[34]   Altschuler M, Mascarenhas JP. Heat shock proteins and effects of heat shock in plants. Plant Mol Biol 1982; 1(2): 103-15.
[http://dx.doi.org/10.1007/BF00024974]

[35]   Kan Y, Mu XR, Gao J, Lin HX, Lin Y. The molecular basis of heat stress responses in plants. Mol Plant 2023; 16(10): 1612-34.
[http://dx.doi.org/10.1016/j.molp.2023.09.013]

[36]   Marchin RM, Backes D, Ossola A, Leishman MR, Tjoelker MG, Ellsworth DS. Extreme heat increases stomatal conductance and drought-induced mortality risk in vulnerable plant species. Glob Change Biol 2022; 28(3): 1133-46.
[http://dx.doi.org/10.1111/gcb.15976]

[37]   Lamaoui M, Jemo M, Datla R, Bekkaoui F. Heat and drought stresses in crops and approaches for their mitigation. Front Chem 2018; 6: 26.
[http://dx.doi.org/10.3389/fchem.2018.00026]

[38]   Sattar A, Sher A, Ijaz M, *et al.* Terminal drought and heat stress alter physiological and biochemical attributes in flag leaf of bread wheat. PLoS One 2020; 15(5): e0232974.
[http://dx.doi.org/10.1371/journal.pone.0232974]

[39]   Duan H, Wu J, Huang G, *et al.* Individual and interactive effects of drought and heat on leaf physiology of seedlings in an economically important crop. AoB Plants 2016; 9(1): plw090.
[http://dx.doi.org/10.1093/aobpla/plw090]

[40]   Hüve K, Bichele I, Rasulov B, Niinemets Ü. When it is too hot for photosynthesis: heat-induced instability of photosynthesis in relation to respiratory burst, cell permeability changes and $H_2O_2$ formation. Plant Cell Environ 2011; 34(1): 113-26.
[http://dx.doi.org/10.1111/j.1365-3040.2010.02229.x]

[41]   Yang D, Peng S, Wang F. Response of photosynthesis to high growth temperature was not related to leaf anatomy plasticity in Rice (*Oryza sativa* L.). Front Plant Sci 2020; 11: 26.
[http://dx.doi.org/10.3389/fpls.2020.00026]

[42]   Quinn PJ. Effects of temperature on cell membranes. Symp Soc Exp Biol 1988; 42: 237-58.

[43]   Raison JK, Pike CS, Berry JA. Growth temperature-induced alterations in the thermotropic properties of Nerium oleander membrane lipids. Plant Physiol 1982; 70(1): 215-8.
[http://dx.doi.org/10.1104/pp.70.1.215]

[44]   Barua D, Heckathorn SA, Coleman JS. Variation in heat-shock proteins and photosynthetic thermotolerance among natural populations of *Chenopodium album* L. from contrasting thermal environments: implications for plant responses to global warming. J Integr Plant Biol 2008; 50(11): 1440-51.
[http://dx.doi.org/10.1111/j.1744-7909.2008.00756.x]

[45]   Ougham HJ, Howarth CJ. Temperature shock proteins in plants. Symp Soc Exp Biol 1988; 42: 259-80.

[46]   Tian F, Hu XL, Yao T, *et al.* Recent advances in the roles of HSFs and HSPs in heat stress response in woody plants. Front Plant Sci 2021; 12: 704905.
[http://dx.doi.org/10.3389/fpls.2021.704905]

[47]   Qu AL, Ding YF, Jiang Q, Zhu C. Molecular mechanisms of the plant heat stress response. Biochem Biophys Res Commun 2013; 432(2): 203-7.
[http://dx.doi.org/10.1016/j.bbrc.2013.01.104]

[48]   Haider S, Raza A, Iqbal J, Shaukat M, Mahmood T. Analyzing the regulatory role of heat shock transcription factors in plant heat stress tolerance: a brief appraisal. Mol Biol Rep 2022; 49(6): 5771-

85.
[http://dx.doi.org/10.1007/s11033-022-07190-x]

[49]    Boinot M, Karakas E, Koehl K, Pagter M, Zuther E. Cold stress and freezing tolerance negatively affect the fitness of *Arabidopsis thaliana* accessions under field and controlled conditions. Planta 2022; 255(2): 39.
[http://dx.doi.org/10.1007/s00425-021-03809-8]

[50]    Eremina M, Rozhon W, Poppenberger B. Hormonal control of cold stress responses in plants. Cell Mol Life Sci 2016; 73(4): 797-810.
[http://dx.doi.org/10.1007/s00018-015-2089-6]

[51]    Ritonga FN, Chen S. Physiological and molecular mechanism involved in cold stress tolerance in plants. Plants 2020; 9(5): 560.
[http://dx.doi.org/10.3390/plants9050560]

[52]    Miller MA, Zachary JF. Mechanisms and morphology of cellular injury, adaptation, and death. Pathologic Basis of Veterinary Disease. 2017; 2–43. e19.
[http://dx.doi.org/10.1016/B978-0-323-35775-3.00001-1]

[53]    Ninagawa T, Eguchi A, Kawamura Y, Konishi T, Narumi A. A study on ice crystal formation behavior at intracellular freezing of plant cells using a high-speed camera. Cryobiology 2016; 73(1): 20-9.
[http://dx.doi.org/10.1016/j.cryobiol.2016.06.003]

[54]    Körber C, Englich S, Rau G. Intracellular ice formation: cryomicroscopical observation and calorimetric measurement. J Microsc 1991; 161(2): 313-25.
[http://dx.doi.org/10.1111/j.1365-2818.1991.tb03092.x]

[55]    Bredow M, Vanderbeld B, Walker VK. Ice-binding proteins confer freezing tolerance in transgenic *Arabidopsis thaliana*. Plant Biotechnol J 2017; 15(1): 68-81.
[http://dx.doi.org/10.1111/pbi.12592]

[56]    Toxopeus J, Sinclair BJ. Mechanisms underlying insect freeze tolerance. Biol Rev Camb Philos Soc 2018; 93(4): 1891-914.
[http://dx.doi.org/10.1111/brv.12425]

[57]    Guy CL, Haskell D. Induction of freezing tolerance in spinach is associated with the synthesis of cold acclimation induced proteins. Plant Physiol 1987; 84(3): 872-8.
[http://dx.doi.org/10.1104/pp.84.3.872]

[58]    Min K, Cho Y, Kim E, Lee M, Lee SR. Exogenous glycine betaine application improves freezing tolerance of Cabbage (*Brassica oleracea* L.) leaves. Plants 2021; 10(12): 2821.
[http://dx.doi.org/10.3390/plants10122821]

[59]    Chen HH, Li PH, Brenner ML. Involvement of abscisic acid in potato cold acclimation. Plant Physiol 1983; 71(2): 362-5.
[http://dx.doi.org/10.1104/pp.71.2.362]

[60]    Minami A, Nagao M, Arakawa K, Fujikawa S, Takezawa D. Abscisic acid-induced freezing tolerance in the moss *Physcomitrella patens* is accompanied by increased expression of stress-related genes. J Plant Physiol 2003; 160(5): 475-83.
[http://dx.doi.org/10.1078/0176-1617-00888]

[61]    Mohapatra SS, Poole RJ, Dhindsa RS. Abscisic acid-regulated gene expression in relation to freezing tolerance in alfalfa. Plant Physiol 1988; 87(2): 468-73.
[http://dx.doi.org/10.1104/pp.87.2.468]

[62]    Shinkawa R, Morishita A, Amikura K, *et al.* Abscisic acid induced freezing tolerance in chilling-sensitive suspension cultures and seedlings of rice. BMC Res Notes 2013; 6(1): 351.
[http://dx.doi.org/10.1186/1756-0500-6-351]

[63]    Jurczyk B, Pociecha E, Janowiak F, Dziurka M, Kościk I, Rapacz M. Changes in ethylene, ABA and

sugars regulate freezing tolerance under low-temperature waterlogging in *Lolium perenne*. Int J Mol Sci 2021; 22(13): 6700.
[http://dx.doi.org/10.3390/ijms22136700]

[64]   Mantyla E, Lang V, Palva ET. Role of abscisic acid in drought-induced freezing tolerance, cold acclimation, and accumulation of LT178 and RAB18 proteins in *Arabidopsis thaliana*. Plant Physiol 1995; 107(1): 141-8.
[http://dx.doi.org/10.1104/pp.107.1.141]

[65]   Byun YJ, Koo MY, Joo HJ, Ha-Lee YM, Lee DH. Comparative analysis of gene expression under cold acclimation, deacclimation and reacclimation in Arabidopsis. Physiol Plant 2014; 152(2): 256-74.
[http://dx.doi.org/10.1111/ppl.12163]

[66]   Guy CL, Niemi KJ, Brambl R. Altered gene expression during cold acclimation of spinach. Proc Natl Acad Sci USA 1985; 82(11): 3673-7.
[http://dx.doi.org/10.1073/pnas.82.11.3673]

[67]   Welch WJ. Heat shock proteins functioning as molecular chaperones: their roles in normal and stressed cells. Philos Trans R Soc Lond B Biol Sci 1993; 339(1289): 327-33.
[http://dx.doi.org/10.1098/rstb.1993.0031]

[68]   Macario AJL. Heat-shock proteins and molecular chaperones: implications for pathogenesis, diagnostics, and therapeutics. Int J Clin Lab Res 1995; 25(2): 59-70.
[http://dx.doi.org/10.1007/BF02592359]

[69]   Barrett J. Thermal hysteresis proteins. Int J Biochem Cell Biol 2001; 33(2): 105-17.
[http://dx.doi.org/10.1016/S1357-2725(00)00083-2]

[70]   Ekpo MD, Xie J, Hu Y, *et al.* Antifreeze proteins: novel applications and navigation towards their clinical application in cryobanking. Int J Mol Sci 2022; 23(5): 2639.
[http://dx.doi.org/10.3390/ijms23052639]

[71]   dos Santos C, Franco OL. Pathogenesis-related proteins (PRs) with enzyme activity activating plant defense responses. Plants 2023; 12(11): 2226.
[http://dx.doi.org/10.3390/plants12112226]

[72]   Wilson SK, Pretorius T, Naidoo S. Mechanisms of systemic resistance to pathogen infection in plants and their potential application in forestry. BMC Plant Biol 2023; 23(1): 404.
[http://dx.doi.org/10.1186/s12870-023-04391-9]

[73]   Satyakam , Zinta G, Singh RK, Kumar R. Cold adaptation strategies in plants—An emerging role of epigenetics and antifreeze proteins to engineer cold resilient plants. Front Genet 2022; 13: 909007.
[http://dx.doi.org/10.3389/fgene.2022.909007]

[74]   Hasanuzzaman M, Fujita M. Plant responses and tolerance to salt stress: physiological and molecular interventions. Int J Mol Sci 2022; 23(9): 4810.
[http://dx.doi.org/10.3390/ijms23094810]

[75]   Zhao S, Zhang Q, Liu M, Zhou H, Ma C, Wang P. Regulation of plant responses to salt stress. Int J Mol Sci 2021; 22(9): 4609.
[http://dx.doi.org/10.3390/ijms22094609]

[76]   Negrão S, Schmöckel SM, Tester M. Evaluating physiological responses of plants to salinity stress. Ann Bot (Lond) 2017; 119(1): 1-11.
[http://dx.doi.org/10.1093/aob/mcw191]

[77]   Balasubramaniam T, Shen G, Esmaeili N, Zhang H. Plants response mechanisms to salinity stress. Plants 2023; 12(12): 2253.
[http://dx.doi.org/10.3390/plants12122253]

[78]   Hnilickova H, Kraus K, Vachova P, Hnilicka F. Salinity stress affects photosynthesis, malondialdehyde formation, and proline content in *Portulaca oleracea* L. Plants 2021; 10(5): 845.
[http://dx.doi.org/10.3390/plants10050845]

[79]  Wani AS, Ahmad A, Hayat S, Fariduddin Q. Salt-induced modulation in growth, photosynthesis and antioxidant system in two varieties of *rassica juncea*. Saudi J Biol Sci 2013; 20(2): 183-93.
[http://dx.doi.org/10.1016/j.sjbs.2013.01.006]

[80]  Alharbi K, Al-Osaimi AA, Alghamdi BA. Sodium chloride (NaCl)-induced physiological alteration and oxidative stress generation in *Pisum sativum* (L.): a toxicity assessment. ACS Omega 2022; 7(24): 20819-32.
[http://dx.doi.org/10.1021/acsomega.2c01427]

[81]  Tester M, Davenport R. Na+ tolerance and Na+ transport in higher plants. Ann Bot (Lond) 2003; 91(5): 503-27.
[http://dx.doi.org/10.1093/aob/mcg058]

[82]  Ondrasek G, Rathod S, Manohara KK, *et al.* Salt stress in plants and mitigation approaches. Plants 2022; 11(6): 717.
[http://dx.doi.org/10.3390/plants11060717]

[83]  Zhao C, Zhang H, Song C, Zhu JK, Shabala S. Mechanisms of plant responses and adaptation to soil salinity. Innovation 2020; 1(1): 100017.
[http://dx.doi.org/10.1016/j.xinn.2020.100017]

[84]  Meng X, Zhou J, Sui N. Mechanisms of salt tolerance in halophytes: current understanding and recent advances. Open Life Sci 2018; 13(1): 149-54.
[http://dx.doi.org/10.1515/biol-2018-0020]

[85]  Queirós F, Fontes N, Silva P, *et al.* Activity of tonoplast proton pumps and Na+/H+ exchange in potato cell cultures is modulated by salt. J Exp Bot 2009; 60(4): 1363-74.
[http://dx.doi.org/10.1093/jxb/erp011]

[86]  Foster KJ, Miklavcic SJ. Toward a biophysical understanding of the salt stress response of individual plant cells. J Theor Biol 2015; 385: 130-42.
[http://dx.doi.org/10.1016/j.jtbi.2015.08.024]

[87]  Krulwich TA. Na+/H+ antiporters. Biochim Biophys Acta Rev Bioenerg 1983; 726(4): 245-64.
[http://dx.doi.org/10.1016/0304-4173(83)90011-3]

[88]  Bassil E, Ohto M, Esumi T, *et al.* The Arabidopsis intracellular Na+/H+ antiporters NHX5 and NHX6 are endosome associated and necessary for plant growth and development. Plant Cell 2011; 23(1): 224-39.
[http://dx.doi.org/10.1105/tpc.110.079426]

[89]  Keisham M, Mukherjee S, Bhatla S. Mechanisms of sodium transport in plants—progresses and challenges. Int J Mol Sci 2018; 19(3): 647.
[http://dx.doi.org/10.3390/ijms19030647]

[90]  Leyval C, Turnau K, Haselwandter K. Effect of heavy metal pollution on mycorrhizal colonization and function: physiological, ecological and applied aspects. Mycorrhiza 1997; 7(3): 139-53.
[http://dx.doi.org/10.1007/s005720050174]

[91]  Nriagu JO. Trace metal pollution of lakes: a global perspective. 2nd International Conference. Trace metals in aquatic environment. 1990.

[92]  Rashid A, Schutte BJ, Ulery A, *et al.* Heavy metal contamination in agricultural soil: environmental pollutants affecting crop health. Agronomy (Basel) 2023; 13(6): 1521.
[http://dx.doi.org/10.3390/agronomy13061521]

[93]  Rascio N, Navari-Izzo F. Heavy metal hyperaccumulating plants: How and why do they do it? And what makes them so interesting? Plant Sci 2011; 180(2): 169-81.
[http://dx.doi.org/10.1016/j.plantsci.2010.08.016]

[94]  Assunção AGL, Cakmak I, Clemens S, González-Guerrero M, Nawrocki A, Thomine S. Micronutrient homeostasis in plants for more sustainable agriculture and healthier human nutrition. J Exp Bot 2022;

73(6): 1789-99.
[http://dx.doi.org/10.1093/jxb/erac014]

[95]   Angulo-Bejarano PI, Puente-Rivera J, Cruz-Ortega R. Metal and metalloid toxicity in plants: an overview on molecular aspects. Plants 2021; 10(4): 635.
[http://dx.doi.org/10.3390/plants10040635]

[96]   Genchi G, Sinicropi MS, Lauria G, Carocci A, Catalano A. The effects of cadmium toxicity. Int J Environ Res Public Health 2020; 17(11): 3782.
[http://dx.doi.org/10.3390/ijerph17113782]

[97]   Alengebawy A, Abdelkhalek ST, Qureshi SR, Wang MQ. Heavy metals and pesticides toxicity in agricultural soil and plants: ecological risks and human health implications. Toxics 2021; 9(3): 42.
[http://dx.doi.org/10.3390/toxics9030042]

[98]   Charkiewicz AE, Omeljaniuk WJ, Nowak K, Garley M, Nikliński J. Cadmium toxicity and health effects—a brief summary. Molecules 2023; 28(18): 6620.
[http://dx.doi.org/10.3390/molecules28186620]

[99]   Schutzendubel A, Polle A. Plant responses to abiotic stresses: heavy metal-induced oxidative stress and protection by mycorrhization. J Exp Bot 2002; 53(372): 1351-65.
[http://dx.doi.org/10.1093/jexbot/53.372.1351]

[100]  Fisher RM, Gupta V. Heavy metals [Internet]. StatPearls Treasure Island (FL): StatPearls Publishing; 2024.
[PMID: 32491738]

[101]  Gür N, Topdemir A, Munzuroğlu Ö, Çobanoğlu D. Heavy metal ions ($Cu+2$, $Pb+2$, $Hg+2$, $Cd+2$) Clivia sp. effects of plant pollen on germination and tube growth (In Turkish). F. U. Journal of Science and Mathematics 2004; 16(2): 177-82.

[102]  Hall JL. Cellular mechanisms for heavy metal detoxification and tolerance. J Exp Bot 2002; 53(366): 1-11.
[http://dx.doi.org/10.1093/jexbot/53.366.1]

[103]  Sharma SS, Dietz KJ. The relationship between metal toxicity and cellular redox imbalance. Trends Plant Sci 2009; 14(1): 43-50.
[http://dx.doi.org/10.1016/j.tplants.2008.10.007]

[104]  Gupta A, Dubey P, Kumar M, *et al.* Consequences of Arsenic Contamination on Plants and Mycoremediation-Mediated Arsenic Stress Tolerance for Sustainable Agriculture. Plants 2022; 11(23): 3220.
[http://dx.doi.org/10.3390/plants11233220]

[105]  Gupta A, Mishra R, Rai S, *et al.* Mechanistic Insights of Plant Growth Promoting Bacteria Mediated Drought and Salt Stress Tolerance in Plants for Sustainable Agriculture. Int J Mol Sci 2022; 23(7): 3741.
[http://dx.doi.org/10.3390/ijms23073741]

[106]  Pinto E, Sigaud-kutner TCS, Leitão MAS, Okamoto OK, Morse D, Colepicolo P. Heavy metal–induced oxidative stress in algae 1. J Phycol 2003; 39(6): 1008-18.
[http://dx.doi.org/10.1111/j.0022-3646.2003.02-193.x]

[107]  Badawy AA, Abdel Rehim MH, Turky GM. Charge transport and heavy metal removal efficacy of graphitic carbon nitride doped with $CeO_2$. RSC Advances 2023; 13(13): 8955-66.
[http://dx.doi.org/10.1039/D3RA00844D]

[108]  Yousef YA, Hvitved-Jacobsen T, Harper HH, Lin LY. Heavy metal accumulation and transport through detention ponds receiving highway runoff. Sci Total Environ 1990; 93: 433-40.
[http://dx.doi.org/10.1016/0048-9697(90)90134-G]

[109]  Singh S, Parihar P, Singh R, Singh VP, Prasad SM. Heavy metal tolerance in plants: role of transcriptomics, proteomics, metabolomics, and ionomics. Front Plant Sci 2016; 6: 1143.

[http://dx.doi.org/10.3389/fpls.2015.01143]

[110]  Noor I, Sohail H, Sun J, *et al.* Heavy metal and metalloid toxicity in horticultural plants: Tolerance mechanism and remediation strategies. Chemosphere 2022; 303(Pt 3): 135196.
[http://dx.doi.org/10.1016/j.chemosphere.2022.135196]

[111]  Blaby-Haas CE, Merchant SS. Lysosome-related organelles as mediators of metal homeostasis. J Biol Chem 2014; 289(41): 28129-36.
[http://dx.doi.org/10.1074/jbc.R114.592618]

[112]  Portnoy M, Schmidt P, Rogers R, Culotta V. Metal transporters that contribute copper to metallochaperones in Saccharomyces cerevisiae. Mol Genet Genomics 2001; 265(5): 873-82.
[http://dx.doi.org/10.1007/s004380100482]

[113]  Emamverdian A, Ding Y, Mokhberdoran F, Xie Y. Heavy metal stress and some mechanisms of plant defense response. ScientificWorldJournal 2015; 2015(1): 756120.
[http://dx.doi.org/10.1155/2015/756120]

[114]  Bartosz G. Reactive oxygen species: Destroyers or messengers? Biochem Pharmacol 2009; 77(8): 1303-15.
[http://dx.doi.org/10.1016/j.bcp.2008.11.009]

[115]  Benavides MP, Gallego SM, Tomaro ML. Cadmium toxicity in plants. Braz J Plant Physiol 2005; 17(1): 21-34.
[http://dx.doi.org/10.1590/S1677-04202005000100003]

[116]  Dvořák P, Krasylenko Y, Zeiner A, Šamaj J, Takáč T. Signaling toward reactive oxygen species-scavenging enzymes in plants. Front Plant Sci 2021; 11: 618835.
[http://dx.doi.org/10.3389/fpls.2020.618835]

[117]  Hasanuzzaman M, Raihan MRH, Masud AAC, *et al.* Regulation of reactive oxygen species and antioxidant defense in plants under salinity. Int J Mol Sci 2021; 22(17): 9326.
[http://dx.doi.org/10.3390/ijms22179326]

[118]  Gill SS, Tuteja N. Reactive oxygen species and antioxidant machinery in abiotic stress tolerance in crop plants. Plant Physiol Biochem 2010; 48(12): 909-30.
[http://dx.doi.org/10.1016/j.plaphy.2010.08.016]

[119]  Devasagayam TP, Kamat JP. Biological significance of singlet oxygen. Indian J Exp Biol 2002; 40(6): 680-92. Available from: http://nopr.niscpr.res.in/handle/123456789/23511

[120]  Urban MV, Rath T, Radtke C. Hydrogen peroxide ($H_2O_2$): a review of its use in surgery. Wien Med Wochenschr 2019; 169(9-10): 222-5.
[http://dx.doi.org/10.1007/s10354-017-0610-2]

[121]  Kehrer JP. The Haber–Weiss reaction and mechanisms of toxicity. Toxicology 2000; 149(1): 43-50.
[http://dx.doi.org/10.1016/S0300-483X(00)00231-6]

[122]  Takagi D, Takumi S, Hashiguchi M, Sejima T, Miyake C. Superoxide and singlet oxygen produced within the thylakoid membranes both cause photosystem I photoinhibition. Plant Physiol 2016; 171(3): 1626-34.
[http://dx.doi.org/10.1104/pp.16.00246]

[123]  Fischbacher A, von Sonntag C, Schmidt TC. Hydroxyl radical yields in the Fenton process under various pH, ligand concentrations and hydrogen peroxide/Fe(II) ratios. Chemosphere 2017; 182: 738-44.
[http://dx.doi.org/10.1016/j.chemosphere.2017.05.039]

[124]  Mansoor S, Ali Wani O, Lone JK, *et al.* Reactive oxygen species in plants: from source to sink. Antioxidants 2022; 11(2): 225.
[http://dx.doi.org/10.3390/antiox11020225]

[125]  Lambert AJ, Brand MD. Reactive oxygen species production by mitochondria. Methods Mol Biol

2009; 554: 165-81.
[http://dx.doi.org/10.1007/978-1-59745-521-3_11]

[126]   Turrens JF. Mitochondrial formation of reactive oxygen species. J Physiol 2003; 552(2): 335-44.
[http://dx.doi.org/10.1113/jphysiol.2003.049478]

[127]   Foyer CH, Hanke G. ROS production and signalling in chloroplasts: cornerstones and evolving
concepts. Plant J 2022; 111(3): 642-61.
[http://dx.doi.org/10.1111/tpj.15856]

[128]   Sandalio LM, Rodríguez-Serrano M, Romero-Puertas MC, del Río LA. Role of peroxisomes as a
source of reactive oxygen species (ROS) signaling molecules. Subcell Biochem 2013; 69: 231-55.
[http://dx.doi.org/10.1007/978-94-007-6889-5_13]

[129]   del Río LA, López-Huertas E. ROS generation in peroxisomes and its role in cell signaling. Plant Cell
Physiol 2016; 57(7): pcw076.
[http://dx.doi.org/10.1093/pcp/pcw076]

[130]   Su LJ, Zhang JH, Gomez H, *et al.* Reactive oxygen species-induced lipid peroxidation in Apoptosis,
autophagy, and ferroptosis. Oxid Med Cell Longev 2019; 1-13.
[http://dx.doi.org/10.1155/2019/5080843]

[131]   Khan MD, Mei L, Ali B, Chen Y, Cheng X, Zhu SJ. Cadmium-induced upregulation of lipid
peroxidation and reactive oxygen species caused physiological, biochemical, and ultrastructural
changes in upland cotton seedlings. BioMed Res Int 2013; 1-10.
[http://dx.doi.org/10.1155/2013/374063]

[132]   Kehm R, Baldensperger T, Raupbach J, Höhn A. Protein oxidation - Formation mechanisms, detection
and relevance as biomarkers in human diseases. Redox Biol 2021; 42: 101901.
[http://dx.doi.org/10.1016/j.redox.2021.101901]

[133]   Hemnani T, Parihar MS. Reactive oxygen species and oxidative DNA damage. Indian J Physiol
Pharmacol 1998; 42(4): 440-52.

[134]   Renaudin X. Reactive oxygen species and DNA damage response in cancer. Int Rev Cell Mol Biol
2021; 364: 139-61.
[http://dx.doi.org/10.1016/bs.ircmb.2021.04.001]

[135]   Kesawat MS, Satheesh N, Kherawat BS, *et al.* Regulation of Reactive Oxygen Species during Salt
Stress in Plants and Their Crosstalk with Other Signaling Molecules—Current Perspectives and Future
Directions. Plants 2023; 12(4): 864.
[http://dx.doi.org/10.3390/plants12040864]

[136]   Hasanuzzaman M, Bhuyan MHM, Zulfiqar F, *et al.* Reactive oxygen species and antioxidant defense
in plants under abiotic stress: revisiting the crucial role of a universal defense regulator. Antioxidants
2020; 9(8): 681.
[http://dx.doi.org/10.3390/antiox9080681]

[137]   Kliebenstein DJ, Dietrich RA, Martin AC, Last RL, Dangl JL. LSD1 regulates salicylic acid induction
of copper zinc superoxide dismutase in *Arabidopsis thaliana*. Mol Plant Microbe Interact 1999;
12(11): 1022-6.
[http://dx.doi.org/10.1094/MPMI.1999.12.11.1022]

[138]   Alscher RG, Erturk N, Heath LS. Role of superoxide dismutases (SODs) in controlling oxidative stress
in plants. J Exp Bot 2002; 53(372): 1331-41.
[http://dx.doi.org/10.1093/jexbot/53.372.1331]

[139]   Mishra N, Jiang C, Chen L, Paul A, Chatterjee A, Shen G. Achieving abiotic stress tolerance in plants
through antioxidative defense mechanisms. Front Plant Sci 2023; 14: 1110622.
[http://dx.doi.org/10.3389/fpls.2023.1110622]

[140]   Sofo A, Scopa A, Nuzzaci M, Vitti A. Ascorbate peroxidase and catalase activities and their genetic
regulation in plants subjected to drought and salinity stresses. Int J Mol Sci 2015; 16(6): 13561-78.

[http://dx.doi.org/10.3390/ijms160613561]

[141] Caverzan A, Passaia G, Rosa SB, Ribeiro CW, Lazzarotto F, Margis-Pinheiro M. Plant responses to stresses: role of ascorbate peroxidase in the antioxidant protection. Genet Mol Biol 2012; 35(4 suppl 1) (Suppl.): 1011-9.
[http://dx.doi.org/10.1590/S1415-47572012000600016]

[142] Davey M, Montagu MV, Inzé D, *et al.* Plant L-ascorbic acid: chemistry, function, metabolism, bioavailability and effects of processing. J Sci Food Agric 2000; 80: 825-60.
[http://dx.doi.org/10.1002/(SICI)1097-0010(20000515)80:7<825::AID-JSFA598>3.0.CO;2-6]

[143] Nandi A, Yan LJ, Jana CK, Das N. Role of catalase in oxidative stress- and age-associated degenerative diseases. Oxid Med Cell Longev 2019; 1-19.
[http://dx.doi.org/10.1155/2019/9613090]

[144] Bailly C, Leymarie J, Lehner A, Rousseau S, Côme D, Corbineau F. Catalase activity and expression in developing sunflower seeds as related to drying. J Exp Bot 2004; 55(396): 475-83. Available from: http://www.jstor.org/stable/24029331
[http://dx.doi.org/10.1093/jxb/erh050]

[145] Folkes LK, Dennis MF, Stratford MRL, Candeias LP, Wardman P. Peroxidase-catalyzed effects of indole-3-acetic acid and analogues on lipid membranes, DNA, and mammalian cells *in vitro*. Biochem Pharmacol 1999; 57(4): 375-82.
[http://dx.doi.org/10.1016/S0006-2952(98)00323-2]

[146] Ferreira AGK, Biasibetti-Brendler H, Sidegum DSV, Loureiro SO, Figueiró F, Wyse ATS. Effect of proline on cell death, cell cycle, and oxidative stress in C6 glioma cell line. Neurotox Res 2021; 39(2): 327-34.
[http://dx.doi.org/10.1007/s12640-020-00311-z]

[147] Hayat S, Hayat Q, Alyemeni MN, Wani AS, Pichtel J, Ahmad A. Role of proline under changing environments. Plant Signal Behav 2012; 7(11): 1456-66.
[http://dx.doi.org/10.4161/psb.21949]

[148] Fichman Y, Gerdes SY, Kovács H, Szabados L, Zilberstein A, Csonka LN. Evolution of proline biosynthesis: enzymology, bioinformatics, genetics, and transcriptional regulation. Biol Rev Camb Philos Soc 2015; 90(4): 1065-99.
[http://dx.doi.org/10.1111/brv.12146]

[149] Turchetto-Zolet AC, Margis-Pinheiro M, Margis R. The evolution of pyrroline-5-carboxylate synthase in plants: a key enzyme in proline synthesis. Mol Genet Genomics 2009; 281(1): 87-97.
[http://dx.doi.org/10.1007/s00438-008-0396-4]

[150] Bilska K, Wojciechowska N, Alipour S, Kalemba EM. Ascorbic acid-the little-known antioxidant in woody plants. Antioxidants 2019; 8(12): 645.
[http://dx.doi.org/10.3390/antiox8120645]

[151] Smirnoff N, Wheeler GL. Ascorbic acid in plants: biosynthesis and function. Crit Rev Biochem Mol Biol 2000; 35(4): 291-314.
[http://dx.doi.org/10.1080/10409230008984166]

[152] Gęgotek A, Skrzydlewska E. Antioxidative and anti-inflammatory activity of ascorbic acid. Antioxidants 2022; 11(10): 1993.
[http://dx.doi.org/10.3390/antiox11101993]

[153] Gallie DR. L-ascorbic acid: a multifunctional molecule supporting plant growth and development. Scientifica (Cairo) 2013; 1-24.
[http://dx.doi.org/10.1155/2013/795964]

[154] Saijo Y, Loo EP. Plant immunity in signal integration between biotic and abiotic stress responses. New Phytol 2020; 225(1): 87-104.
[http://dx.doi.org/10.1111/nph.15989]

[155] Lamers J, van der Meer T, Testerink C. How plants sense and respond to stressful environments. Plant Physiol 2020; 182(4): 1624-35.
[http://dx.doi.org/10.1104/pp.19.01464]

[156] Hammond-Kosack K, Jones JDG. Responses to plant pathogens. Biochem Mol Biol Plants 2000; 1: 1102-56.

[157] Peck S, Mittler R. Plant signaling in biotic and abiotic stress. J Exp Bot 2020; 71(5): 1649-51.
[http://dx.doi.org/10.1093/jxb/eraa051]

[158] Wang Z, Ma LY, Cao J, *et al.* Recent advances in mechanisms of plant defense to Sclerotinia sclerotiorum. Front Plant Sci 2019; 10: 1314.
[http://dx.doi.org/10.3389/fpls.2019.01314]

[159] Sobiczewski P, Iakimova ET, Mikiciński A, Węgrzynowicz-Lesiak E, Dyki B. Necrotrophic behaviour of *Erwinia amylovora* in apple and tobacco leaf tissue. Plant Pathol 2017; 66(5): 842-55.
[http://dx.doi.org/10.1111/ppa.12631]

[160] Osman HA, Ameen HH, Mohamed M, Elkelany US. Efficacy of integrated microorganisms in controlling root-knot nematode Meloidogyne javanica infecting peanut plants under field conditions. Bull Natl Res Cent 2020; 44(1): 134.
[http://dx.doi.org/10.1186/s42269-020-00366-0]

[161] Pallas V, García JA. How do plant viruses induce disease? Interactions and interference with host components. J Gen Virol 2011; 92(12): 2691-705.
[http://dx.doi.org/10.1099/vir.0.034603-0]

[162] Taiz L, Zeiger E. Secondary metabolites and plant defense. Plant Physiol 2006; 4: 315-44.

*Current and Future Developments in Physiology, Vol. 2,* 2024, 51-88

# CHAPTER 3

# New-Generation Plant Growth Regulators

**Ergun Kaya**[1,*] and **Damla Ekin Özkaya**[1,2]

[1] *Muğla Sıtkı Koçman University, Faculty of Science, Molecular Biology and Genetics Department, 48000, Menteşe, Muğla, Türkiye*

[2] *Okan University, Vocational School of Health Services, Medical Laboratory Techniques Department, 34959, Tuzla, İstanbul, Türkiye*

**Abstract:** It is known that metabolic conditions such as differentiation, growth, flower and fruit formation, and development in plants are mostly organized by the plant growth regulators. These organic substances that can be made naturally in plants, control growth and other metabolic conditions related to it. They can be carried from where they occur to other parts of the plant. They can be efficient even at very small volumes and are called plant growth regulators. These are the most significant molecules affecting the subsequent plant growth and development and the internal formation of different metabolic reactions. Growth regulators were initially used only for germination of seeds and rooting of cuttings. Later, it has also been used to increase yield, product quality, and the resistance of plants against pests and diseases in the period from seed to harvest. Plant growth regulators can contribute to increasing plant resistance against diseases by stimulating the plant defense system through various physiological or biochemical reactions that occur as a result of host-pathogen interaction.

**Keywords:** Abscisic acid auxin, Cytokinin, Ethylene, Gibberellic acid.

## INTRODUCTION

The main internal factors that regulate the growth and development of the plant are chemicals. Plant growth regulating substances can be produced by plants or given externally to the plant in very small amounts. In addition, they are organic substances that can affect growth, development and other physiological events in the plant positively or negatively. Plants themselves produce these basic substances that they need for growth, development and change. These substances that are formed in the plant and regulate growth and development (physiological events) are called plant growth regulators (PGR) or phytohormones (plant hormones) [1 - 3].

---

* **Corresponding author Ergun Kaya:** Muğla Sıtkı Koçman University, Faculty of Science, Molecular Biology and Genetics Department, 48000, Menteşe, Muğla, Türkiye; E-mail: ergunkaya@mu.edu.tr

**Ergun Kaya (Ed.)**

The importance of PGR was understood for the first time in the 1930s, and from this date on, their function in agricultural products began to be investigated. Danish botanist Went has experimentally demonstrated that a growth agent flows from these cut ends by fitting the cut coleoptile ends onto agar blocks. When this researcher cut the coleoptile apex, he saw that the growth rate of the coleoptile was greatly reduced. He observed that when he reattached the cut end to the top of the cut coleoptile, the growth of the coleoptile began again [4, 5]. Studies on plant physiology have revealed the roles of PGR in plant growth and development, and it has been understood that not only growth-promoting substances but also growth-inhibiting substances are synthesized in the plant over time [6 - 8]. Today, the effects of PGRs, which are widely used to affect the growth rate and development of the plant from germination to harvest and post-harvest storage, on yield are generally indirect [9, 10]. In the external application of these substances to plants, the selection of the appropriate chemical, the determination of the appropriate concentration, and the application time are very important for the desired effect [11, 12].

The main purposes of using PGRs in agriculture can be listed as follows: To ensure propagation by cuttings, to increase the germination power of seeds, to encourage or delay flowering, to increase cold resistance, to increase seed formation in fruits, to increase fruit size, to extend the storage time of fruit, to increase resistance of plants to diseases and pests, to control weeds, to prevent lodging in cotton and cereals [1, 7, 13, 14], to prevent fruit shedding before harvest, to enable all plants to mature at the same time to facilitate machine harvesting, to reduce labor at harvest, to prevent dormancy by accelerating ripening, to break dormancy, to encourage root-shoot and tuber formation especially in tissue culture studies [15 - 17]. The nature, formation and effects of growth regulators found naturally in plants are given in Table **1**.

**Table 1. The effects of the growth regulators naturally found in plants [1, 10, 13, 18].**

| Plant Growth Regulator | The Part of the Plant where the Growth Regulator is Synthesized | Effect of Growth Regulator on the Plant |
|---|---|---|
| Auxins | Synthesized in leaf primordia, young leaves, and developing seeds. | Promote stem elongation, root growth, cell differentiation, and branching and regulate phototropism and fruit development. |
| Cytokinins | Synthesized in root tips. | Promote root growth and differentiation, cell division and growth, and seed germination. |
| Gibberellins | Synthesized in young tissues of shoots and developing seeds. | Promote seed and bud germination, stem elongation, and leaf development. It affects root growth and differentiation. |

*(Table 1) cont.....*

| Plant Growth Regulator | The Part of the Plant where the Growth Regulator is Synthesized | Effect of Growth Regulator on the Plant |
|---|---|---|
| Abscisic Acid | Synthesized in leaves due to water stress. | Inhibits growth, closes stomata during water stress, and prevents dormancy from breaking. |
| Ethylene | Synthesized in many tissues of the plant depending on the stress. | Increases fruit ripening, and suppresses some auxin effects. |
| Brassinosteroids | Synthesized in seeds, fruits, stems, leaves. | Inhibits root growth, inhibits leaf abscission, and increases xylem differentiation. |

In order for a compound to be qualified as PGR, it must be formed in the plant, be transported from the place where it is formed to another place, manage or regulate different life events in the place where it is transported, and show these effects even at very low concentrations [18 - 20].

Among natural PGRs, ethylene is the most widely used growth regulator in the world with 23%, while auxin is in the second place with 20% and Gibberellins is in the third place with 17%. Of these, auxins, cytokinins and gibberellins are growth promoters; While dormins can be grouped as inhibitors, ethylene mostly plays a regulatory role in fruit ripening [21 - 23].

## PLANT GROWTH REGULATORS

The terms "Plant Growth Regulators" or "Growth Regulatory Substances" include natural plant hormones as well as chemical substances that have been found to be effective on plant growth and development, the number of which has been increasing in recent years and can be obtained synthetically. While every "Plant Hormone" is a "Growth Regulatory Substance", not every "Growth Regulatory Substance" is a plant hormone [6, 24].

### Auxins

Auxins are one of the oldest used plant growth regulators in agriculture. These are substances that provide cell expansion and growth, and cell elongation, tissue development and root formation are promoted by them. Auxins, which are plant growth regulators, are synthesized by all higher plants and the most abundant auxin form is Indole-3-acetic acid (IAA) [25, 26].

Researchers have stated that IAA is the only naturally occurring auxin. Natural auxins occur mostly in the apical buds and leaves and descend from the apex in the plant. IAA is abundant in the growing tips of the plant (coleoptile tip, bud, leaf and root tip). After clarification of the chemical structure of auxin (Fig. 1), it was determined that many chemical substances that are more or less similar in structure to IAA have effects like auxin in plants [27, 28]. The most common auxins other than IAA are; indole butyric acid (IBA), naphthalene acetic acid (NAA), naphthoxy acetic acid (NOAA), phenoxy acetic acid (FOAA), 2,4-D, phenyl acetic acid (FAA), parachlorophenoxy acetic acid (4-CPA) and 2,4,5-trichlorophenoxy acetic acid (2,4,5-T). In various sources, it is mentioned that auxin causes necrosis by accumulating in tissues containing boron element in plants and that there is a close relationship between auxin metabolism and boron deficiency [29, 30].

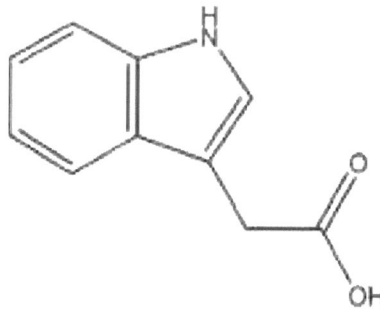

**Fig. (1).** Chemical structure of Indole-3-acetic acid (IAA).

We can list some functions of auxins in plants as follows: (*i*) Since auxins can be transported omnidirectionally or polarly, it is possible to transfer them from cell to cell. It is effective in accelerating cell division, growth and development in plants [26, 30]. (*ii*) It has been found that it is effective in cell growth by increasing osmosis in the cell, facilitating the permeability of the cell against water, increasing the synthesis of specific RNA and protein-structured enzymes that increase the flexibility and width of the cell wall. This can be a solution to seed germination problems in plants with hard seed coats [31, 32]. (*iii*) Providing adventitious root development (Fig. **2**); Significant commercial results have been obtained, especially from the treatment of the cut surfaces of vegetatively propagated woody plants with IAA.

In other words, success has been achieved in rooting from stem segments with auxin application [33, 34]. (*iv*) Obtaining parthenocarpic fruits; In the absence of pollination, flowers and fruits generally do not form. In some plants, fertilization of an egg cell is sufficient for normal fruit development. However, some plants, such as apples or melons, require a large number of seeds to be fertilized in order for the ovary wall to mature and become fleshy. It is possible to obtain parthenocarpic fruits without fertilization (pollination) by applying auxin (Naphthoxy Acetic Acid = NOA) to some plant species. In most plants of the Solanaceae family, they promote fruit set in unfertilized ovaries. These include seedless cucumber, tomato, eggplant, melon, and watermelon. Moreover, it increases the amount of dry matter in many plant species of agricultural importance such as potatoes, beans, corn, sugar beet, *etc.* [35, 36]. (*v*) Prevention of leaf and fruit shedding; Controlling the shedding of leaves, flowers and fruits is of great importance in agriculture. In particular, auxin is applied to transport ornamental plants that are desired to remain evergreen, and to prevent shedding of citrus fruits before harvest. In addition, excessive application of auxin may accelerate fruit shedding.

**Fig. (2).** *In vitro* adventitious root development of endemic *Thymus cariensis* Hub.-Mor. et Jalas on MS medium [43] supplemented with 4.56 µML$^{-1}$ indole-3-butyric acid (IBA), pH 5.8, after four weeks incubation at growth room conditions (23 ± 2°C, 16 h photoperiod, 36 µmol m$^{-2}$ s$^{-1}$). Bar 1 cm in size [44].

In fruits such as apples or olives, auxin (NAA) is applied to make the fruit thinner and to cause better ripening of the remaining fruits. As a result of this application, the shedding of raw fruit at the end of the season is prevented [37, 38]. (*vi*) High auxin concentration in plants, during periods of rapid growth in the tips, buds in the lower parts are prevented from waking up and they cannot last. This is called apical dominance (top bud pressure). Auxins are light sensitive, and as a result of their inactivation in light, they slow down cell growth and cause a tendency towards light in unilateral illumination of plants, known as phototropism. NAA also promotes flowering in cotton, if the wounds opened in fruit trees are covered with NAA-added wound paste, re-shoot formation can be seen from the cut areas, and it is also thought that auxin plays a role as a morphogen (form-former) in the cell, and it is reported that it also works in determining the leaf position [39, 40]. (*vii*) It is effective in controlling weeds. Some synthetic auxins, such as 2,4-D and picloram, are widely used in agricultural weed control. Compounds with 2,4-D cause disorders in many plant functions such as phloem transport, absorption and photosynthesis. In studies conducted in recent years, the use of 2,4-D compounds in tomato and eggplant has been prohibited due to the carcinogenic effect. The use of these compounds in very high doses resulted in darkening, enlargement of the lobes and formation of a hollow structure in sensitive fruit varieties [41, 42].

## Cytokinins

Unlike other plant growth regulators, which appear in plant tissues, especially during cell divisions, are organic substances in the structure of quinine found in both plants and animals. Cytokinins are mainly divided into two groups: synthetic phenylurea derivatives, 1-phenyl-3-(1,2,3-hiadiazol-5-yl) urea, known as thidiazuron (TDZ) and N-(2-chloro-4-pyridyl)-N'-phenylurea (CPPU), and naturally occurring the adenine derivatives kinetin (KIN) and 6-benzyladenine (BA). Synthetic phenyl urea derivatives have a higher potency than TDZ adenine derivatives [45].

Kinetin, 6-benzyl adenine, and zeatin (ZEA) are the most common of the cytokinins (Fig. **3**). They are usually found in young tissues. Many kinetin-like substances have been isolated from germinating seeds, running sap, and young fruit [46, 47]. Zeatin, which is a natural cytokinin, is obtained from corn grains and is also found in high levels in coconut endosperm and horse chestnut fruit [48]. All tissues with active cell division contain sufficient amounts of cytokinins. It is especially synthesized in root meristems and then transported to the green parts of the plant *via* xylem. They are hormones that are effective in cell division and delay aging. While auxins promote root formation, cytokinins promote shoot formation. They contribute to organ formation and development in tissue culture media. It is thought that cytokinins prevent protein degradation by inhibiting nuclease and protease formation in leaves and thus delay senescence. Cytokinins are also effective in breaking dormancy, accelerating carbohydrate transfer, and inhibiting apex shoot dominance. Kinetin (6-furfurylamino purine) maintains protein and nucleic acid synthesis, making cut flowers last for a long time. 6-Benzylamino (BA) purine makes green vegetables last longer after harvest [49 - 51].

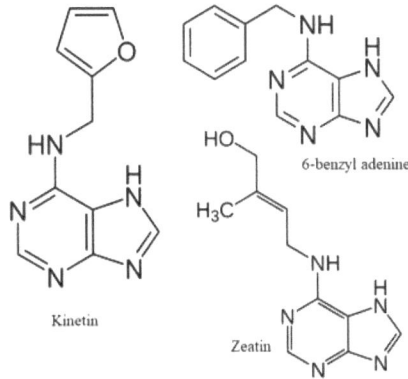

**Fig. (3).** Chemical structure of Kinetin (KIN), 6-benzyl adenine (BA), and zeatin (ZEA).

Although the practical application of cytokinin is not as widespread as auxin, cytokinin is used in plant breeding studies and tissue culture applications (Fig. **4**). The most important organic compounds added to nutrient media in tissue culture studies are auxin and cytokinins. Depending on the balance of these two compounds, root and shoot formation can be controlled [52]. The administration of IAA and Kinetin causes rapid cell division and the cells to remain in a continuous meristematic state. In the case of roughly equal concentrations of auxin and stockin, unorganized new cells continue to form in the callus tissue. Delaying the yellowing (chlorophyll loss) in the leaves is also possible with the application of cytokinin. Synthesis of all cytokinins based on simple amino purines is easy. Purine-based cytokinins are also used to increase fruit sets, to obtain large fruits, and to reduce post-harvest deterioration of cut flowers and vegetables such as lettuce and parsley, whose green parts are edible. It is also used for germination of seeds [53, 54].

## Gibberellins

It was isolated for the first time from Gibberella fujikuroi mushrooms in Japan and was noticed because this fungus causes excessive height growth in rice [56]. Today, about 76 different gibberellins have been isolated from plant species and Gibberella fungus and their properties have been determined. Gibberellin growth regulator (GA), which is used in agricultural production and especially in horticultural crops, is mostly obtained from the Gibberella fungus by fermentation. There are nearly 100 GA series known today, of which more than 50 have been found in plant seeds. However, $GA_3$ (Gibberellic Acid) (Fig. **5**), $GA_5$, $GA_7$, $GA_2$ and $GA_8$, are the most widely used for commercial purposes [57, 58].

**Fig. (4).** *In vitro* shoot formation of *Eucalyptus camaldulensis* Dehnh. on MS medium [43] supplemented with 1.76 μM 6-benzyl adenine (BA), pH 5.8, after four weeks incubation at the growth room conditions (23 ± 2°C, 16 h photoperiod, 36 μmol m$^{-2}$ s$^{-1}$). Bar 1 cm in size [55].

**Fig. (5).** Chemical structure of gibberellic acid (GA$_3$).

They are used in the malting industry to take advantage of production and time, to increase harvest yield in sugarcane, celery and artichoke, to control sex characteristics in cucurbits and to maintain parent lines in hybrid seed production. They are effective substances in the growth and ultimately division of the cell (meristematic formation). For this reason, they are effective in growth and development. They are also used to terminate dormancy in some plants [59, 60].

We can list some functions of gibberellins in plants as follows: (*i*) it provides elongation in genetically stunted plants, reverses the inhibitory effect of red light on stem growth, and promotes stem elongation (Fig. **6**). Since it encourages early flowering in some plant species, flowering, and hybridization can be delayed and the breeding period can be shortened. Since the growth will accelerate with the application of Gibberellin, the probability of the parts to be taken in these regions will be very high [61, 62]. (*ii*) It breaks dormancy in some buds or tubers and increases germination by breaking dormancy in apical seeds [63]. (*iii*) If gibberellin is applied in plants that require long day conditions and chilling, flowering can be achieved even if these conditions are not met [64]. (*iv*) As with auxins, it provides parthenocarpic fruit development in some fruit species (especially effective in species where auxins are not effective). Gibberellin provides seedlessness when given a certain time before flowering, and grain coarsening when given after flowering. It is used to increase grain and cluster size in grapes [65, 66]. (*v*) It promotes germination in light-sensitive seeds and extends the large period of growth, keeping the plants green for a long time. It is very important in floriculture, especially with vegetables whose green parts are edible [67, 68].

**Fig. (6).** *In vitro* shoot elongation of *Thymus longicaulis* C. Presl. on MS medium supplemented with 0.87 µM $GA_3$ and 4.7 µM KIN, pH 5.8, after four weeks of incubation at the growth room conditions ($23 \pm 2°C$, 16 h photoperiod, 36 µmol $m^{-2}s^{-1}$). Bar 1 cm in size [69].

## Ethylene

It has been known for many years that ethylene ($C_2H_4$), a simple compound, is a highly effective plant growth regulator in gas form produced by the plant itself. Ethylene can be produced in all tissues. Under normal conditions, they are gaseous (Fig. **7**), volatile and partially inactive. It is a growth regulator that can be produced at every stage of plant growth and development [70]. It is also known as ripening gas. Due to ethylene secretion in fruits and vegetables in warehouses that are not adequately ventilated, products ripen more quickly, loosen and deteriorate. Bananas, which are plucked before they are ripe, are placed in the same environment with a substance (carbide) that produces ethylene secretion and matured. The synthetically produced names for commercial ethylene are etephone or ethrel [71, 72].

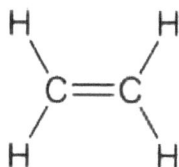

$$\begin{array}{c} H \\ \diagdown \\ C = C \\ \diagup \qquad \diagdown \\ H \qquad\qquad H \end{array}$$

**Fig. (7).** Chemical structure of Ethylene.

The most important functions of ethylene in the plant under natural conditions are as follows: (*i*) To ensure fruit ripening; Under the influence of ethylene, the chlorophyll in the fruit is broken down and turns into pigments, the natural color of the fruit. During ripening, starch, organic acids or oils (as in avocado) are converted into sugars. Although it is generally used in fleshy fruits such as tomatoes and bananas, it has also been used in grapes and walnuts [71, 73]. (*ii*) It promotes aging [74]. (*iii*) Ethylene facilitates separation by increasing the activities of enzymes that provide shedding, making it suitable for machine harvesting. It encourages yellowing of leaves and easy separation of leaves, flowers and fruit stems. In cherries, raspberries, citrus, grapes and mulberries, it provides the loosening of the stems to which the fruits are attached, thus facilitating mechanical harvesting [75, 76]. (*iv*) It stimulates adventitious root formation [77]. (*v*) Due to its regulating effect on blooming, ethylene is used especially in some ornamental plants to provide flower formation at the same time [78]. (*vi*) It is an effective factor in determining the sex of plants, and it is used to direct the sex in plants where male and female organs are in the same individual. The high use of ethylene causes the male flowers to drop and encourages the formation of females [79, 80].

## *Signaling Mechanism of the Ethylene*

Ethylene, which is known as the maturation hormone in plants, differs from other plant hormones because it is a gas slightly lighter than air and well soluble in oil. While it plays a role in many stages of plant life such as seed germination, leaf rupture, fruit rupture, and fruit ripening, it has also been determined that it plays a role in the vegetative response to stress conditions such as low/high temperature, drought, and some chemicals. Although the ethylene synthesis capacity in the cell varies, the synthesis is continuous in almost every cell. Although synthesized ethylene has many precursors in the cell, methionine, as its main precursor, can be easily converted into ethylene in the plant [81, 82].

With the retention of the methanethiolate ($CH_3S$) group released by the cycle, it is converted back to methionine by the plants, and thanks to this cycle mechanism, the amount of ethylene can be kept constant even if the methionine level in the cell decreases. With the retention of the $CH_3S$ group released by the cycle, it is converted back to methionine by the plants, and thanks to this cycle mechanism, the amount of ethylene can be kept constant even if the methionine level in the cell decreases [83, 84].

If we look briefly at the ethylene cycle, S-adenosyl methionine, which is an intermediate, turns into methylthioadenosize form by fermentation and hydrolyzes to form methylthioribozu and then methionine. The $CH_3S$ group of S-denosylmethionine forms methylthioadenosine, while the remaining amount is transformed into 1-aminocyclopropane-1 carboxylic acid. The precursor of ethylene is 1-aminocyclopropane-1 carboxylic acid, which transforms into ethylene when the enzyme's oxygen is present [85, 86]. In addition to auxin that increases rooting and cytokine hormones that prevent rooting, there are different opinions that ethylene may also play a role in adventitious root formation. While one group of researchers explained that ethylene inhibits cytokinins by reducing them and promotes rooting, another group of researchers reported that ethylene inhibits rooting by reducing the amount of auxin available [77, 87].

## Abscisic Acid

Abscisic acid (ABA) is a growth inhibitor that is naturally synthesized in the plant and plays a role in the regulation of plant development. ABA, which has a closed chemical formula of $C_{15}H_{20}O_4$ (Fig. **8**) and a molecular weight of 264.32 g $mol^{-1}$, has a sesquiterpene structure. ABA, which contains asymmetric carbon atoms in its chemical structure, has optical (enantiomer) and geometric isomerism properties [88, 89].

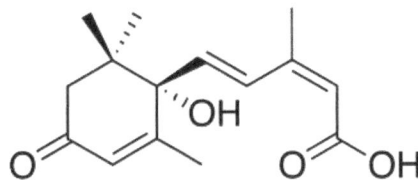

**Fig. (8).** Chemical structure of Abscisic acid (ABA).

It is a natural antagonist of growth promoters such as auxin, gibberellin, and cytokinin. It is found almost everywhere in plants and at all times. It only increases or decreases as environmental conditions change. Accordingly, its effect on physiological events also changes. It is normally found in high amounts in dormant seeds and buds and is thought to have a dormancy-maintaining effect. However, it is also found in leaves, stems, and fruits [90, 91].

Although it is not used commercially, it is sometimes used as a growth inhibitor. Plants send the ABA they produce in their bodies to the regions where there will be leaf fall, and in this way, leaf shedding is observed in autumn. In plant production, ABA can be produced both naturally and synthetically [92]. Its functions in plants are: (*i*) ABA promotes closure of stomata in most plant species. It is known that ABA slows down RNA and protein synthesis and causes the closure of stomata together with $CO_2$ in plants under water stress. Since ABA synthesis increases with water deficiency, its effect on the regulation of stomata during sweating is highly likely [93]. (*ii*) They prevent growth in storage organs such as seeds in annual plants and buds and tubers in biennial and perennial plants. It stimulates the production of storage protein in the grain and is also responsible for inhibiting the early germination of seeds. Breaking dormancy in most seeds is associated with reduced ABA levels in the grain [94, 95].

## Signaling Mechanism of the Abscisic Acid

Under normal conditions, the amount of ABA in plant cells is low and this amount of ABA is used for the normal development of the plant. In addition, ABA triggers the increase of reactive oxygen species (superoxide, hydrogen peroxide, hydroxyl radical) formed as a result of metabolic activity in plant tissues, thus stimulating the antioxidant defense system in the plant. It has been proven by physiological studies that ABA, also known as the stress hormone, plays an active role as a regulator in the response of the plant to stress and increases due to the molecular changes that occur in the plant under abiotic stress conditions, especially drought and salt stress [96, 97].

ABA, which is synthesized in the root under the low water potential that occurs with stress, is transferred to the xylem by the roots and transported to the stomatal cells, where it binds to the ABA receptor, causing the closure of stomata and the stimulation of many related genes. ABA causes an increase in the $Ca^{+2}$ ion in the guard cells in the stomata and the increase plays an active role in the closure of the stomata. Many phospholipid-derived intracellular messengers (such as phospholipase C, and phospholipase D) play a role in the response of ABA to stress. With the activation of these intracellular messengers, secondary messengers (phosphatidic acid) are formed [98, 99].

The production of ABA in the root and its transport mechanism to the leaves and its role in stomal conductivity have been explained in many studies. In some studies, it has been tried to show that ABA reduces transpiration (sweating) using the bioassay approach [100]. It is now known that ABA plays a direct role in guard cell closure. In many plant species, the ABA ratio in the roots varies in relation to soil moisture and water content in the roots. Studies conducted on tobacco (*Nicotiana plumbaginifolia*) under drought have revealed that ABA is completely rooted [101]. Although there are no definite findings about the synthesis site in the root, studies conducted in Pea (*Pisum sativum*) and Asian Sunflower (*Commenlina comunis*) explained that ABA synthesis occurs only at the root tips, while some studies have stated that it occurs at a distance of 3 cm from the root tip to the root tip [102, 103].

It has also been proven that the dominant role of ABA between root and shoot is that the concentration of ABA measured in xylem in drought-stressed plants is lower than the concentrations of exogenous ABA (necessary for the closure of stomata) in the leaf. Moreover; the locations of enzymes associated with ABA biosynthesis (such as aldehyde oxidase) have been studied, but these studies have focused more on ABA production under different nitrogen amounts than drought stress. Although there are many studies and approaches supporting the role of ABA in stomatal control, some substances in xylem (jasmonic acid) have been reported to play a role in drought-induced stomatal closure in accordance with ABA. In the study performed in Arabidopsis, protein phosphatases (ABI1 and ABI2) and transcription factors (ABI3-5) associated with the ABA signaling pathway were isolated [104, 105].

ABA biosynthesis starts with the zeaxanthin epoxidase (ZEP) enzyme in roots and leaves. The key regulatory step in the synthesis is the catalysis of the 9-cis-epoxycarotenoid dioxygenase (NCED) enzyme (Fig. **9**). The NCED enzyme converts the precursor epoxy carotenoid molecule to zeaxanthin (xanthoxin) in plastids, and then it is converted to zeaxanthin ABA by two cytosolic enzymes, accompanied by abscisic aldehyde. In studies on the NCED gene in plants such as tomatoes and corn, it has been observed that the NCED promoter is affected by drought stress in both roots and leaves [106, 107].

In the last stage of ABA biosynthesis, ABA is catalyzed by abscisic aldehyde oxidase enzyme ($AAO_3$). The control of this enzyme is at the transcriptional or post-transcriptional levels, but the precise functions and translational regulation of this regulation are still unclear. It has been proven that both signaling pathways, ABA-dependent and ABA-independent, interact with each other. It has been proven that DRE and ABRE elements can be found simultaneously in the promoters of genes associated with these signaling pathways, and that some genes (such as rd29A) are involved in the ABA-independent signaling pathway in the first hours of stress, while they act in an ABA-dependent role in the later stages of gene expression [108, 109].

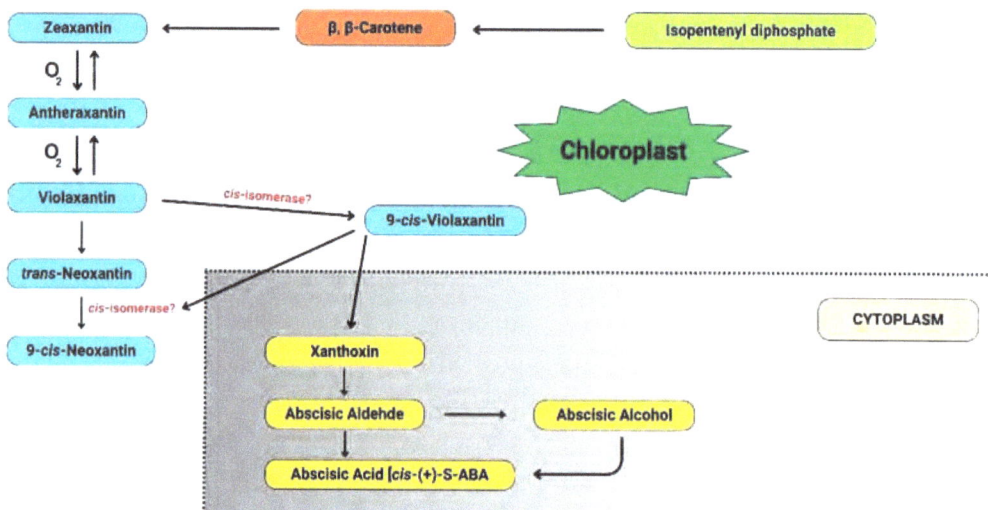

**Fig. (9).** The ABA synthesis pathway begins with zeaxanthin.

## Jasmonic Acid

Jasmonic acid [3-oxo-2-(2'-cis-pentenyl)-cyclopentane-1-acetic-acid] is synthesized from α-linolenic acid. About 20 years ago, the plant growth inhibitory role of jasmonic acid (Fig. **10a**) and its fragrant ester, methyl jasmonate (Fig. **10b**), was determined. It is a fragrant compound obtained from flowers (*e.g.* jasmine, *Jasminum grandiflorum* L. and *Rosmarinus Officinalis* L.) and various fruits. Jasmonic acid has been found in approximately 206 plant species, including ferns, mosses, some fungi, and algae [110]. Recently, it has been determined that methyl jasmonate is important in signaling molecules in plant genes, and it has been found to significantly increase the manifestation of some specific plant genes. It is especially effective in the formation of response genes that occur in case of damage to the plant [111].

Methyl jasmonate has most characteristics of plant growth regulators. Some of the effects seen in plants are: (*i*) It causes yellowing of leaves in plants, and breakage in leaf stems, and prevents growth and development [112]. In studies conducted in tissue culture media, it has been found to inhibit growth and development. Jasmonic acid was isolated from three *Camellia* species, but it was observed that using jasmonic acid alone was more effective in inhibiting the germination of pollen of *C. sinensis* L in the medium. It has been determined that jasmonic acid and methyl jasmonate can strongly inhibit kinetin-like substances that promote the development of soybean callus, even at very low concentrations, and a similar inhibitory effect on callus formation in potato meristem culture [113]. (*ii*) It promotes root formation [112]. (*iii*) If it is sprayed on the potato plant in gaseous state, it increases tuber formation and also promotes tuber formation in tissue culture (*in vitro*) environments [114, 115].

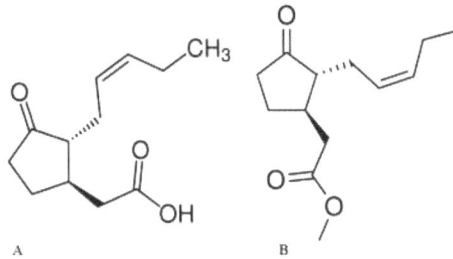

**Fig. (10).** Chemical structure of jasmonic acid (**A**) and methyl jasmonate (**B**).

An extract with tuber formation promoting properties was obtained from potato leaves and old tubers, and it was observed that it promoted tuber formation when added to the *in vitro* environment. Since it is very similar to jasmonic acid in terms of chemical structure, it has been determined that this extract jasmonic acid promotes tuber formation even at very low concentrations such as 7- 10 M. The discovery of this phenomenon has led to important developments in plant physiology as well as its practical importance in the production of mini and micro tubers in potato seeding programs [114, 116 - 118]. (*iv*) It increases ethylene synthesis and thus fruit ripening [119]. Methyl jasmonate also strongly stimulates ethylene production at all stages of fruit ripening. The effects of methyl jasmonate on ripening of tomato fruit are similar to ethylene; Chlorophyll is broken down, and the accumulation of lycopene, which gives the tomato its red color, is stimulated [120, 121]. (*v*) it causes β-carotene synthesis [111]. (*vi*) it causes curls in vine shoots [122]. (*vii*) It has also been reported to inhibit seed germination, callus formation, root growth, chlorophyll production and pollen grain germination. Therefore, it is also effective in photosynthesis [112].

It has been found that externally applied jasmonic acid causes closure of stomata in barley seedlings, reducing water loss by evaporation and thus increasing salinity resistance. The effect of jasmonic acid on photosynthesis was determined not only by the closure of stomata, but also by reducing oxygen transfer [123]. (*viii*) In case of any injury to the plant, it inhibits the effect of enzymes that cause the proteins in the plant (especially the vegetative storage proteins in soybean) to break down. The activities of these genes are also promoted by ABA [111, 115]. (*ix*) It has been determined that externally applied methyl jasmonate contributes to the long-term preservation of stored potatoes and other fruits and vegetables with high water content without losing their quality. While low concentrations prevent germination in storage, high concentrations increase reducing sugar accumulation [124]. Due to these effects, they also affect the final processing quality of potatoes. Numerous studies have shown that jasmonic acid application extends the shelf life of fruits and vegetables up to two weeks. It has been stated that the external application of methyl jasmonate is important for the long-term preservation of quality and other fresh nutritional properties in stored potatoes and other foods with high water content. It has been determined that when applied at low concentrations as a liquid spray or vapor, it limits shoot formation and sugar accumulation [114]. Gaseous spraying has been shown to significantly reduce bacterial density in celery and pepper [124]. (*x*) It has effects on aging. It has been determined by many studies that jasmonic acid accelerates the death event due to increasing respiration in the dark, and is effective in the breakdown of chlorophyll in the leaves and yellowing. Researchers have shown that jasmonic acid is among the death (plant growth inhibitory) hormones and they have determined that it is transferred from the developing seeds and fruits to the leaves, thus promoting yellowing [115].

Some of the effects of jasmonic acid identified so far are that it promotes yellowing of leaves in plants, causes rupture of leaves and flower stalks and shedding of fruits, promotes root formation, increases tuber formation in potatoes, causes curling in vine shoots, and increases β-carotene and ethylene synthesis. In contrast, jasmonic acid has been reported to inhibit seed germination, callus development, root growth, chlorophyll production, and pollen germination [125 - 129]. Numerous proteins have been identified that promote the formation of jasmonic acid in plants. Proteins whose functions are promoted by jasmonic acid and methyl jasmonate are; (*i*) Soybean VDP [130, 131]. (*ii*) Proteinase inhibitory proteins in tomato and potato [132]. (*iii*) Fat skeleton membrane proteins with DDP in rapeseed embryos [133].

It has also been determined that external application of methyl jasmonate promotes the manifestation of VDP genes and protein accumulation in leaf cells [131], while jasmonic acid and ABA are effective in the accumulation of hydroxy-cinnamic acid amides and tryptophan essential amino acids, which are effective in osmotic stress in barley [123]. Methyl jasmonate is also an active component of the pheromone secreted by male fruit moths, which is attractive to females. It has been reported that exogenous jasmonic acid application can affect microorganism growth and development and pest attack, since jasmonic acid is found in α-linolenic acid, which is in the structure of fungi that may or may not cause diseases in plants [111].

## Brassinosteroids

Brassinosteroids (BRs) (Fig. **11**) have various regulatory activities on growth and development in plants [134, 135]. The main effects of BRs on plant development can be listed as cell division and expansion, cellular differentiation, lateral root development, vascular differentiation, pollen tube development, maintenance of apical dominance, flowering, senescence, and increased stress tolerance [136, 137].

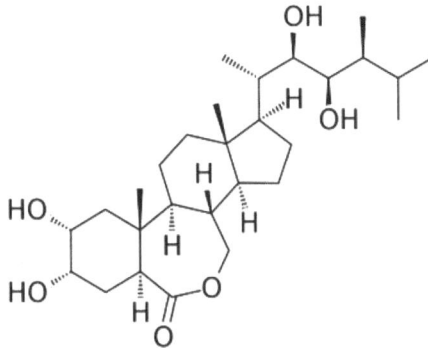

**Fig. (11).** Chemical structure of brassinosteroids.

The most important functions of BRs in the plant under natural conditions are as follows: (*i*) They promote cell expansion and cell division [138]. The growth-promoting effects of BRs occur by accelerating cell elongation and division. This physiological effect of BRs was demonstrated for the first time with the bean second internode experiment [139]. The other experiment is the leaf-bending biological experiment in rice. The bending of the leaf blade in rice was caused by BR-induced cell expansion. Leaf blade bending is a condition similar to epinasty caused by ethylene. However, in this experiment, BRs caused the cells on the upper surface where the rice lamina touched the leaf sheath to elongate more than the cells on the lower surface, thus causing the leaf to bend vertically towards the outer part [140]. Microscopic study of BR-deficient mutants and leaves of wild-type plants showed that BRs are an important class of plant growth hormones in the stem, as the cell size of mutants was smaller than that of wild-type. In addition, the high expression of the DWF4 gene related to BR biosynthesis caused an increase in plant growth. Thus, it was concluded that the most important and early recognition feature of BRs is to promote cell division and elongation [141].

Cell wall relaxation for cell elongation, and water intake into the cell by osmotic transport to maintain turgor pressure and the cell wall thickness are achieved by synthesis. Both are similarly adjusted by BRs. BRs, aquaporins, and vacuolar $H^+$-

ATPase affect water uptake. The Arabidopsis mutant det3 was obtained by mutation in one of the subunits of V-ATPase. The det3 mutant showed a similar phenotype to the det2 mutant that does not express BR, but the det3 mutant was found to be less sensitive to exogenous administration [142]. BRs also increase cell wall relaxation and promote the expression of cell wall-modifying enzymes such as xyloglucan endo-glycosylase/hydroxylase (XTHs) and non-enzymatic proteins found in the cell wall. Cell elongation also occurs with the control of microtubule organization. Microtubule alignment helps organize the cellulose microfibrils during synthesis, and in normal cell elongation, the microfibrils are strung across the cell wall [143].

As a result of the microscopic analysis, it was found that the microtubules of Arabidopsis mutants that do not synthesize BRs were very few and they were arranged parallel to each other. Treatment of the mutant with BRs showed that normal microtubule number and organization were restored. It has been found that BRs do not increase the total amount of tubulin protein in the cell, but play a role in regulating microtubule formation and organization [139]. In addition to cell elongation, BRs also stimulate cell proliferation. Cytokinin promotes cell division depending on the expression of D-type cyclin (CYCD3). Studies have shown that 24-epiBL application also increases the expression of the CYCD3 gene [144]. (*ii*) BRs both inhibit and promote root growth: mutants that do not express BR have typically been found to have reduced root growth and BRs are essential for normal root elongation. However, similar to auxin, exogenously applied BRs can affect root growth negatively or positively depending on the concentration. When exogenous BR treatment was applied to mutants not expressing BR, it was observed that root growth increased at low concentrations and inhibited at high concentrations. The concentration limit for inhibition varies depending on the activity of the BR analog used. As a result, it has been revealed that BRs have strong effects on root morphology such as root elongation and branching [145]. (*iii*) BRs promote xylem differentiation throughout vascular development: BRs play an important role in vascular development by promoting xylem and suppressing phloem differentiation. Evidence for this was demonstrated by the disruption of vascular systems in det2 mutants, with a high phloem/xylem ratio, unlike wild-type plants. In mutants not expressing BR, a decrease in the number of vascular bundles and irregular gaps between bundles were observed. On the other hand, it was found that more xylem production occurred in mutant plants as a result of overexpression of the BR receptor protein compared to wild-type plants. However, the fact that BR receptor-protein-like encoding genes are only expressed in vascular tissues supports that the BR signal transduction pathway is specific for the vascular system [146]. (*iv*) BRs are necessary for pollen tube development. Pollens are rich sources of BRs, demonstrating the importance of BRs in male fertility. BRs promote pollen tube development (from the stigma to the embryo sac). For example, in a study with the Arabidopsis cpd mutant that does not express BR, it has been shown that post-germination pollen tube elongation on the stigma fails and pollen tube elongation is partially dependent on BR application. Similarly, in the study with BR-non-responsive mutants, the mutants were self-pollinated, pollen tube development failed, and ultimately no seeds were obtained. However, fertile seeds were produced by hand pollination of mutants and wild-type plants. Thus, it was concluded that both BR and BR signaling are required for normal pollen tube development [147].

The stimulating effects of BRs on development occur especially in young and developing body tissues. The kinetics of cell expansion, which is the effect of brassinolide at nanomolar concentrations, differs from that of auxin. For example, brassinolide was applied to epicotyl parts of soybean and the elongation rate increased by brassinolide started after 45 minutes, and the maximum rate was reached a few hours after the application. In contrast, auxin began to promote elongation after 15 minutes and reached its maximum rate within 45 minutes. The researchers thought that while BRs carry out growth with a slower pathway within the scope of gene transcription, auxin may be faster, perhaps without causing gene transcription, or that the auxin-induced stimulatory gene expression may be higher than the gene expression induced by brassinolide. In fact, auxin and BRs increase body development synergistically and independently of each other, and both hormones are needed for optimum activity. Moreover, auxins and BRs are associated with each other in many of the same developmental processes, including vascular differentiation, flower, fruit, and root development [148, 149]. It was found that auxin increased BR biosynthesis in Arabidopsis by promoting the expression of the DWARF4 gene. In their previous studies, the researchers found that among the BR biosynthesis enzymes in Arabidopsis, DWARF4 is the enzyme that determines the rate of BR synthesized synthesis. In these studies, GUS (DWF4pro: GUS) was inserted into the Arabidopsis DWF4 gene as a promoter. Auxin was applied to DWF4pro: GUS plant at different doses and time intervals, and as a result of the study, it was determined that the biosynthesis of BR was increased by auxin. Another analysis performed in the study was microarray analysis, and at the end of this study, they revealed that auxin regulates BR biosynthesis in Arabidopsis and that the growth-promoting effects of BRs are maintained by auxins [150, 151].

The effects of BRs on root growth are independent of the effects of auxin and gibberellin. In the study, an inhibitor (2,3,5-triiodobenzoic acid-TIBA) that inhibits polar auxin transport was used, but root growth promoted by BR could not be prevented. When BR and auxins are applied at the same time, both should add supportive and inhibitory effects on root growth. On the other hand, mutants with reduced root growth not expressing BR could not be normalized by gibberellin application. Studies have shown that BRs inhibit root growth but do not interact with auxin or gibberellin. However, high concentrations of BR, such as auxin, stimulate ethylene production, thus making it possible for several of the inhibitory effects of BR on root growth to be mediated by ethylene. BRs can promote lateral root formation at low concentrations. In this case, BRs and auxin play a synergistic role. In the model supported today, BRs increase lateral root development and some of this is due to the effect of polar auxin transport [145].

Seeds, like pollen grains, contain high levels of brassinosteroids, thereby promoting seed germination. Although the molecular basis of the interactions is not known, it is known that this event occurs as a result of interaction with other plant hormones [152]. It has been found that gibberellic acid has positive effects on seed germination and abscisic acid has negative effects on seed germination. In the study, it was found that BRs increased germination in tobacco independent of gibberellic acid signaling. In addition, BRs abolished delayed germination in gibberellic acid mutants and BR mutants were more sensitive to inhibition by ABA than wild-type plants. Thus, it has been demonstrated that BRs are necessary to overcome the inhibitory effect of ABA and promote germination [153].

The roles of plant growth regulators (ABA, ethylene, jasmonic acid, and salicylic acid) in the adaptation of the plant to biotic and abiotic stresses have been defined. While ethylene, jasmonic acid, and salicylic acid are important in defense against pathogens and insect attacks, ABA has been found to be the key hormone in drought and salt stress. ABA, ethylene, jasmonic acid, and salicylic acid are growth regulators associated with heat stress, and as a result of studies with mutants, it has been revealed that ethylene, salicylic acid, and ABA are effective in gaining heat tolerance to plants. It has been stated that ABA inhibits the effects of BRs in the plant's response to stress, but there may be a mutual signal exchange in the signaling pathway of BRs, ethylene, salicylic acid and ABA in the formation of the opposite response in the stress environment [154, 155].

## OTHER PLANT GROWTH REGULATORS AND GROWTH INHIBITORS

Some compounds with positive or opposite effects, naturally occurring in many plants or given externally, can reduce or increase plant growth as part of a normal defense mechanism. These substances such as, tuberonic acid, salicylates, chlormequat chlorur and daminozid, ancymidol, maleic hydrazide, phosphon-do and Amo-1618, and Paclobutrazol can accelerate growth, development, flower and fruit formation by inhibiting or promoting, and accelerating the physiological events that develop depending on them.

## Tuberonic Acid

Obtained from potato leaves and aged potato tubers, this compound (Fig. **12**) is chemically similar to jasmonic acid. Since it is known to contribute to tuber accumulation in the potato plant, tuberonic acid is used to promote tuber production *in vitro* [156]. A compound called tuberonic acid, called "aglycone," has been shown to promote tuber formation in single-nosed stem shoots when added to an agar medium on which shoots develop [157].

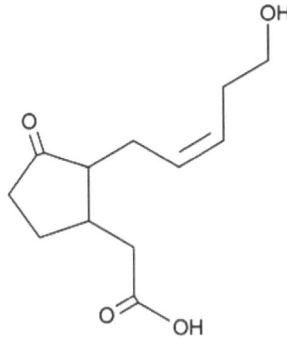

**Fig. (12).** Chemical structure of tuberonic acid.

## Salicylates

It is a plant phenol compound (Fig. **13**) that shows the same activity as salicylic acid. Although it is known to occur naturally in many plants, it has been isolated from around 34 plant species. It has been determined that it promotes flowering in plants, increases heat production in thermogenic plants, and therefore increases resistance to pests [158]. Lopez-Delgado and Scott [159] determined that by adding acetylsalicylic acid (Aspirin=ASA) to the medium where potato microplants are present, 100% tuber formation was obtained in these environments.

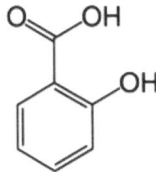

**Fig. (13).** Chemical structure of salicylates.

## Chlormequat Chloride

Chlormequat chloride (Fig. **14**) is a plant growth retardant and also a low-toxic plant growth regulator. It can be found in plant seeds and absorbed through leaves, branches, buds, and root systems. By suppressing the plant's node development, they can control its overgrowth and increase its resistance. Because it also increases the chlorophyll content, it can make leaves greener and thicker. Thus, photosynthesis will be strengthened and it will increase the quality and yield of the fruit. This growth regulator can also improve the environmental compatibility of plants, such as increasing drought, frost, disease, pest, and salinization resistances [160, 161]. In a study on soybeans, it was determined that it increased the number of beans per stem, flower bud formation, the number of grains in the pod and the total beans yield [162]. It is also used to inhibit the development of potted ornamental plants such as azalea, geranium and poinsettia, to encourage early flowering, to increase flowering, and to increase the number of buds and flowers per plant. In addition, it is applied to increase fruit set in grapes, and anthocyanin synthesis in apples, reduce fruit shedding before harvest and storage damage. It is also used in the fruit industry to reduce vegetative growth, change fruit shape and color, and improve fruit quality [163].

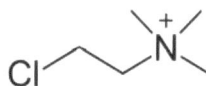

**Fig. (14).** Chemical structure of chlormequat chloride.

## Ancymidol

Ancymidol (Fig. **15**) is a plant growth inhibitor used to reduce internode length in plants [164]. It is especially applied in the cultivation of ornamental plants. Ancymidol has a growth retarding effect and reduces the gibberellic acid content in the plant. The effect of ancymidol on higher plants can be prevented by the simultaneous application of gibberellic acid, but this is only effective at low (100 µM) concentrations of ancymidol [165]. It has been reported that Ancymidol has an inhibitory effect on cellulose synthesis in plants, and inhibits cell growth and new cell wall formation in high-dose applications [166].

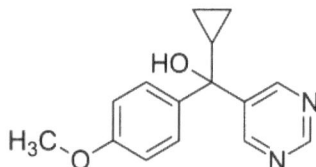

**Fig. (15).** Chemical structure of Ancymidol.

## Maleic hydrazide

Maleic Hydrazide (Fig. **16**) is a plant growth inhibitor that inhibits cell division and bud formation in woody plants [167]. It is used to control germination in onions and tubers. The synthetic compound is applied to the leaves before harvest and is rapidly transported to the storage organs. Its use has been limited to lawn crops, as it often causes much damage [168].

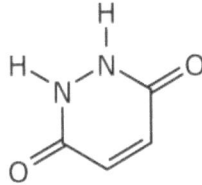

**Fig. (16).** Chemical structure of maleic hydrazide.

## Phosphon-D and Amo-1618

The natural growth regulator gibberellin, which is known to be responsible for longitudinal growth in plants, is thought to control mitotic activity in the subapical meristem area [169]. Artificial growth inhibitors such as Phosphon-D (tributyl-2, 4-dichlorobenzylphosphonium chloride) (Fig. **17A**) and Amo-1618 (2-isopropyl -5-methyl-4-trimethylammonium chloride) (Fig. **17B**), on the other hand, affect the synthesis of gibberellin in the plants to which they are applied, reduce the level of endogenous gibberellin, meristematic activity and cell expansion, and strongly inhibit longitudinal growth. For example, as a result of the treatment of chrysanthemum stems with Amo-1618, it was determined that the mitotic activities of the plants decreased and they showed limited height growth [170].

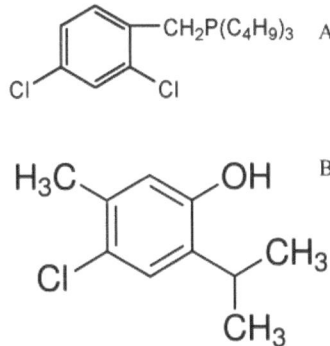

**Fig. (17).** Chemical structure of Phosphon-D (**A**) and Amo-1618 (**B**).

## Paclobutrazol

The same plant growth regulator may give different responses in different tissues of a plant or may be effective in different development stages of the same tissue. Some plant growth regulators have a stimulating effect on plants, while others have an inhibitory effect. Paclobutrazol, [(2RS, 3RS)-1-(4-chlorophenyl)- 4, 4-

dimethyl-2-(1H-1, 2, 4-trizol-1-yl)-pentan-3-ol], (Fig. **18**) is a synthetic plant growth inhibitor that decreases vegetative growth and increases generative growth. It is used to prevent excessive vegetative growth especially in fruit growing and ornamental plants [171]. Previously, plants and standard fruits are obtained. It has been noted that when Paclobutrazol, which inhibits Gibberellin biosynthesis, is applied in some plants, the plant becomes resistant to cold. While paclobutrazol controls vegetative growth, it can also have positive effects on fruit yield and quality. For example, in apple, cherry and peach, increased fruit yield was observed in Paclobutrazol applications by spraying from the soil and leaves compared to the control [172, 173].

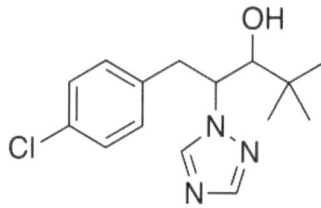

**Fig. (18).** Chemical structure of Paclobutrazol.

## Strigolactones

Plants produce signaling molecules called strigolactones. They serve two primary purposes: as endogenous hormones that regulate plant growth, and as constituents of root exudates that facilitate symbiotic relationships between soil microorganisms and plants. Certain plants that feed on other plants have developed a third purpose: when their seeds are near the roots of an appropriate host plant, they encourage the germination of their seeds. This third role was responsible for the initial identification and discovery of strigolactones [174 - 176].

Strigolactones have an impact on various facets of plant growth. This was first noticed in 2008 when it was found that a strigolactone is actually the chemical signal that was previously unknown and that was transferred from roots to shoots to inhibit the growth of secondary branches. The fact that plant mutants deficient in strigolactone synthesis generated a large number of secondary branches, which could be stopped from growing by applying the synthetic strigolactone GR24, provided the proof. It has since been found that strigolactones cause the stem to thicken secondary times and can encourage the growth of lateral roots and root hairs [177 - 179].

## COCLUSION

In order for plants to grow and develop continuously, it is imperative that they regularly take in water and minerals from the soil and gases such as $CO_2$ and $O_2$ from the air. As vegetation progresses, the plant grows, develops and some cells, tissues and organs are formed and take a unique shape. In order for normal growth and development to occur, a number of internal and external factors play a role together [3, 10]. Plants need chemical communication between cells in order to grow and develop. Plant growth regulators (PGRs) are the basic molecules that provide cell-to-cell communication in plants. Plants produce these basic substances they need in order to continue their growth and development. In general, organic molecules that are naturally synthesized in plants, that control growth and other related physiological events, that can be effective in the regions where they are carried to other parts of the plants, and that can show their effect even at very low concentrations are called hormones (phytohormone-plant growth regulators). These substances are obtained from various organs of higher plants and some fungi [15, 17]. Most of the physiological events that take place in plants are controlled by plant growth regulators.

Plant growth regulators come in two forms, natural and synthetic. While natural hormones are synthesized by the plant, synthetic hormones are substances with different structures developed by the chemical industry. Synthetic hormones have similar effects with natural hormones, and in some cases they may have more effects [13, 15].

The use of plant growth regulators in practice can be summarized as follows; accelerate maturation, provide cuttings, increase the germination power of seeds, encourage or delay flowering, increase cold resistance, seed formation in fruits, and fruit size, extend fruit storage period, increase plant resistance to diseases and pests, provide weed control, pre-harvest to prevent fruit shedding, facilitate machine harvesting, break dormancy, and encourage root-shoot and tuber formation in tissue culture studies [1, 10, 13].

The importance of growth regulators in plant production is quite significant. With the use of growth regulators, plant development can be directed as desired and successful results can be obtained. For a quality production, growth regulators should be used consciously and there should be no consequences that would pose a risk to health. Growth regulators, which can be effective even at very low doses, are being studied in the same group as pesticides. Although the effects of plant growth regulators on the environment are so low that they cannot be compared with pesticides, studies have shown that adverse effects can occur as a result of misuse. Knowing the properties and transport routes of the hormones synthesized by plants provides information on how to use growth regulators. Likewise, starting from the information about phytohormones, the internal hormone concentrations can be increased at certain points or in the whole plant by applications such as branch angle expansion, wounding, bending, use of microorganisms, etiolation and low gamma radiation without using synthetic growth regulators. In addition, with such applications, the effects of plant growth regulators can be increased or it may be necessary to use them at lower levels. Therefore, knowing the applications that increase the internal hormone biosynthesis and putting these applications into use in practice will contribute to a more successful production.

It has been proven by studies in the fields of plant physiology that the roles of PGRs in growth and development are very important. However, the effects of some of them on metabolic processes still need to be clarified. In this context, technological developments and molecular-level approaches such as transcriptomics-proteomics will provide great benefits in fully elucidating these functions in the future and this will be a guide for more effective use of these molecules.

## REFERENCES

[1]     George EF, Hall MA, Klerk GJD. Plant Growth Regulators I: Introduction; Auxins, their Analogues and Inhibitors. In: George EF, Hall MA, Klerk GJD, Eds. Plant Propagation by Tissue Culture. Dordrecht: Springer 2008; pp. 175-204.
        [http://dx.doi.org/10.1007/978-1-4020-5005-3_5]

[2]     Singh A, Kumar J, Kumar P. Effects of plant growth regulators and sucrose on post harvest physiology, membrane stability and vase life of cut spikes of gladiolus. Plant Growth Regul 2008; 55(3): 221-9.
        [http://dx.doi.org/10.1007/s10725-008-9278-3]

[3]     Rademacher W. Plant growth regulators: backgrounds and uses in plant production. J Plant Growth Regul 2015; 34(4): 845-72.
        [http://dx.doi.org/10.1007/s00344-015-9541-6]

[4]     Went FW. Over stoffen, die den groei in het coleoptiel van Avena sativa versnellen. Kon Akad Wet Amst 1926; 35: 723.

[5]     Van Overbeek J. Auxins. Bot Rev 1959; 25(2): 269-350. Available from: https://www.jstor.org/stable/4353596
        [http://dx.doi.org/10.1007/BF02860041]

[6]     Sun S, Zhou X, Cui X, *et al.* Exogenous plant growth regulators improved phytoextraction efficiency by *Amaranths hypochondriacus* L. in cadmium contaminated soil. Plant Growth Regul 2020; 90(1): 29-40.
        [http://dx.doi.org/10.1007/s10725-019-00548-5]

[7]     Jiménez VM. Involvement of plant hormones and plant growth regulators on *in vitro* somatic embryogenesis. Plant Growth Regul 2005; 47(2-3): 91-110.
        [http://dx.doi.org/10.1007/s10725-005-3478-x]

[8]     Meng H, Hua S, Shamsi IH, Jilani G, Li Y, Jiang L. Cadmium-induced stress on the seed germination and seedling growth of *Brassica napus* L., and its alleviation through exogenous plant growth

regulators. Plant Growth Regul 2009; 58(1): 47-59.
[http://dx.doi.org/10.1007/s10725-008-9351-y]

[9]    Egamberdieva D. Alleviation of salt stress by plant growth regulators and IAA producing bacteria in wheat. Acta Physiol Plant 2009; 31(4): 861-4.
[http://dx.doi.org/10.1007/s11738-009-0297-0]

[10]    Li W, Liu X, Ajmal Khan M, Yamaguchi S. The effect of plant growth regulators, nitric oxide, nitrate, nitrite and light on the germination of dimorphic seeds of *Suaeda salsa* under saline conditions. J Plant Res 2005; 118(3): 207-14.
[http://dx.doi.org/10.1007/s10265-005-0212-8] [PMID: 15937723]

[11]    Shahbaz M, Ashraf M, Athar H-R. Does exogenous application of 24-epibrassinolide ameliorate salt induced growth inhibition in wheat (*Triticum aestivum* L.)? Plant Growth Regul 2008; 55(1): 51-64.
[http://dx.doi.org/10.1007/s10725-008-9262-y]

[12]    de Freitas ST, Martinelli F, Feng B, Reitz NF, Mitcham EJ. Transcriptome approach to understand the potential mechanisms inhibiting or triggering blossom-end rot development in tomato fruit in response to plant growth regulators. J Plant Growth Regul 2018; 37(1): 183-98.
[http://dx.doi.org/10.1007/s00344-017-9718-2]

[13]    Dawar K, Rahman U, Alam SS, *et al.* Nitrification inhibitor and plant growth regulators improve wheat yield and nitrogen use efficiency. J Plant Growth Regul 2022; 41(1): 216-26.
[http://dx.doi.org/10.1007/s00344-020-10295-x]

[14]    Gemrotová M, Kulkarni MG, Stirk WA, Strnad M, Van Staden J, Spíchal L. Seedlings of medicinal plants treated with either a cytokinin antagonist (PI-55) or an inhibitor of cytokinin degradation (INCYDE) are protected against the negative effects of cadmium. Plant Growth Regul 2013; 71(2): 137-45.
[http://dx.doi.org/10.1007/s10725-013-9813-8]

[15]    Kaur A, Kaur N, Singh H, Murria S, Jawanda SK. Efficacy of plant growth regulators and mineral nutrients on fruit drop and quality attributes of plum cv. Satluj purple. Plant Physiology Reports 2021; 26(3): 541-7.
[http://dx.doi.org/10.1007/s40502-021-00609-w]

[16]    Kermode AR. Role of abscisic acid in seed dormancy. J Plant Growth Regul 2005; 24(4): 319-44.
[http://dx.doi.org/10.1007/s00344-005-0110-2]

[17]    Malladi A, Burns JK. Communication by plant growth regulators in roots and shoots of horticultural crops. Hortic Sci (Prague) 2007; 42(5): 1113-7. Available from: https://journals.ashs.org/hortsci/view/journals/hortsci/42/5/article-p1113.xml

[18]    Wally OSD, Critchley AT, Hiltz D, *et al.* Regulation of phytohormone biosynthesis and accumulation in Arabidopsis following treatment with commercial extract from the marine macroalga *Ascophyllum nodosum.* J Plant Growth Regul 2013; 32(2): 324-39.
[http://dx.doi.org/10.1007/s00344-012-9301-9]

[19]    Vlahoviček-Kahlina K, Jurić S, Marijan M, *et al.* Synthesis, characterization, and encapsulation of novel plant growth regulators (PGRs) in biopolymer matrices. Int J Mol Sci 2021; 22(4): 1847.
[http://dx.doi.org/10.3390/ijms22041847] [PMID: 33673329]

[20]    Upreti KK, Sharma M. Role of Plant Growth Regulators in Abiotic Stress Tolerance. In: Rao N, Shivashankara K, Laxman R, Eds. Abiotic Stress Physiology of Horticultural Crops. New Delhi: Springer 2016; pp. 19-46.
[http://dx.doi.org/10.1007/978-81-322-2725-0_2]

[21]    Jan S, Singh R, Bhardwaj R, *et al.* Plant growth regulators: a sustainable approach to combat pesticide toxicity. 3 Biotech. 2020; 10: 466.
[http://dx.doi.org/10.1007/s13205-020-02454-4]

[22]    Shi X, Jin F, Huang Y, *et al.* Simultaneous determination of five plant growth regulators in fruits by

modified quick, easy, cheap, effective, rugged, and safe (QuEChERS) extraction and liquid chromatography-tandem mass spectrometry. J Agric Food Chem 2012; 60(1): 60-5.
[http://dx.doi.org/10.1021/jf204183d] [PMID: 22148585]

[23]     Yan H, Yang Z, Chen S, Wu J. Exploration and development of artificially synthesized plant growth regulators. Advanced Agrochem 2024; 3(1): 47-56.
[http://dx.doi.org/10.1016/j.aac.2023.07.008]

[24]     Gross D, Parthier B. Novel natural substances acting in plant growth regulation. J Plant Growth Regul 1994; 13(2): 93-114.
[http://dx.doi.org/10.1007/BF00210953]

[25]     Sauer M, Robert S, Kleine-Vehn J. Auxin: simply complicated. J Exp Bot 2013; 64(9): 2565-77.
[http://dx.doi.org/10.1093/jxb/ert139] [PMID: 23669571]

[26]     Gomes GLB, Scortecci KC. Auxin and its role in plant development: structure, signalling, regulation and response mechanisms. Plant Biol 2021; 23(6): 894-904.
[http://dx.doi.org/10.1111/plb.13303] [PMID: 34396657]

[27]     Lavy M, Estelle M. Mechanisms of auxin signaling. Development 2016; 143(18): 3226-9.
[http://dx.doi.org/10.1242/dev.131870] [PMID: 27624827]

[28]     Perrot-Rechenmann C, Napier RM. Auxins. Vitam Horm 2005; 72: 203-33.
[http://dx.doi.org/10.1016/S0083-6729(04)72006-3] [PMID: 16492472]

[29]     Majda M, Robert S. The role of auxin in cell wall expansion. Int J Mol Sci 2018; 19(4): 951.
[http://dx.doi.org/10.3390/ijms19040951] [PMID: 29565829]

[30]     Mockaitis K, Estelle M. Auxin receptors and plant development: a new signaling paradigm. Annu Rev Cell Dev Biol 2008; 24(1): 55-80.
[http://dx.doi.org/10.1146/annurev.cellbio.23.090506.123214] [PMID: 18631113]

[31]     Naser V, Shani E. Auxin response under osmotic stress. Plant Mol Biol 2016; 91(6): 661-72.
[http://dx.doi.org/10.1007/s11103-016-0476-5] [PMID: 27052306]

[32]     Kalve S, Sizani BL, Markakis MN, *et al.* Osmotic stress inhibits leaf growth of *Arabidopsis thaliana* by enhancing ARF-mediated auxin responses. New Phytol 2020; 226(6): 1766-80.
[http://dx.doi.org/10.1111/nph.16490] [PMID: 32077108]

[33]     Agulló-Antón MÁ, Sánchez-Bravo J, Acosta M, Druege U. Auxins or sugars: What makes the difference in the adventitious rooting of stored carnation cuttings? J Plant Growth Regul 2011; 30(1): 100-13.
[http://dx.doi.org/10.1007/s00344-010-9174-8]

[34]     Guan L, Tayengwa R, Cheng ZM, Peer WA, Murphy AS, Zhao M. Auxin regulates adventitious root formation in tomato cuttings. BMC Plant Biol 2019; 19(1): 435.
[http://dx.doi.org/10.1186/s12870-019-2002-9] [PMID: 31638898]

[35]     Serrani JC, Fos M, Atarés A, García-Martínez JL. Effect of gibberellin and auxin on parthenocarpic fruit growth induction in the cv micro-tom of tomato. J Plant Growth Regul 2007; 26(3): 211-21.
[http://dx.doi.org/10.1007/s00344-007-9014-7]

[36]     Molesini B, Rotino GL, Spena A, Pandolfini T. Expression profile analysis of early fruit development in iaaM-parthenocarpic tomato plants. BMC Res Notes 2009; 2(1): 143.
[http://dx.doi.org/10.1186/1756-0500-2-143] [PMID: 19619340]

[37]     Dal Cin V, Velasco R, Ramina A. Dominance induction of fruitlet shedding in Malus × domestica (L. Borkh): molecular changes associated with polar auxin transport. BMC Plant Biol 2009; 9(1): 139.
[http://dx.doi.org/10.1186/1471-2229-9-139] [PMID: 19941659]

[38]     Basu MM, González-Carranza ZH, Azam-Ali S, Tang S, Shahid AA, Roberts JA. The manipulation of auxin in the abscission zone cells of Arabidopsis flowers reveals that indoleacetic acid signaling is a prerequisite for organ shedding. Plant Physiol 2013; 162(1): 96-106.

[http://dx.doi.org/10.1104/pp.113.216234] [PMID: 23509178]

[39]   Aloni R, Aloni E, Langhans M, Ullrich CI. Role of cytokinin and auxin in shaping root architecture: regulating vascular differentiation, lateral root initiation, root apical dominance and root gravitropism. Ann Bot (Lond) 2006; 97(5): 883-93.
[http://dx.doi.org/10.1093/aob/mcl027] [PMID: 16473866]

[40]   Balla J, Medveďová Z, Kalousek P, *et al.* Auxin flow-mediated competition between axillary buds to restore apical dominance. Sci Rep 2016; 6(1): 35955.
[http://dx.doi.org/10.1038/srep35955] [PMID: 27824063]

[41]   Sobiech Ł, Joniec A, Loryś B, Rogulski J, Grzanka M, Idziak R. Autumn application of synthetic auxin herbicide for weed control in cereals in Poland and Germany. Agriculture 2022; 13(1): 32.
[http://dx.doi.org/10.3390/agriculture13010032]

[42]   Xu J, Liu X, Napier R, Dong L, Li J. Mode of action of a novel synthetic auxin herbicide halauxifen-methyl. Agronomy (Basel) 2022; 12(7): 1659.
[http://dx.doi.org/10.3390/agronomy12071659]

[43]   Murashige T, Skoog F. A revised medium for rapid growth bio assays with tobacco tissue cultures. Physiol Plant 1962; 15(3): 473-97.
[http://dx.doi.org/10.1111/j.1399-3054.1962.tb08052.x]

[44]   Ozudogru EA, Kaya E. Cryopreservation of *Thymus cariensis* and *T. vulgaris* shoot tips: comparison of three vitrification-based methods. Cryo Lett 2012; 33(5): 363-75.
[PMID: 23224369]

[45]   Te-chato S, Hilae A, In-peuy K. Effects of cytokinin types and concentrations on growth and development of cell suspension culture of oil palm. J Agric Sci Technol 2008; 4(2): 157-63.

[46]   Prasad R. Cytokinin and its key role to enrich the plant nutrients and growth under adverse conditions-an update. Front Genet 2022; 13: 883924.
[http://dx.doi.org/10.3389/fgene.2022.883924] [PMID: 35795201]

[47]   Barciszewski J, Massino F, Clark BFC. Kinetin—A multiactive molecule. Int J Biol Macromol 2007; 40(3): 182-92.
[http://dx.doi.org/10.1016/j.ijbiomac.2006.06.024] [PMID: 16899291]

[48]   Schäfer M, Brütting C, Meza-Canales ID, *et al.* The role of *cis* -zeatin-type cytokinins in plant growth regulation and mediating responses to environmental interactions. J Exp Bot 2015; 66(16): 4873-84.
[http://dx.doi.org/10.1093/jxb/erv214] [PMID: 25998904]

[49]   Werner T, Schmülling T. Cytokinin action in plant development. Curr Opin Plant Biol 2009; 12(5): 527-38.
[http://dx.doi.org/10.1016/j.pbi.2009.07.002] [PMID: 19740698]

[50]   Kieber JJ, Schaller GE. Cytokinin signaling in plant development. Development 2018; 145(4): dev149344.
[http://dx.doi.org/10.1242/dev.149344] [PMID: 29487105]

[51]   Werner T, Motyka V, Strnad M, Schmülling T. Regulation of plant growth by cytokinin. Proc Natl Acad Sci USA 2001; 98(18): 10487-92.
[http://dx.doi.org/10.1073/pnas.171304098] [PMID: 11504909]

[52]   Jing H, Strader LC. Interplay of auxin and cytokinin in lateral root development. Int J Mol Sci 2019; 20(3): 486.
[http://dx.doi.org/10.3390/ijms20030486] [PMID: 30678102]

[53]   Danilova MN, Kudryakova NV, Doroshenko AS, Daminova AG, Oelmüller R, Kusnetsov VV. Versatile effect of cytokinin on detached senescing leaves of Arabidopsis in the light. Plant Growth Regul 2023; 99(2): 313-22.
[http://dx.doi.org/10.1007/s10725-022-00909-7]

[54]     Zhang H, Zhou C. Signal transduction in leaf senescence. Plant Mol Biol 2013; 82(6): 539-45.
         [http://dx.doi.org/10.1007/s11103-012-9980-4] [PMID: 23096425]

[55]     Kaya E, Alves A, Rodrigues L, *et al.* Cryopreservation of eucalyptus genetic resources. Cryo Lett
         2013; 34(6): 608-18.
         [PMID: 24441371]

[56]     Hori S. Some observations on 'Bakanae' disease of the rice plant. Mem Agric Res Sta (Tokyo) 1898;
         12: 110-9.

[57]     Hedden P, Sponsel V. A century of gibberellin research. J Plant Growth Regul 2015; 34(4): 740-60.
         [http://dx.doi.org/10.1007/s00344-015-9546-1] [PMID: 26523085]

[58]     Yamaguchi S. Gibberellin metabolism and its regulation. Annu Rev Plant Biol 2008; 59(1): 225-51.
         [http://dx.doi.org/10.1146/annurev.arplant.59.032607.092804] [PMID: 18173378]

[59]     Castro-Camba R, Sánchez C, Vidal N, Vielba JM. Plant development and crop yield: the role of
         gibberellins. Plants 2022; 11(19): 2650.
         [http://dx.doi.org/10.3390/plants11192650] [PMID: 36235516]

[60]     Chen SY, Kuo SR, Chien CT. Roles of gibberellins and abscisic acid in dormancy and germination of
         red bayberry (*Myrica rubra*) seeds. Tree Physiol 2008; 28(9): 1431-9.
         [http://dx.doi.org/10.1093/treephys/28.9.1431] [PMID: 18595855]

[61]     Zhang N, Xie YD, Guo HJ, *et al.* Gibberellins regulate the stem elongation rate without affecting the
         mature plant height of a quick development mutant of winter wheat (*Triticum aestivum* L.). Plant
         Physiol Biochem 2016; 107: 228-36.
         [http://dx.doi.org/10.1016/j.plaphy.2016.06.008] [PMID: 27317908]

[62]     Yamaguchi N, Winter CM, Wu MF, *et al.* Gibberellin acts positively then negatively to control onset
         of flower formation in Arabidopsis. Science 2014; 344(6184): 638-41.
         [http://dx.doi.org/10.1126/science.1250498] [PMID: 24812402]

[63]     Deligios PA, Rapposelli E, Mameli MG, Baghino L, Mallica GM, Ledda L. Effects of physical,
         mechanical and hormonal treatments of seed-tubers on bud dormancy and plant productivity.
         Agronomy (Basel) 2019; 10(1): 33.
         [http://dx.doi.org/10.3390/agronomy10010033]

[64]     Skalicky M, Kubes J, Vachova P, Hajihashemi S, Martinkova J, Hejnak V. Effect of gibberellic acid
         on growing-point development of non-vernalized wheat plants under long-day conditions. Plants 2020;
         9(12): 1735.
         [http://dx.doi.org/10.3390/plants9121735] [PMID: 33316881]

[65]     Sharif R, Su L, Chen X, Qi X. Hormonal interactions underlying parthenocarpic fruit formation in
         horticultural crops. Hortic Res 2022; 9: uhab024.
         [http://dx.doi.org/10.1093/hr/uhab024]

[66]     Wang X, Zhao M, Wu W, Korir NK, Qian Y, Wang Z. Comparative transcriptome analysis of berry-
         sizing effects of gibberellin (GA3) on seedless *Vitis vinifera* L. Genes Genomics 2017; 39(5): 493-507.
         [http://dx.doi.org/10.1007/s13258-016-0500-9]

[67]     Chen SY, Kuo SR, Chien CT. Roles of gibberellins and abscisic acid in dormancy and germination of
         red bayberry (Myrica rubra) seeds. Tree Physiol. 2008; 28(9).

[68]     Rademacher W. Inhibitors of Gibberellin Biosynthesis: Applications in Agriculture and Horticulture.
         In: Takahashi N, Phinney BO, MacMillan J, Eds. Gibberellins. New York: Springer 1991; pp. 296-
         310.
         [http://dx.doi.org/10.1007/978-1-4612-3002-1_29]

[69]     Ozudogru EA, Kaya E, Kirdok E, Issever-Ozturk S. *In vitro* propagation from young and mature
         explants of thyme (*Thymus vulgaris* and *T. longicaulis*) resulting in genetically stable shoots. *In Vitro*
         Cell. Dev Biol Plant 2011; 47(2): 309-20.

[http://dx.doi.org/10.1007/s11627-011-9347-6]

[70]    Wang F, Cui X, Sun Y, Dong CH. Ethylene signaling and regulation in plant growth and stress responses. Plant Cell Rep 2013; 32(7): 1099-109.
[http://dx.doi.org/10.1007/s00299-013-1421-6] [PMID: 23525746]

[71]    Barry CS, Giovannoni JJ. Ethylene and fruit ripening. J Plant Growth Regul 2007; 26(2): 143-59.
[http://dx.doi.org/10.1007/s00344-007-9002-y]

[72]    Lakhwani D, Sanchita , Pandey A, Sharma D, Asif MH, Trivedi PK. Novel microRNAs regulating ripening-associated processes in banana fruit. Plant Growth Regul 2020; 90(2): 223-35.
[http://dx.doi.org/10.1007/s10725-020-00572-w]

[73]    Christ B, Hörtensteiner S. Mechanism and significance of chlorophyll breakdown. J Plant Growth Regul 2014; 33(1): 4-20.
[http://dx.doi.org/10.1007/s00344-013-9392-y]

[74]    Schaller GE. Ethylene and the regulation of plant development. BMC Biol 2012; 10(1): 9.
[http://dx.doi.org/10.1186/1741-7007-10-9] [PMID: 22348804]

[75]    Holm RE, Wilson WC. Ethylene and fruit loosening from combinations of citrus abscission chemicals. J   Am   Soc   Hortic   Sci   1977;   102(5):   576-9.   Available   from:
https://journals.ashs.org/jashs/view/journals/jashs/102/5/article-p576.xml
[http://dx.doi.org/10.21273/JASHS.102.5.576]

[76]    Martin GC. Mechanical olive harvest: use of fruit loosening agents. Acta Hortic 1994; (356): 284-91.
[http://dx.doi.org/10.17660/ActaHortic.1994.356.60]

[77]    Clark DG, Gubrium EK, Barrett JE, Nell TA, Klee HJ. Root formation in ethylene-insensitive plants. Plant Physiol 1999; 121(1): 53-60.
[http://dx.doi.org/10.1104/pp.121.1.53] [PMID: 10482660]

[78]    Li Z, Wang J, Zhang X, *et al.* The genome of *Aechmea fasciata* provides insights into the evolution of tank epiphytic habits and ethylene-induced flowering. Commun Biol 2022; 5(1): 920.
[http://dx.doi.org/10.1038/s42003-022-03918-4] [PMID: 36071139]

[79]    Ramos MJN, Coito JL, Silva HG, Cunha J, Costa MMR, Rocheta M. Flower development and sex specification in wild grapevine. BMC Genomics 2014; 15(1): 1095.
[http://dx.doi.org/10.1186/1471-2164-15-1095] [PMID: 25495781]

[80]    Kempe K, Gils M. Pollination control technologies for hybrid breeding. Mol Breed 2011; 27(4): 417-37.
[http://dx.doi.org/10.1007/s11032-011-9555-0]

[81]    Zhang M, Smith JAC, Harberd NP, Jiang C. The regulatory roles of ethylene and reactive oxygen species (ROS) in plant salt stress responses. Plant Mol Biol 2016; 91(6): 651-9.
[http://dx.doi.org/10.1007/s11103-016-0488-1] [PMID: 27233644]

[82]    Vanderstraeten L, Depaepe T, Bertrand S, Van Der Straeten D. The ethylene precursor ACC affects early vegetative development independently of ethylene signaling. Front Plant Sci 2019; 10: 1591.
[http://dx.doi.org/10.3389/fpls.2019.01591] [PMID: 31867034]

[83]    Rodrigues MA, Bianchetti RE, Freschi L. Shedding light on ethylene metabolism in higher plants. Front Plant Sci 2014; 5: 665.
[http://dx.doi.org/10.3389/fpls.2014.00665] [PMID: 25520728]

[84]    North JA, Narrowe AB, Xiong W, *et al.* A nitrogenase-like enzyme system catalyzes methionine, ethylene, and methane biogenesis. Science 2020; 369(6507): 1094-8.
[http://dx.doi.org/10.1126/science.abb6310] [PMID: 32855335]

[85]    Van de Poel B, Van Der Straeten D. 1-aminocyclopropane-1-carboxylic acid (ACC) in plants: more than just the precursor of ethylene! Front Plant Sci 2014; 5: 640.
[http://dx.doi.org/10.3389/fpls.2014.00640] [PMID: 25426135]

[86]   Yoon GM, Kieber JJ. 1-Aminocyclopropane-1-carboxylic acid as a signalling molecule in plants. AoB Plants 2013; 5(0): plt017.
[http://dx.doi.org/10.1093/aobpla/plt017]

[87]   Yu J, Niu L, Yu J, *et al.* The involvement of ethylene in calcium-induced adventitious root formation in cucumber under salt stress. Int J Mol Sci 2019; 20(5): 1047.
[http://dx.doi.org/10.3390/ijms20051047] [PMID: 30823363]

[88]   Miyazono K, Miyakawa T, Sawano Y, *et al.* Structural basis of abscisic acid signalling. Nature 2009; 462(7273): 609-14.
[http://dx.doi.org/10.1038/nature08583] [PMID: 19855379]

[89]   Zaharia LI, Walker-Simmon MK, Rodríguez CN, Abrams SR. Chemistry of abscisic acid, abscisic acid catabolites and analogs. J Plant Growth Regul 2005; 24(4): 274-84.
[http://dx.doi.org/10.1007/s00344-005-0066-2]

[90]   Hauser F, Waadt R, Schroeder JI. Evolution of abscisic acid synthesis and signaling mechanisms. Curr Biol 2011; 21(9): R346-55.
[http://dx.doi.org/10.1016/j.cub.2011.03.015] [PMID: 21549957]

[91]   Chen K, Li GJ, Bressan RA, Song CP, Zhu JK, Zhao Y. Abscisic acid dynamics, signaling, and functions in plants. J Integr Plant Biol 2020; 62(1): 25-54.
[http://dx.doi.org/10.1111/jipb.12899] [PMID: 31850654]

[92]   Milborrow BV. The chemistry and physiology of abscisic-acid. Annu Rev Plant Physiol 1974; 25(1): 259-307.
[http://dx.doi.org/10.1146/annurev.pp.25.060174.001355]

[93]   Bharath P, Gahir S, Raghavendra AS. Abscisic acid-induced stomatal closure: an important component of plant defense against abiotic and biotic stress. Front Plant Sci 2021; 12: 615114.
[http://dx.doi.org/10.3389/fpls.2021.615114] [PMID: 33746999]

[94]   Nakashima K, Yamaguchi-Shinozaki K. ABA signaling in stress-response and seed development. Plant Cell Rep 2013; 32(7): 959-70.
[http://dx.doi.org/10.1007/s00299-013-1418-1] [PMID: 23535869]

[95]   Kang J, Yim S, Choi H, *et al.* Abscisic acid transporters cooperate to control seed germination. Nat Commun 2015; 6(1): 8113.
[http://dx.doi.org/10.1038/ncomms9113] [PMID: 26334616]

[96]   Jiang M, Zhang J. Water stress-induced abscisic acid accumulation triggers the increased generation of reactive oxygen species and up-regulates the activities of antioxidant enzymes in maize leaves. J Exp Bot 2002; 53(379): 2401-10.
[http://dx.doi.org/10.1093/jxb/erf090] [PMID: 12432032]

[97]   Sah SK, Reddy KR, Li J. Abscisic acid and abiotic stress tolerance in crop plants. Front Plant Sci 2016; 7: 571.
[http://dx.doi.org/10.3389/fpls.2016.00571] [PMID: 27200044]

[98]   Hu B, Cao J, Ge K, Li L. The site of water stress governs the pattern of ABA synthesis and transport in peanut. Sci Rep 2016; 6(1): 32143.
[http://dx.doi.org/10.1038/srep32143] [PMID: 27694957]

[99]   Manzi M, Lado J, Rodrigo MJ, Zacarías L, Arbona V, Gómez-Cadenas A. Root ABA accumulation in long-term water-stressed plants is sustained by hormone transport from aerial organs. Plant Cell Physiol 2015; 56(12): 2457-66.
[http://dx.doi.org/10.1093/pcp/pcv161] [PMID: 26542111]

[100]  Lovisolo C, Perrone I, Hartung W, Schubert A. An abscisic acid-related reduced transpiration promotes gradual embolism repair when grapevines are rehydrated after drought. New Phytol 2008; 180(3): 642-51.
[http://dx.doi.org/10.1111/j.1469-0137.2008.02592.n] [PMID: 18700860]

[101]  Wood NT, Allan AC, Haley A, Viry-Moussaïd M, Trewavas AJ. The characterization of differential calcium signalling in tobacco guard cells. Plant J 2000; 24(3): 335-44.
[http://dx.doi.org/10.1046/j.1365-313x.2000.00881.x] [PMID: 11069707]

[102]  Falik O, Novoplansky A. Is ABA the exogenous vector of interplant drought cuing? Plant Signal Behav 2022; 17(1): 2129295.
[http://dx.doi.org/10.1080/15592324.2022.2129295] [PMID: 36200554]

[103]  MacRobbie EAC. Effects of ABA in 'isolated' guard cells of *Commelina communis* L. J Exp Bot 1981; 32(3): 563-72.
[http://dx.doi.org/10.1093/jxb/32.3.563]

[104]  Lv X, Ding Y, Long M, *et al.* Effect of foliar application of various nitrogen forms on starch accumulation and grain filling of wheat (*Triticum aestivum* L.) under drought stress. Front Plant Sci 2021; 12: 645379.
[http://dx.doi.org/10.3389/fpls.2021.645379] [PMID: 33841473]

[105]  Leung J, Merlot S, Giraudat J. The arabidopsis abscisic acid-insensitive2 (ABI2) and ABI1 genes encode homologous protein phosphatases 2C involved in abscisic acid signal transduction. Plant Cell 1997; 9(5): 759-71.
[http://dx.doi.org/10.1105/tpc.9.5.759] [PMID: 9165752]

[106]  Schwartz SH, Qin X, Zeevaart JAD. Elucidation of the indirect pathway of abscisic acid biosynthesis by mutants, genes, and enzymes. Plant Physiol 2003; 131(4): 1591-601.
[http://dx.doi.org/10.1104/pp.102.017921] [PMID: 12692318]

[107]  Endo A, Okamoto M, Koshiba T. ABA Biosynthetic and Catabolic Pathways. In: Zhang DP, Ed. Abscisic Acid: Metabolism, Transport and Signaling. Dordrecht: Springer 2014; pp. 21-45.
[http://dx.doi.org/10.1007/978-94-017-9424-4_2]

[108]  Yoshida T, Mogami J, Yamaguchi-Shinozaki K. ABA-dependent and ABA-independent signaling in response to osmotic stress in plants. Curr Opin Plant Biol 2014; 21: 133-9.
[http://dx.doi.org/10.1016/j.pbi.2014.07.009] [PMID: 25104049]

[109]  Ndathe R, Dale R, Kato N. Dynamic modeling of ABA-dependent expression of the *Arabidopsis RD29A* gene. Front Plant Sci 2022; 13: 928718.
[http://dx.doi.org/10.3389/fpls.2022.928718] [PMID: 36092424]

[110]  Meyer A, Miersch O, Büttner C, Dathe W, Sembdner G. Occurrence of the plant growth regulator jasmonic acid in plants. J Plant Growth Regul 1984; 3(1-4): 1-8.
[http://dx.doi.org/10.1007/BF02041987]

[111]  Staswick PE, Su W, Howell SH. Methyl jasmonate inhibition of root growth and induction of a leaf protein are decreased in an *Arabidopsis thaliana* mutant. Proc Natl Acad Sci USA 1992; 89(15): 6837-40.
[http://dx.doi.org/10.1073/pnas.89.15.6837] [PMID: 11607311]

[112]  Sembdner G, Parthier B. Biochemistry, physiological and molecular actions of Jasmonates. Annu Rev Plant Physiol Plant Mol Biol 1993; 44(1): 569-89.
[http://dx.doi.org/10.1146/annurev.pp.44.060193.003033]

[113]  Zhang J, Zhang X, Ye M, Li XW, Lin SB, Sun XL. The jasmonic acid pathway positively regulates the polyphenol oxidase-based defense against tea geometrid caterpillars in the tea plant (*Camellia sinensis*). J Chem Ecol 2020; 46(3): 308-16.
[http://dx.doi.org/10.1007/s10886-020-01158-6] [PMID: 32016775]

[114]  Koda Y, Kikuta Y, Tazaki H, Tsujino Y, Sakamura S, Yoshihara T. Potato tuber-inducing activities of jasmonic acid and related compounds. Phytochemistry 1991; 30(5): 1435-8.
[http://dx.doi.org/10.1016/0031-9422(91)84180-Z]

[115]  van den Berg JH, Ewing EE. Jasmonates and their role in plant growth and development, with special reference to the control of potato tuberization: A review. Am Potato J 1991; 68(11): 781-94.

[http://dx.doi.org/10.1007/BF02853808]

[116]   Koda Y, Okazawa Y. Detection of potato tuber-inducing activity in potato leaves and old tubers. Plant Cell Physiol 1988; 29: 1047-51.
[http://dx.doi.org/10.1093/oxfordjournals.pcp.a077602]

[117]   Abdala G, Castro G, Guiñazú R, Tizio R, Miersch O. Occurrence of jasmonic acid in organs of *Solanum tuberosum* L. and its effect on tuberization. Plant Growth Regul 1996; 19: 139-43.
[http://dx.doi.org/10.1007/BF00024580]

[118]   Koda Y, Kikuta Y. Effects of jasmonates on *in vitro* tuberization in several potato cultivars that differ greatly in maturity. Plant Prod Sci 2001; 4(1): 66-70.
[http://dx.doi.org/10.1626/pps.4.66]

[119]   Fan X, Mattheis JP, Fellman JK, Patterson ME. Effect of methyl jasmonate on ethylene and volatile production by summered apple depends on fruit developmental stage. J Agric Food Chem 1997; 45(1): 208-11.
[http://dx.doi.org/10.1021/jf9603846]

[120]   Wang J, Song L, Gong X, Xu J, Li M. Functions of jasmonic acid in plant regulation and response to abiotic stress. Int J Mol Sci 2020; 21(4): 1446.
[http://dx.doi.org/10.3390/ijms21041446]

[121]   Thaler JS. Induced resistance in agricultural crops: effects of jasmonic acid on herbivory and yield in tomato plants. Environ Entomol 1999; 28(1): 30-7.
[http://dx.doi.org/10.1093/ee/28.1.30]

[122]   Falkenstein E, Groth B, Mithöfer A, Weiler EW. Methyljasmonate and α-linolenic acid are potent inducers of tendril coiling. Planta 1991; 185(3): 316-22.
[http://dx.doi.org/10.1007/BF00201050] [PMID: 24186412]

[123]   Ogura Y, Ishihara A, Iwamura H. Induction of hydroxycinnamic acid amides and tryptophan by jasmonic acid, abscisic acid and osmotic stress in barley leaves. Z Naturforsch C J Biosci 2001; 56(3-4): 193-202.
[http://dx.doi.org/10.1515/znc-2001-3-405] [PMID: 11371008]

[124]   Buta JG, Moline HE. Methyl jasmonate extends shelf life and reduces microbial contamination of fresh-cut celer and peppers. J Agric Food Chem 1998; 46(4): 1253-6.
[http://dx.doi.org/10.1021/jf9707492]

[125]   Parthier B. Jasmonates: Hormonal regulators or stress factors in leaf senescence? J Plant Growth Regul 1990; 9(1-4): 57-63.
[http://dx.doi.org/10.1007/BF02041942]

[126]   Koda Y, Kikuta Y. Wound-induced accumulatin of jasmonic acid in tissues of potato tubers. Plant Cell Physiol 1994; 35(5): 751-6.
[http://dx.doi.org/10.1093/oxfordjournals.pcp.a078653]

[127]   Turner JG, Ellis C, Devoto A. The jasmonate signal pathway. Plant Cell 2002; 14(Suppl) (Suppl. 1): S153-64.
[http://dx.doi.org/10.1105/tpc.000679] [PMID: 12045275]

[128]   Schaller F, Schaller A, Stintzi A. Biosynthesis and metabolism of jasmonates. J Plant Growth Regul 2004; 23(3): 179-99.
[http://dx.doi.org/10.1007/s00344-004-0047-x]

[129]   Rohwer CL, Erwin JE. Horticultural applications of jasmonates. J Hortic Sci Biotechnol 2008; 83(3): 283-304.
[http://dx.doi.org/10.1080/14620316.2008.11512381]

[130]   Mason HS, Mullet JE. Expression of two soybean vegetative storage protein genes during development and in response to water deficit, wounding, and jasmonic acid. Plant Cell 1990; 2(6): 569-79.

[http://dx.doi.org/10.1105/tpc.2.6.569] [PMID: 2152178]

[131]  Huang JF, Bantroch DJ, Greenwood JS, Staswick PE. Methyl jasmonate treatment eliminates cell-specific expression of vegetative storage protein genes in soybean leaves. Plant Physiol 1991; 97(4): 1512-20.
       [http://dx.doi.org/10.1104/pp.97.4.1512] [PMID: 16668578]

[132]  Farmer EE, Johnson RR, Ryan CA. Regulation of expression of proteinase inhibitor genes by methyl jasmonate and jasmonic Acid. Plant Physiol 1992; 98(3): 995-1002.
       [http://dx.doi.org/10.1104/pp.98.3.995] [PMID: 16668777]

[133]  Wilen RW, van Rooijen GJH, Pearce DW, Pharis RP, Holbrook LA, Moloney MM. Effects of jasmonic Acid on embryo-specific processes in brassica and linum oilseeds. Plant Physiol 1991; 95(2): 399-405.
       [http://dx.doi.org/10.1104/pp.95.2.399] [PMID: 16667997]

[134]  Cheon J, Fujioka S, Dilkes BP, Choe S. Brassinosteroids regulate plant growth through distinct signaling pathways in Selaginella and Arabidopsis. PLoS One 2013; 8(12): e81938.
       [http://dx.doi.org/10.1371/journal.pone.0081938] [PMID: 24349155]

[135]  Li Z, He Y. Roles of brassinosteroids in plant reproduction. Int J Mol Sci 2020; 21(3): 872.
       [http://dx.doi.org/10.3390/ijms21030872] [PMID: 32013254]

[136]  Bajguz A. Metabolism of brassinosteroids in plants. Plant Physiol Biochem 2007; 45(2): 95-107.
       [http://dx.doi.org/10.1016/j.plaphy.2007.01.002] [PMID: 17346983]

[137]  Clouse SD. Brassinosteroids. Arabidopsis Book 2011; 9: e0151.
       [http://dx.doi.org/10.1199/tab.0151] [PMID: 22303275]

[138]  Oh MH, Honey SH, Tax FE. The control of cell expansion, cell division, and vascular development by brassinosteroids: a historical perspective. Int J Mol Sci 2020; 21(5): 1743.
       [http://dx.doi.org/10.3390/ijms21051743] [PMID: 32143305]

[139]  Catterou M, Dubois F, Schaller H, *et al.* Brassinosteroids, microtubules and cell elongation in *Arabidopsis thaliana*. II. Effects of brassinosteroids on microtubules and cell elongation in the bul1 mutant. Planta 2001; 212(5-6): 673-83.
       [http://dx.doi.org/10.1007/s004250000467] [PMID: 11346940]

[140]  Yamamuro C, Ihara Y, Wu X, *et al.* Loss of function of a rice brassinosteroid insensitive1 homolog prevents internode elongation and bending of the lamina joint. Plant Cell 2000; 12(9): 1591-605.
       [http://dx.doi.org/10.1105/tpc.12.9.1591] [PMID: 11006334]

[141]  Hong Z, Ueguchi-Tanaka M, Fujioka S, *et al.* The Rice brassinosteroid-deficient dwarf2 mutant, defective in the rice homolog of Arabidopsis DIMINUTO/DWARF1, is rescued by the endogenously accumulated alternative bioactive brassinosteroid, dolichosterone. Plant Cell 2005; 17(8): 2243-54.
       [http://dx.doi.org/10.1105/tpc.105.030973] [PMID: 15994910]

[142]  Fleurat-Lessard P, Frangne N, Maeshima M, Ratajczak R, Bonnemain JL, Martinoia E. Increased expression of vacuolar aquaporin and H+-ATPase related to motor cell function in *Mimosa pudica* L. Plant Physiol 1997; 114(3): 827-34.
       [http://dx.doi.org/10.1104/pp.114.3.827] [PMID: 12223745]

[143]  Kushwah S, Banasiak A, Nishikubo N, *et al.* Arabidopsis XTH4 and XTH9 contribute to wood cell expansion and secondary wall formation. Plant Physiol 2020; 182(4): 1946-65.
       [http://dx.doi.org/10.1104/pp.19.01529] [PMID: 32005783]

[144]  Collins C, Maruthi NM, Jahn CE. CYCD3 D-type cyclins regulate cambial cell proliferation and secondary growth in *Arabidopsis*. J Exp Bot 2015; 66(15): 4595-606.
       [http://dx.doi.org/10.1093/jxb/erv218] [PMID: 26022252]

[145]  Müssig C, Shin GH, Altmann T. Brassinosteroids promote root growth in Arabidopsis. Plant Physiol 2003; 133(3): 1261-71.
       [http://dx.doi.org/10.1104/pp.103.028662] [PMID: 14526105]

[146] Lee J, Han S, Lee HY, *et al.* Brassinosteroids facilitate xylem differentiation and wood formation in tomato. Planta 2019; 249(5): 1391-403.
[http://dx.doi.org/10.1007/s00425-019-03094-6] [PMID: 30673841]

[147] Vogler F, Schmalzl C, Englhart M, Bircheneder M, Sprunck S. Brassinosteroids promote Arabidopsis pollen germination and growth. Plant Reprod 2014; 27(3): 153-67.
[http://dx.doi.org/10.1007/s00497-014-0247-x] [PMID: 25077683]

[148] Taiz L. Plant cell expansion: regulation of cell wall mechanical properties. Annu Rev Plant Physiol 1984; 35(1): 585-657.
[http://dx.doi.org/10.1146/annurev.pp.35.060184.003101]

[149] Groenewald EG, Van der Westhuizen AJ. The roles of prostaglandins and brassinosteroids as plant growth regulators: review article. S Afr J Sci 2005; 101(1): 67-74. Available from: https://hdl.handle.net/10520/EJC96343

[150] Chung Y, Maharjan PM, Lee O, *et al.* Auxin stimulates DWARF4 expression and brassinosteroid biosynthesis in Arabidopsis. Plant J 2011; 66(4): 564-78.
[http://dx.doi.org/10.1111/j.1365-313X.2011.04513.x] [PMID: 21284753]

[151] Kim HB, Kwon M, Ryu H, *et al.* The regulation of DWARF4 expression is likely a critical mechanism in maintaining the homeostasis of bioactive brassinosteroids in Arabidopsis. Plant Physiol 2006; 140(2): 548-57.
[http://dx.doi.org/10.1104/pp.105.067918] [PMID: 16407451]

[152] Steber CM, McCourt P. A role for brassinosteroids in germination in Arabidopsis. Plant Physiol 2001; 125(2): 763-9.
[http://dx.doi.org/10.1104/pp.125.2.763] [PMID: 11161033]

[153] Leubner-Metzger G. Brassinosteroids and gibberellins promote tobacco seed germination by distinct pathways. Planta 2001; 213(5): 758-63.
[http://dx.doi.org/10.1007/s004250100542] [PMID: 11678280]

[154] Divi UK, Rahman T, Krishna P. Brassinosteroid-mediated stress tolerance in Arabidopsis shows interactions with abscisic acid, ethylene and salicylic acid pathways. BMC Plant Biol 2010; 10(1): 151.
[http://dx.doi.org/10.1186/1471-2229-10-151] [PMID: 20642851]

[155] Kour J, Kohli SK, Khanna K, *et al.* Brassinosteroid signaling, crosstalk and, physiological functions in plants under heavy metal stress. Front Plant Sci 2021; 12: 608061.
[http://dx.doi.org/10.3389/fpls.2021.608061] [PMID: 33841453]

[156] Koda Y, Takahashi K, Kikuta Y. Potato tuber-inducing activities of salicylic acid and related compounds. J Plant Growth Regul 1992; 11: 215-9.
[http://dx.doi.org/10.1007/BF02115480]

[157] Miyawaki K, Inoue S, Kitaoka N, Matsuura H. Potato tuber-inducing activities of jasmonic acid and related-compounds (II). Biosci Biotechnol Biochem 2021; 85(12): 2378-82.
[http://dx.doi.org/10.1093/bbb/zbab161]

[158] Raskin I. Role of salicylic acid in plants. Annu Rev Plant Physiol Plant Mol Biol 1992; 43(1): 439-63.
[http://dx.doi.org/10.1146/annurev.pp.43.060192.002255]

[159] López-Delgado H, Scott IM. Acetylsalicylic acid: its effects on a highly expressed phosphatase from Solanum cardiophyllum. Biotecnol Apl 1996; 13: 186-9.

[160] Haque S, Farooqi AHA, Gupta MM, Sangwan RS, Khan A. Effect of ethrel, chlormequat chloride and paclobutrazol on growth and pyrethrins accumulation in *Chrysanthemum cinerariaefolium* Vis. Plant Growth Regul 2007; 51(3): 263-9.
[http://dx.doi.org/10.1007/s10725-007-9170-6]

[161] Zhang M, Yang J, Pan H, Pearson BJ. Dwarfing effects of chlormequat chloride and uniconazole on

potted baby primrose. Horttechnology 2020; 30(5): 536-43.
[http://dx.doi.org/10.21273/HORTTECH04646-20]

[162] Ramesh R, Ramprasad E. Effect of Plant Growth Regulators on Morphological, Physiological and Biochemical Parameters of Soybean (*Glycine max* L. Merrill). In: Kumar A, Ed. Biotechnology and Bioforensics. Singapore: Springer 2015; pp. 61-71.
[http://dx.doi.org/10.1007/978-981-287-050-6_7]

[163] Moniruzzaman M. Effect of cycocel (CCC) on the growth and yield manipulation of vegetable soybean. Agric. Res. Center Report. 2000; 1-16.

[164] Hernández-Altamirano JM, Largo-Gosens A, Martínez-Rubio R, *et al.* Effect of ancymidol on cell wall metabolism in growing maize cells. Planta 2018; 247(4): 987-99.
[http://dx.doi.org/10.1007/s00425-018-2840-y] [PMID: 29330614]

[165] Hofmannová J, Schwarzerová K, Havelková L, Boríková P, Petrásek J, Opatrný Z. A novel, cellulose synthesis inhibitory action of ancymidol impairs plant cell expansion. J Exp Bot 2008; 59(14): 3963-74.
[http://dx.doi.org/10.1093/jxb/ern250] [PMID: 18832186]

[166] Thakur R, Sood A, Nagar PK, Pandey S, Sobti RC, Ahuja PS. Regulation of growth of Lilium plantlets in liquid medium by application of paclobutrazol or ancymidol, for its amenability in a bioreactor system: growth parameters. Plant Cell Rep 2006; 25(5): 382-91.
[http://dx.doi.org/10.1007/s00299-005-0094-1] [PMID: 16369766]

[167] Hossain ABMS, Mizutani F. Dwarf peach trees by using abscisic acid hormone, maleic hydrazide and cycocel as growth inhibitors in phloemic stress conditions grafted on vigorous rootstock. Can J Pure Appl Sci 2008; 2: 335-42.

[168] Cutrubinis M, Delincée H, Bayram G, *et al.* Germination test for identification of irradiated garlic. Eur Food Res Technol 2004; 219: 178-83.
[http://dx.doi.org/10.1007/s00217-004-0935-0]

[169] Bhatla SC, Lal MA. Gibberellins. In: Bhatla SC, Lal MA, Eds. Plant Physiology, Development and Metabolism. Singapore: Springer 2023; pp. 431-41.
[http://dx.doi.org/10.1007/978-981-99-5736-1_17]

[170] Lendzian KJ, Ziegler H, Sankhla N. Effect of phosphon-D on photosynthetic light reactions and on reactions of the oxidative and reductive pentose phosphate cycle in a reconstituted spinach (*Spinacia oleracea* L.) chloroplast system. Planta 1978; 141(2): 199-204.
[http://dx.doi.org/10.1007/BF00387889] [PMID: 24414777]

[171] Santos Filho FB, Silva TI, Dias MG, Alves ACL, Grossi JAS. Paclobutrazol reduces growth and increases chlorophyll indices and gas exchanges of basil (*Ocimum basilicum*). Braz J Biol 2022; 82: e262364.
[http://dx.doi.org/10.1590/1519-6984.262364] [PMID: 35857950]

[172] Wang SY, Sun T, Faust M. Translocation of paclobutrazol, a gibberellin biosynthesis inhibitor, in apple seedlings. Plant Physiol 1986; 82(1): 11-4.
[http://dx.doi.org/10.1104/pp.82.1.11] [PMID: 16664976]

[173] Hamid M, Williams R. Translocation of paclobutrazol and gibberellic acid in Sturt's Desert Pea (*Swainsona formosa*). Plant Growth Regul 1997; 23: 167-71.
[http://dx.doi.org/10.1023/A:1005982002914]

[174] Bouwmeester HJ, Fonne-Pfister R, Screpanti C, De Mesmaeker A. Strigolactones: Plant Hormones with Promising Features. Angew Chem Int Ed 2019; 58(37): 12778-86.
[http://dx.doi.org/10.1002/anie.201901626] [PMID: 31282086]

[175] Wang X, Li Z, Shi Y, *et al.* Strigolactones promote plant freezing tolerance by releasing the WRKY41 -mediated inhibition of *CBF / DREB1* expression. EMBO J 2023; 42(19): e112999.
[http://dx.doi.org/10.15252/embj.2022112999] [PMID: 37622245]

[176] Dun EA, Brewer PB, Gillam EMJ, Beveridge CA. Strigolactones and Shoot Branching: What Is the Real Hormone and How Does It Work? Plant Cell Physiol 2023; 64(9): 967-83.
[http://dx.doi.org/10.1093/pcp/pcad088] [PMID: 37526426]

[177] Umehara M, Hanada A, Yoshida S, *et al.* Inhibition of shoot branching by new terpenoid plant hormones. Nature 2008; 455(7210): 195-200.
[http://dx.doi.org/10.1038/nature07272] [PMID: 18690207]

[178] Gomez-Roldan V, Fermas S, Brewer PB, *et al.* Strigolactone inhibition of shoot branching. Nature 2008; 455(7210): 189-94.
[http://dx.doi.org/10.1038/nature07271] [PMID: 18690209]

[179] Brewer PB, Koltai H, Beveridge CA. Diverse roles of strigolactones in plant development. Mol Plant 2013; 6(1): 18-28.
[http://dx.doi.org/10.1093/mp/sss130] [PMID: 23155045]

# CHAPTER 4

# Current Assessment and Future Perspectives of Secondary Metabolites

**Nurdan Saraç[1,*], Aysel Uğur[2], Tuba Baygar[3] and İrem Demir[4]**

[1] *Muğla Sıtkı Koçman University, Faculty of Science, Biology Department, Menteşe, Muğla, Türkiye*

[2] *Gazi University, Faculty of Dentistry, Basic Science Department, Ankara, Türkiye*

[3] *Mugla Sitki Kocman University, Research Laboratories Center, Menteşe, Mugla, Türkiye*

[4] *Mugla Sitki Kocman University, Institute of Science, Biology Department, Menteşe, Muğla, Türkiye*

**Abstract:** Herbal secondary metabolites have become more and more significant in recent years because of the negative side effects of synthetic medications used to treat a variety of illnesses and the growing demand for natural industrial products. Green chemicals are becoming more and more popular, particularly as a result of the negative environmental impacts of synthetic chemicals. Among these green chemicals, especially herbal products are used in the pharmaceutical industry, in the production of natural dyes, and in the production of herbal fragrances and flavor substances. The importance of these secondary metabolites has led in recent years to investigate the possibilities of increasing their production using tissue culture technology. Thanks to plant tissue culture, it is possible to obtain secondary metabolites more cheaply and efficiently, without being affected by seasonal fluctuations. Secondary metabolites are obtained from medicinal plants and are also named as phytochemicals, natural products, or plant constituents. The studies about the plant secondary metabolites have been increasing in the last 50 years. These molecules have a major role in the adaptation of plants to their environment and also in the defense system of predators; and response to environmental stresses.

**Keywords:** Green compounds, Medicinal use, Plant, Secondary metabolite.

## INTRODUCTION

Although humans have used various plant secondary metabolites in medicine and pharmacology for thousands of years, the chemical analysis began nearly 200 years ago with the first isolation of morphine from Papaver somniferum [1].

---

* **Corresponding author Nurdan Saraç:** Muğla Sıtkı Koçman University, Faculty of Science, Biology Department, Menteşe, Muğla, Türkiye; E-mail: nsarac@mu.edu.tr

**Ergun Kaya (Ed.)**

There have been significant developments in plant analysis in the last 20-30 years. Natural products isolated from plants are used for the discovery and development of modern medicines [2]. Chromatography, electrophoresis, isotope techniques, and enzymology have been effective in elucidating the exact regime formulas and the most important biosynthetic pathways [3, 4].

The primary metabolites (nucleotides, amino acids, and organic acids) are directly effective in the vital functions of plants. Although the secondary metabolites are not directly effective in the vital functions of the plant. In the early years when secondary compounds were studied, it was argued that these substances were 'functionless and waste substances' [5, 6]. In the following years, it was explained by many scientists that these components are very important for the defense system of plants [7 - 13].

The metabolite name is usually limited to these small molecules, which are products of metabolism. The metabolites have various functions in plants such as fuel, structural, signaling, stimulatory and inhibitory effects on enzymes, catalytic activity of their own (a cofactor), defense system, and interactions with other organisms. Plants produce a wide variety of organic compounds, the vast majority of which are not directly involved in growth and development [14, 15].

Secondary metabolism refers to metabolic pathways and distribution of minor process products that, unlike primary metabolism, are not necessary for the growth and reproduction of the organism [16]. The secondary metabolites are produced in the secondary metabolic pathways of plants (Fig. **1**). Plant secondary metabolites have a huge group of various structure compounds [17].

## Classification of Plant Secondary Metabolites

Plant secondary metabolites are generally classified according to their biosynthetic pathways [19]. A basic classification of secondary metabolites includes three main groups: terpenes (such as plant volatiles, cardiac glycosides, carotenoids, and sterols), phenolics (such as phenolic acids, coumarins, lignans, stilbenes, flavonoids, tannins, and lignin) and nitrogen-containing compounds (such as alkaloids and glucosinolates) (Fig. **2**) [20].

Terpenes are one of the largest classes of natural products and more than 22,000 compounds belonging to this group have been identified [21]. While there may be terpenes that contain only hydrocarbons, there are also terpenes that contain oxygen. Oxygen-containing terpenes are also called "terpenoids" [22].

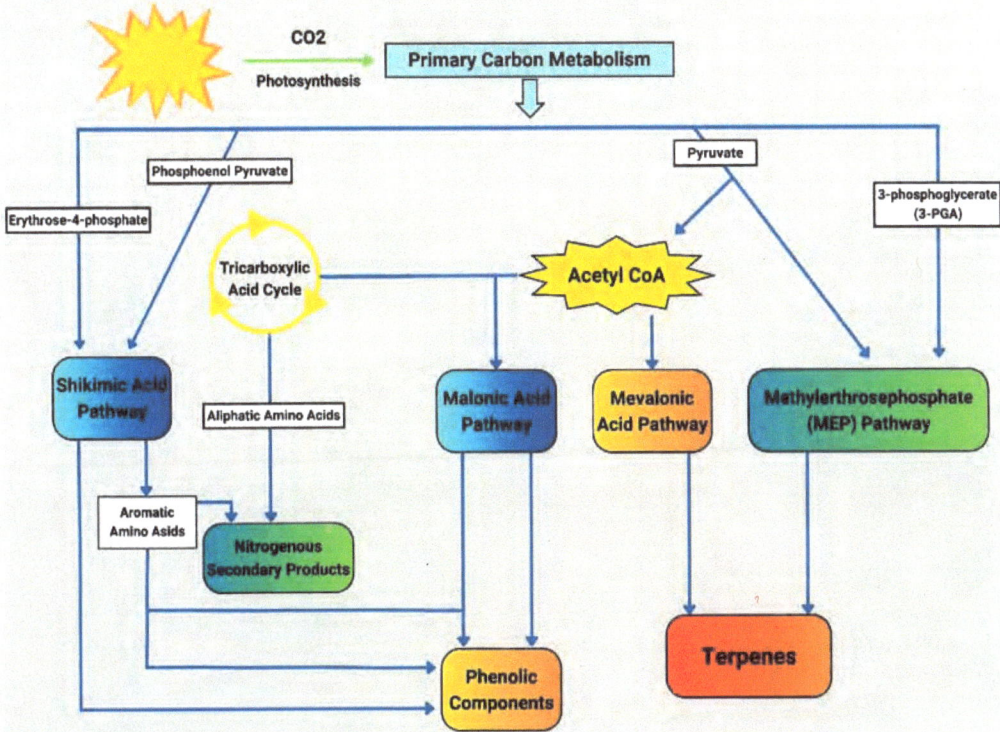

**Fig. (1).** Simplified view of the main pathways and primary metabolism of the secondary metabolites biosynthesis [18].

Terpenoids have various functional roles in plants. Examples of these are phytol and carotenoids as photosynthetic pigments, ubiquinone and plastoquinone as electron carriers, abscisic acid as hormones, and sterols as structural components of cell membranes [24].

Most terpenoids are lipophilic and readily interact with biomembranes and membrane proteins. They increase the fluidity and permeability of membranes, leading to uncontrolled flow of ions and metabolites and even cell leakage, which can ultimately cause necrotic or apoptotic cell death [25]. In general, terpenes are cytotoxic to bacteria and fungi, insects, and vertebrates and are used against infections [26]. Compounds in this group are very important in terms of odor potential [27]. The recent biological activities of some terpenes are given in Table **1**.

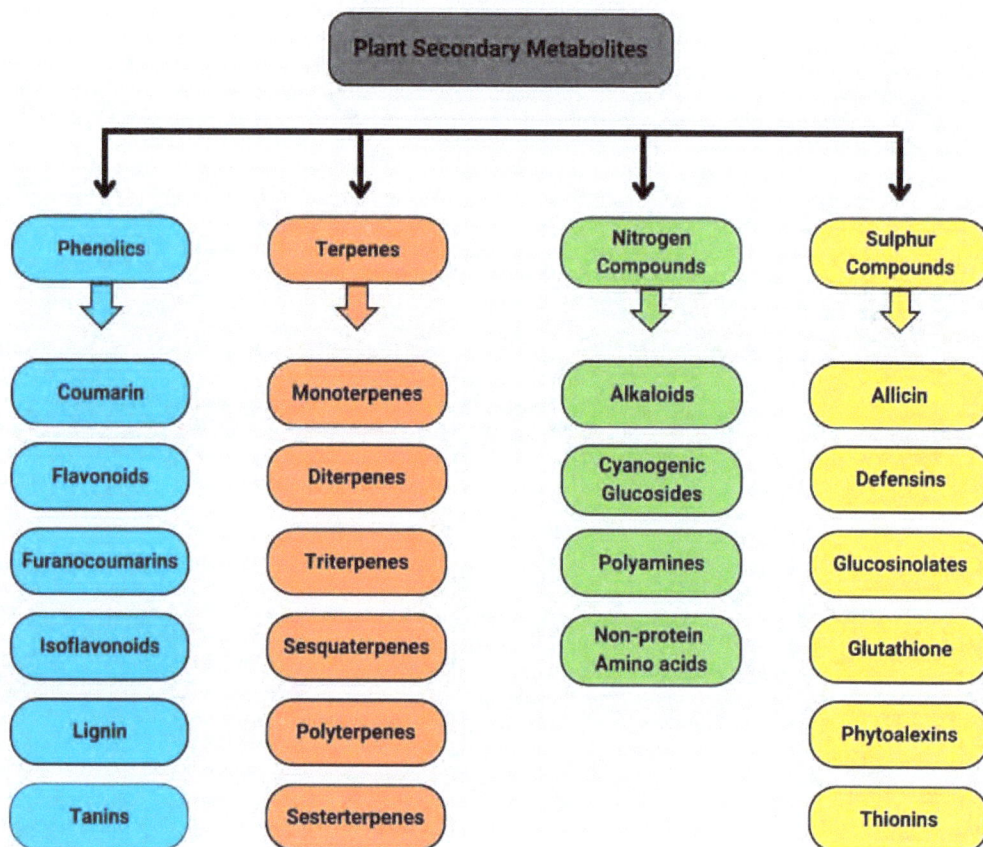

**Fig. (2).** Classification of some plant secondary metabolites [23].

**Table 1. The recent biological activities of some terpenes.**

| Terpene | Source | Activity | References |
|---|---|---|---|
| Limonene | Essential oils of various plants | Pain control | [28] |
| | *Lavandula luisieri* | Anti-inflammatory activity | [29] |
| *p*-cymene | Commercially obtained | Anti-hyperalgesic and anti-inflammatory activity (*in vivo*) | [30] |
| | | Antioxidant and neuroprotective activity | [31] |
| (−)-linalool | Lamiaceae species | Antinociceptive effect in an animal model of chronic non-inflammatory muscle pain | [32] |
| | - | Antibacterial activity against *Pseudomonas aeruginosa* | [33] |

| Terpene | Source | Activity | References |
|---------|--------|----------|-----------|
| Terpinene | *Melaleuca alternifolia* | Antiprotozoal activity | [34] |
| | *Bursera morelensis* | Antifungal activity against *Candida albicans* | [35] |
| Sesquiterpenes | *Centaurea drabifolia* | Cytotoxic activity | [36] |
| | *Scorzonera divaricata* | Cytotoxic activity against HepG2 (human liver cancer), K562 (human chronic myelogenous leukemia), and Hela (human cervical carcinoma) cells, Antibacterial activity against various bacteria | [37] |
| Diterpene | *Copaifera* spp. | Treating acute injuries such as inflammation or skin disorders. | [38] |
| | *Euphorbia connata* (Euphorbiaceae) | Cytotoxic activity against MDA-MB (breast cancer) and MCF-7 (breast cancer) cells. | [39] |

Phenolic compounds protect plants from herbivores, pathogen attacks, and other animals with their deterrent effects. High concentrations are effective in resistance against fungi [40]. Phenolic compounds are often found bound to sugars, and this property reduces their endogenous toxicity. These compounds also protect the plant from UV radiation and cold stress [41].

One of the biggest categories of secondary metabolites is made up of phenolic compounds. These compounds are synthesized by fruits, vegetables, teas, cocoa, and other plants. They have various biological activities such as antioxidant, anti-inflammatory, anti-carcinogenic, *etc,* and may protect from oxidative stress and some diseases [42]. Simple phenolics are bactericidal, antiseptic, and anthelmintic [43]. The recent biological activities of some phenolic compounds are given in Table **2**.

**Table 2. The recent biological activities of some phenolic compounds.**

| Phenolic Name | Source | Activity | References |
|---------------|--------|----------|-----------|
| Catechol | - | *In vitro* Carbonic anhydrase inhibitory capacity against hCA I, II, IX, and XII. | [44] |
| Ferulic acid | - | Anti-inflammatory activity on $H_2O_2$-induced HEK293 (Human Embryonic Kidney) cells. | [45] |
| | - | Anti-inflammatory activity on LPS-induced BEECs (bovine endometrial epithelial cells). | [46] |
| | | Antibacterial activity on *Shigella flexneri* | [47] |
| Scopoletin | - | Cytotoxic activity on HeLa (*in vitro*) | [48] |
| | | Cytotoxic activity on U937 (Histiocytic Lymphoma) | [49] |

*(Table 2) cont.....*

| Phenolic Name | Source | Activity | References |
|---|---|---|---|
| Luteolin | - | Reducing oxidative stress (*in vivo*) | [50] |
| | | Antioxidant activity | [51] |
| | | Antioxidant activity on HT29 (colon cancer cell) | [52] |
| Kaempferol | - | Anti-inflammatory and antioxidant activity | [53] |
| | | Antibiofilm activity against *Candida* species | [54] |
| | | Inhibitory effect on the proliferation activity of HCT116 (human colon cancer cells) | [55] |
| Gallic acid | - | Chondroprotective activity (*in vitro* and *in vivo*) | [56] |
| | | Cytotoxicity on HCT-15 (colon cancer) cells | [57] |
| | | Antioxidant activity | [58] |
| Pinoresinol | - | The osteoblastic proliferation and differentiation activity | [59] |
| | - | Antiinflammatory activity | [60] |
| | *Cinnamomum Camphora* | Antibacterial activity | [61] |

(-): Not specified in the literature.

Alkaloids are nitrogen-containing compounds. These are derived from amino acids such as tyrosine, lysine, tryptophan, and aspartic acid [62]. These compounds, found in approximately 20% of plant species, play a defensive role against herbivores and pathogen attacks [63 - 65]. They have highly variable effects at the cellular level. Some of them act on the nervous system, some affect protein synthesis, and others affect membrane transport and enzyme activities [66].

Approximately 12000 alkaloids are used as narcotics, pharmaceuticals, and poisons due to their various biological activities [67]. Plant alkaloids with widespread use as medicines include vincristine, vinblastine, and camptothecin used as anticancers, colchicine as a gout suppressant, morphine and codeine as analgesics, and scopolamine as a sedative [68]. The recent biological activities of some alkaloids are given in Table **3**.

**The Plant Secondary Metabolites, their Roles and Usage**

The plants produce a widespread diversity of secondary metabolites [83]. More importantly, most of these secondary metabolites are used in the pharmaceutical industry (for their pharmacological activities or toxicologies on humans and animals). They are also used in cosmetic products, nutrition, drugs, dyes, fragrances, flavors, and dietary products. Therefore, the scientific and industrial importance of secondary metabolites is high [84].

**Table 3. The recent biological activities of some alkaloids.**

| Alkaloid | Source | Activity | References |
|---|---|---|---|
| Berberine | - | Cytotoxic activity on T47D (breast cancer) and MCF-7 cells. | [69] |
| | | Cytotoxic activity on Tca8113 (oral squamous cell carcinoma), CNE2 (nasopharyngeal carcinoma cell), MCF-7 (breast cancer), Hela (cervical carcinoma), and HT29 (colon cancer). | [70] |
| Sanguinarine | - | Antibiofilm activity on *Providencia rettgeri* | [71] |
| | | Antibiofilm activity on *Candida albicans* | [72] |
| Allocryptopine | *Chelidonium majus* | Antibacterial activity on *Staphylococcus aureus* and cytotoxicity on L929 cells | [73] |
| | *Glaucium corniculatum* | Neuroprotective effect | [74] |
| Magnoflorine | Chinese herb Magnolia or Aristolochia | Cytotoxic activity on MCF-7 and MDA-MB-231 cells | [75] |
| | - | Inhibition of alpha-glucosidase activity, which is necessary for normal cell wall composition and virulence of *Candida albicans*. | [76] |
| | | Anti-inflammatory activity in acute lung injury | [77] |
| Coptisine | *Coptidis rhizoma* | Cytotoxic activity on MDA-MB-231 cells | [78] |
| | *Coptis chinensis* | Cytotoxic activity on gastric cancer cell lines ACC-201 and NCI-N87 | [79] |
| | - | Anti- *Helicobacter pylori* activity | [80] |
| Scoulerine | - | Cytotoxic activity against OVCAR3 (ovarian cancer) | [81] |
| | | Cytotoxic activity on colorectal cancer cells | [82] |

Secondary metabolites are at least as important as primary metabolites for the vital functions of the plants, as they provide plants with resistance to pests, diseases, and adverse environmental conditions and have an effect against some weeds [85, 86]. Secondary metabolites are also used in industrial areas in paint, fiber, glue, oil, flavor, perfume, and medicines. Determining the biological properties of the secondary metabolite has stimulated the search for new drugs, antibiotics, insecticides, and herbicides [87, 88].

Today, herbal formulations are popular because of their effect on various diseases and disorders. These formulations do not cause any significant adverse effects and are available at reasonable prices [89]. The World Health Organization reports that herbal medicines are popular in developing countries and more than 80% of the world's population use herbal medicines for their basic health needs. Therefore, there is a need to research plants with medicinal value [14].

In recent years, the traditional use of medicines has been a matter of global debate. Many of the plant species recognized as medicinal plants have been scientifically evaluated for their possible medicinal effects. Obtaining plant-derived compounds is becoming increasingly difficult due to the rapid disappearance of the natural habitats of medicinal plants along with environmental and geopolitical instability. This situation has led the pharmaceutical industry and scientists to research cell cultures as an alternative method to classical medicinal plant production [14].

## Plant Tissue Culture in Secondary Metabolite Production

The research and commercial importance of secondary metabolites resulted in great interest, particularly in the production of bioactive plant metabolites using tissue culture technology [14].

Plant cell and tissue culture technology can be used routinely for both the propagation of plants in a sterile environment and the extraction of secondary metabolites from leaves, stems, roots, and meristems. There are various studies on obtaining secondary metabolites from cell suspension cultures of commercial and medicinal plants [14, 90].

The ability to produce secondary metabolites from plant cell culture has been understood recently in the history of *in-vitro* culture. It was thought that undifferentiated cells, such as callus or cell suspension cultures, were unable to produce secondary metabolites, unlike differentiated cells and specialized organs [91]. Zenk and his colleagues evidenced experimentally that this idea was incorrect by observing that the undifferentiated cell culture of *Morinda citrifolia* yielded 2.5 g of anthraquinone per liter [92]. This discovery enabled the possible use of plant cultures for the production of secondary compounds [93] and demonstrated the possible use of plant cell cultures for the specific preparation of natural compounds [94, 95].

The cell culture system has several advantages over plants grown using classical methods. These useful metabolites can be produced under controlled conditions, regardless of climate changes or soil conditions; the cultured plant cells are free of microbes and insects; cells of any plant in the tropical or alpine region can be easily multiplied to yield their specific metabolites. Rational regulation of metabolite processes with automatic control of cell growth reduces labor costs and increases productivity and the ability to extract organic substances from callus cultures. For these advantages, tissue culture technology has made great progress in the production of phytochemicals [96].

The guidelines for the production of secondary metabolites from plant cell cultivation are given in Fig. (**3**) [97].

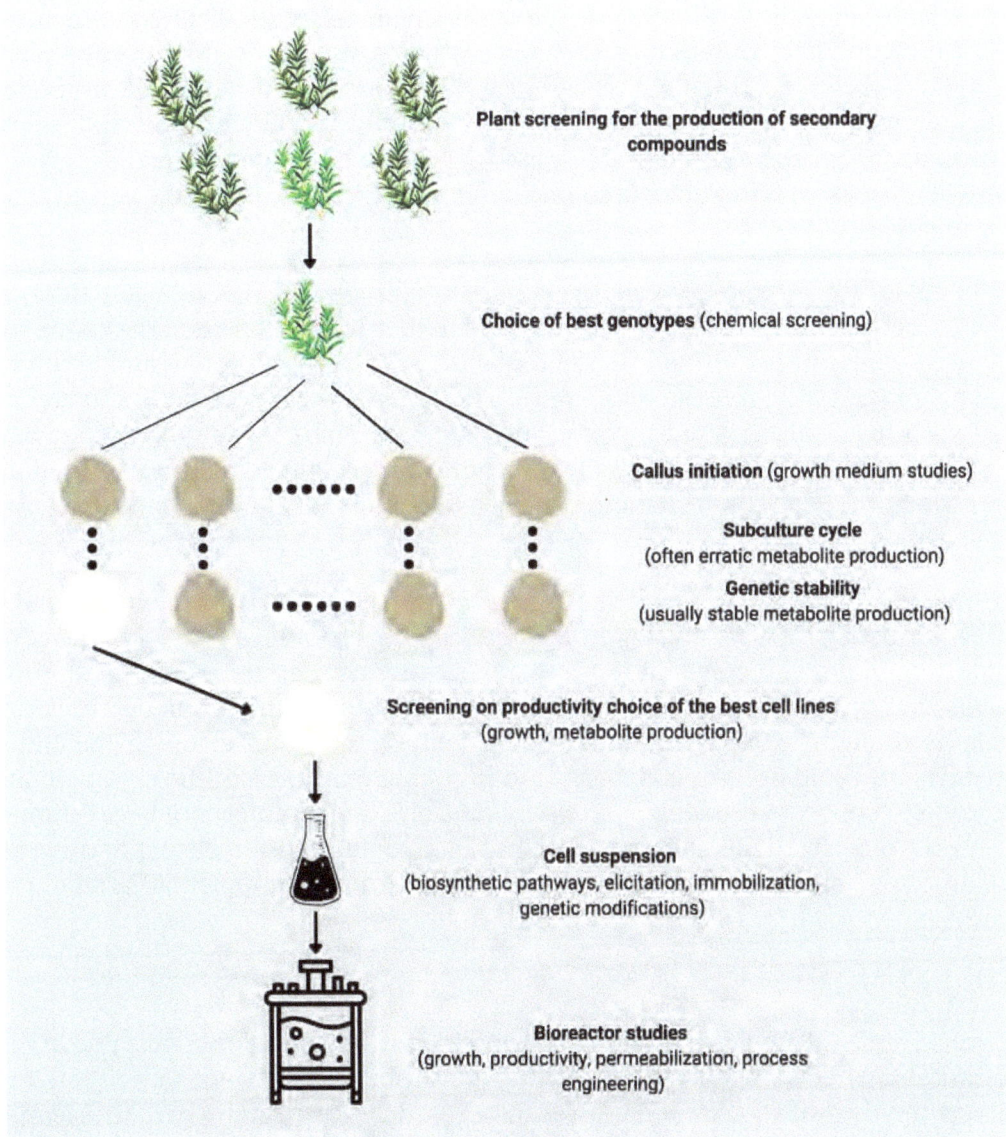

**Fig. (3).** The production stages of secondary metabolites from plant cells [95].

## FUTURE PROSPECTS

The majority of prescription drugs used around the world are of plant origin. Among the raw drug materials produced in the plant-derived cell, tissue or organ cultures, diosgenin, codeine, morphine, atropine, hyoscyamine, scopolamine, digoxin, digitoxin, quinine, reserpine, artemisinin and taxol are the main secondary metabolites. Plants produce many chemicals with different structures and functions. It is known that very few of the secondary metabolites of plant origin have been identified so far, and many of them are known to have structures and biological activities and have been used in a wide variety of industrial fields. Especially in medicine, there are some that are used as painkillers, sedatives, muscle relaxants, antimicrobials against infectious diseases, insecticides, and food preservatives.

Although many countries in the world are very rich in terms of plant flora, it cannot be said that research on the isolation, purification, determination of chemical structures of these important compounds, cell culture, identification of coding genes, use of relevant cloning techniques, *etc.* is sufficient. As a matter of fact, as a result of the studies conducted, lack of resources, and inadequate and faulty policies, an impression that the phenomenon examined does not fall within the scope of modern biotechnology and remains more of a preliminary study has been gained.

In order to be affected by all these negativities as little as possible, to make effective use of newly developed techniques and technologies in the fields of molecular biology, biotechnology, and genetic engineering in the future, to grow plants that are resistant to various diseases agents, pests, environmental factors such as drought, excessive humidity and temperature, and to obtain healthier products, it would be extremely useful to investigate the possibilities, provide the necessary support to research and researchers, and direct biotechnological studies to obtain plant varieties that are more productive and more resistant to adverse conditions in order to make the best use of the agricultural potential around the world.

## CONCLUSION

In recent years, there has been an increasing interest in natural products due to the undesirable effects of synthetic drugs and the antibiotic resistance that negatively affects the treatments. Among these natural products, especially herbal essential oils and extracts and their various metabolites are widely researched. Many pharmaceutical raw material studies have been carried out, especially on phenolics, terpenes and alkaloids, which are the main secondary metabolite groups, and some of these compounds are used as components of some important drugs. Recently, cultivation and especially tissue culture techniques have been used instead of collecting them from nature to obtain these plants and their secondary metabolites, which are candidates for pharmaceutical raw materials. Plant tissue culture studies, in which differences in soil structure and composition can be eliminated and mass production can be carried out, without being affected by seasonal changes, offer great advantages to both researchers and industrial producers. Introducing new active secondary metabolites to the pharmaceutical industry, which will be obtained as a result of advanced purification, characterization, and biological activity research, will not only prevent the undesirable effects and environmental problems caused by synthetic drugs but also significantly prevent treatment failures.

## REFERENCES

[1]     Hartmann T. From waste products to ecochemicals: Fifty years research of plant secondary metabolism. Phytochemistry 2007; 68(22-24): 2831-46.
[http://dx.doi.org/10.1016/j.phytochem.2007.09.017] [PMID: 17980895]

[2]     Okada T, Mochamad Afendi F, Altaf-Ul-Amin M, Takahashi H, Nakamura K, Kanaya S. Metabolomics of medicinal plants: the importance of multivariate analysis of analytical chemistry data. Curr Computeraided Drug Des 2010; 6(3): 179-96.
[http://dx.doi.org/10.2174/157340910791760055] [PMID: 20550511]

[3]     Balunas MJ, Douglas Kinghorn A. Drug discovery from medicinal plants. 2005; 78(5): 431-441.
[http://dx.doi.org/10.1016/j.lfs.2005.09.012]

[4]     Harborne JB. Phytochemical Methods: A Guide to Modern Techniques of Plant Analysis. 3rd ed., UK: Chapman and Hall 1998. Available from: https://link.springer.com/book/9780412572609

[5]     Mothes K. Physiology of Alkaloids. Annu Rev Plant Physiol 1955; 6(1): 393-432.
[http://dx.doi.org/10.1146/annurev.pp.06.060155.002141]

[6]     Paech K. Die stickstoffhaltigen sekundären Pflanzenstoffe. In: Paech K, Ed. Biochemie und Physiologie Der Sekundären Pflanzenstoffe. Berlin, Heidelberg: Springer 1950; pp. 203-54.
[http://dx.doi.org/10.1007/978-3-662-29290-7_7]

[7]     Stahl E. Pflanzen und schnecken. Jena Z Naturwiss 1888; 22: 557.

[8]     Fraenkel GS. The raison d'etre of secondary substances. Science 1959; 129(3361): 1466-70.
[http://dx.doi.org/10.1126/science.129.3361.1466] [PMID: 13658975]

[9]     Levin DA. The chemical defenses of plants to pathogens and herbivores. Annu Rev Ecol Syst 1976; 7(1): 121-59.
[http://dx.doi.org/10.1146/annurev.es.07.110176.001005]

[10]    Levinson HZ. The defensive role of alkaloids in insects and plants. Experientia 1976; 32(4): 408-11.
[http://dx.doi.org/10.1007/BF01920763]

[11]    Schildknecht H. Protective substances of arthropods and plants. Pontif Accad Sci 1977; 3: 59-107.

[12]    Janzen DH. New horizons in the biology of plant defenses Herbivores: their interactions with secondary plant metabolites. New York, USA: Academic Press 1979; pp. 331-48.

[13]    Harborne JB. Recent advances in chemical ecology. Nat Prod Rep 1986; 3(4): 323-44.
[http://dx.doi.org/10.1039/np9860300323] [PMID: 3547185]

[14]    Jain C, Khatana S, Vijayvergia R. Bioactivity of secondary metabolites of various plants: a review. Int J Pharm Sci Res 2019; 10(2): 494-504.

[http://dx.doi.org/10.13040/IJPSR.0975-8232.10(2).494-04]

[15]   Zhao J, Davis LC, Verpoorte R. Elicitor signal transduction leading to production of plant secondary metabolites. Biotechnol Adv 2005; 23(4): 283-333.
[http://dx.doi.org/10.1016/j.biotechadv.2005.01.003] [PMID: 15848039]

[16]   Yang L, Wen KS, Ruan X, Zhao YX, Wei F, Wang Q. Response of plant secondary metabolites to environmental factors. Molecules 2018; 23(4): 762.
[http://dx.doi.org/10.3390/molecules23040762] [PMID: 29584636]

[17]   Piasecka A, Jedrzejczak-Rey N, Bednarek P. Secondary metabolites in plant innate immunity: conserved function of divergent chemicals. New Phytol 2015; 206(3): 948-64.
[http://dx.doi.org/10.1111/nph.13325] [PMID: 25659829]

[18]   Tien L. Root cultures for secondary products. In: Yildirim E, Turan M, Ekinci M, Eds. Plant Roots. London: IntechOpen 2020; pp. 1-21.

[19]   Walton NJ, Brown DE. Chemicals from plants : perspectives on plant secondary products. London: World Scientific 1999; p. 425.
[http://dx.doi.org/10.1142/3203]

[20]   Agostini-Costa TDS, Vieira RF, Bizzo HR, Silveira D, Gimenes MA. Secondary metabolites. In: Dhanarasu S, Ed. Chromatography and its applications. InTech 2012; pp. 131-64.
[http://dx.doi.org/10.5772/35705]

[21]   Connolly JD, Hill RA. Dictionary of terpenoids. (London: Chapman and Hall). Cordell GA. (1976). Biosynthesis of sesquiterpenes. Chem Rev 1991; 76(4): 425-60.

[22]   Boiteau P, Pasich B, Ratsımamanga R. Triterpenoides. Gauthier-Villars Paris 1969; pp. 3-5.

[23]   Garg P, Awasthi S, Horne D, Salgia R, Singhal SS. The innate effects of plant secondary metabolites in preclusion of gynecologic cancers: Inflammatory response and therapeutic action. Biochim Biophys Acta Rev Cancer 2023; 1878(4): 188929.
[http://dx.doi.org/10.1016/j.bbcan.2023.188929] [PMID: 37286146]

[24]   Mahmoud SS, Croteau RB. Strategies for transgenic manipulation of monoterpene biosynthesis in plants. Trends Plant Sci 2002; 7(8): 366-73.
[http://dx.doi.org/10.1016/S1360-1385(02)02303-8] [PMID: 12167332]

[25]   Wink M. Molecular modes of action of cytotoxic alkaloids: from DNA intercalation, spindle poisoning, topoisomerase inhibition to apoptosis and multiple drug resistance. Alkaloids: Chemistry and Biology, 2007; 64, 1–47.
[http://dx.doi.org/10.1016/S1099-4831(07)64001-2]

[26]   Goetz Francis Hadji-Minaglou P, Phytothérapie CE. Guide for prescribers.

[27]   Özkaya O, Şen K, Aubert C, Dündar Ö, Gunata Z. Characterization of the free and glycosidically bound aroma potential of two important tomato cultivars grown in Turkey. J Food Sci Technol 2018; 55(11): 4440-9.
[http://dx.doi.org/10.1007/s13197-018-3362-0] [PMID: 30333640]

[28]   Kaimoto T, Hatakeyama Y, Takahashi K, Imagawa T, Tominaga M, Ohta T. Involvement of transient receptor potential A1 channel in algesic and analgesic actions of the organic compound limonene. Eur J Pain 2016; 20(7): 1155-65.
[http://dx.doi.org/10.1002/ejp.840] [PMID: 27030509]

[29]   Rufino AT, Ribeiro M, Sousa C, *et al.* Evaluation of the anti-inflammatory, anti-catabolic and pro-anabolic effects of E-caryophyllene, myrcene and limonene in a cell model of osteoarthritis. Eur J Pharmacol 2015; 750: 141-50.
[http://dx.doi.org/10.1016/j.ejphar.2015.01.018] [PMID: 25622554]

[30]   de Santana MF, Guimarães AG, Chaves DO, *et al.* The anti-hyperalgesic and anti-inflammatory profiles of *p*-cymene: Evidence for the involvement of opioid system and cytokines. Pharm Biol 2015;

53(11): 1583-90.
[http://dx.doi.org/10.3109/13880209.2014.993040] [PMID: 25856703]

[31]    de Oliveira TM, de Carvalho RBF, da Costa IHF, *et al.* Evaluation of *p* -cymene, a natural antioxidant. Pharm Biol 2015; 53(3): 423-8.
[http://dx.doi.org/10.3109/13880209.2014.923003] [PMID: 25471840]

[32]    Nascimento SS, Camargo EA, DeSantana JM, *et al.* Linalool and linalool complexed in β-cyclodextrin produce anti-hyperalgesic activity and increase Fos protein expression in animal model for fibromyalgia. Naunyn Schmiedebergs Arch Pharmacol 2014; 387(10): 935-42.
[http://dx.doi.org/10.1007/s00210-014-1007-z] [PMID: 24958161]

[33]    Liu X, Cai J, Chen H, *et al.* Antibacterial activity and mechanism of linalool against *Pseudomonas aeruginosa.* Microb Pathog 2020; 141: 103980.
[http://dx.doi.org/10.1016/j.micpath.2020.103980] [PMID: 31962183]

[34]    Baldissera MD, Grando TH, Souza CF, *et al. In vitro* and *in vivo* action of terpinen-4-ol, γ-terpinene, and α-terpinene against Trypanosoma evansi. Exp Parasitol 2016; 162: 43-8.
[http://dx.doi.org/10.1016/j.exppara.2016.01.004] [PMID: 26773165]

[35]    Rivera-Yañez C, Terrazas L, Jimenez-Estrada M, *et al.* Anti-candida activity of *Bursera morelensis* Ramirez essential oil and two compounds, α-pinene and γ-terpinene—an *in vitro* study. Molecules 2017; 22(12): 2095.
[http://dx.doi.org/10.3390/molecules22122095] [PMID: 29206158]

[36]    Formisano C, Sirignano C, Rigano D, *et al.* Antiproliferative activity against leukemia cells of sesquiterpene lactones from the Turkish endemic plant *Centaurea drabifolia* subsp. detonsa. Fitoterapia 2017; 120: 98-102.
[http://dx.doi.org/10.1016/j.fitote.2017.05.016] [PMID: 28579551]

[37]    Wu QX, He XF, Jiang CX, *et al.* Two novel bioactive sulfated guaiane sesquiterpenoid salt alkaloids from the aerial parts of *Scorzonera divaricata.* Fitoterapia 2018; 124: 113-9.
[http://dx.doi.org/10.1016/j.fitote.2017.10.011] [PMID: 29066296]

[38]    de S Vargas F, D O de Almeida P, Aranha ES, *et al.* Biological activities and cytotoxicity of diterpenes from Copaifera spp. Oleoresins. Molecules 2015; 20(4): 6194-210.
[http://dx.doi.org/10.3390/molecules20046194] [PMID: 25859778]

[39]    Shadi S, Saeidi H, Ghanadian M, *et al.* New macrocyclic diterpenes from *Euphorbia connata* Boiss. with cytotoxic activities on human breast cancer cell lines. Nat Prod Res 2015; 29(7): 607-14.
[http://dx.doi.org/10.1080/14786419.2014.979418] [PMID: 25426544]

[40]    Parr AJ, Bolwell GP. Phenols in the plant and in man. The potential for possible nutritional enhancement of the diet by modifying the phenols content or profile. J Sci Food Agric 2000; 80(7): 985-1012.
[http://dx.doi.org/10.1002/(SICI)1097-0010(20000515)80:7<985::AID-JSFA572>3.0.CO;2-7]

[41]    Harborne JB, Mabry TJ. The Flavonoids: advances in research. London: Springer 1982.
[http://dx.doi.org/10.1007/978-1-4899-2915-0]

[42]    Park ES, Moon WS, Song MJ, Kim MN, Chung KH, Yoon JS. Antimicrobial activity of phenol and benzoic acid derivatives. Int Biodeterior Biodegradation 2001; 47(4): 209-14.
[http://dx.doi.org/10.1016/S0964-8305(01)00058-0]

[43]    Pengelly A. The constituents of medicinal plants: An introduction to the chemistry and therapeutics of herbal medicine. 2nd ed. London: Routledge 2020; pp. 1-172.
[http://dx.doi.org/10.4324/9781003117964]

[44]    Scozzafava A, Passaponti M, Supuran CT, Gülçin İ. Carbonic anhydrase inhibitors: guaiacol and catechol derivatives effectively inhibit certain human carbonic anhydrase isoenzymes (hCA I, II, IX and XII). J Enzyme Inhib Med Chem 2015; 30(4): 586-91.
[http://dx.doi.org/10.3109/14756366.2014.956310] [PMID: 25373500]

[45]   Bian YY, Guo J, Majeed H, *et al.* Ferulic acid renders protection to HEK293 cells against oxidative damage and apoptosis induced by hydrogen peroxide. *In Vitro* Cell. Dev Biol Anim 2015; 51(7): 722-9.
[http://dx.doi.org/10.1007/s11626-015-9876-0] [PMID: 25678463]

[46]   Yin P, Zhang Z, Li J, *et al.* Ferulic acid inhibits bovine endometrial epithelial cells against LPS-induced inflammation *via* suppressing NK-κB and MAPK pathway. Res Vet Sci 2019; 126: 164-9.
[http://dx.doi.org/10.1016/j.rvsc.2019.08.018] [PMID: 31499425]

[47]   Kang J, Liu L, Liu Y, Wang X. Ferulic acid inactivates *Shigella flexneri* through cell membrane destruction, biofilm retardation, and altered gene expression. J Agric Food Chem 2020; 68(27): 7121-31.
[http://dx.doi.org/10.1021/acs.jafc.0c01901] [PMID: 32588628]

[48]   Liu F, Xu J, Yang R, Liu S, Hu S, Yan M, Han F. New light on treatment of cervical cancer: Chinese medicine monomers can be effective for cervical cancer by inhibiting the PI3K/Akt signaling pathway. 2022; 157: 114084.
[http://dx.doi.org/10.1016/j.biopha.2022.114084]

[49]   Alkorashy AI, Doghish AS, Abulsoud AI, *et al.* Effect of scopoletin on phagocytic activity of U937-derived human macrophages: Insights from transcriptomic analysis. Genomics 2020; 112(5): 3518-24.
[http://dx.doi.org/10.1016/j.ygeno.2020.03.022] [PMID: 32243896]

[50]   Zhang R, Li D, Xu T, *et al.* Antioxidative effect of luteolin pretreatment on simulated ischemia/reperfusion injury in cardiomyocyte and perfused rat heart. Chin J Integr Med 2017; 23(7): 518-27.
[http://dx.doi.org/10.1007/s11655-015-2296-x] [PMID: 26956461]

[51]   Wang H, Wang H, Cheng H, Che Z. Ameliorating effect of luteolin on memory impairment in an Alzheimer's disease model. Mol Med Rep 2016; 13(5): 4215-20.
[http://dx.doi.org/10.3892/mmr.2016.5052] [PMID: 27035793]

[52]   Kang KA, Piao MJ, Ryu YS, *et al.* Luteolin induces apoptotic cell death *via* antioxidant activity in human colon cancer cells. Int J Oncol 2017; 51(4): 1169-78.
[http://dx.doi.org/10.3892/ijo.2017.4091] [PMID: 28791416]

[53]   Tian C, Liu X, Chang Y, *et al.* Investigation of the anti-inflammatory and antioxidant activities of luteolin, kaempferol, apigenin and quercetin. S Afr J Bot 2021; 137: 257-64.
[http://dx.doi.org/10.1016/j.sajb.2020.10.022]

[54]   Rocha MFG, Sales JA, da Rocha MG, *et al.* Antifungal effects of the flavonoids kaempferol and quercetin: a possible alternative for the control of fungal biofilms. Biofouling 2019; 35(3): 320-8.
[http://dx.doi.org/10.1080/08927014.2019.1604948]

[55]   Wu H, Cui M, Li C, *et al.* Kaempferol reverses aerobic glycolysis *via* miR-339-5p-mediated PKM alternative splicing in colon cancer cells. cite this. J Agric Food Chem 2021; 69(10): 3060-8.
[http://dx.doi.org/10.1021/acs.jafc.0c07640] [PMID: 33663206]

[56]   Wen L, Qu TB, Zhai K, Ding J, Hai Y, Zhou JL. Gallic acid can play a chondroprotective role against AGE-induced osteoarthritis progression. J Orthop Sci 2015; 20(4): 734-41.
[http://dx.doi.org/10.1007/s00776-015-0718-4] [PMID: 25824985]

[57]   Subramanian AP, Jaganathan SK, Mandal M, Supriyanto E, Muhamad II. Gallic acid induced apoptotic events in HCT-15 colon cancer cells. World J Gastroenterol 2016; 22(15): 3952-61.
[http://dx.doi.org/10.3748/wjg.v22.i15.3952] [PMID: 27099438]

[58]   Yigitturk G, Acara AC, Erbas O, *et al.* The antioxidant role of agomelatine and gallic acid on oxidative stress in STZ induced type I diabetic rat testes. Biomed Pharmacother 2017; 87: 240-6.
[http://dx.doi.org/10.1016/j.biopha.2016.12.102] [PMID: 28061407]

[59]   Jiang X, Chen W, Shen F, *et al.* Pinoresinol promotes MC3T3-E1 cell proliferation and differentiation *via* the cyclic AMP/protein kinase-A signaling pathway. Mol Med Rep 2019; 20(3): 2143-50.

[http://dx.doi.org/10.3892/mmr.2019.10468] [PMID: 31322181]

[60] Dutta A, Das D, Baruah BJ, *et al.* Pinoresinol targets NF-kB alongside STAT3 pathway to attenuate IL-6-induced inflammation. Res Sq. 2024.
[http://dx.doi.org/10.21203/rs.3.rs-3900297/v1]

[61] Zhou H, Ren J, Li Z. Antibacterial activity and mechanism of pinoresinol from *Cinnamomum Camphora* leaves against food-related bacteria. Food Control 2017; 79: 192-9.
[http://dx.doi.org/10.1016/j.foodcont.2017.03.041]

[62] Loomis WD, Croteau R. Biochemistry of terpenoids 1980. Available from: https://www.sciencedirect.com/science/article/pii/B9780126754049500199
[http://dx.doi.org/10.1016/B978-0-12-675404-9.50019-9]

[63] Hegnauer R. Biochemistry, distribution and taxonomic relevance of higher plant alkaloids. Phytochemistry 1988; 27(8): 2423-7. Available from: https://www.sciencedirect.com/science/article/pii/0031942288870067
[http://dx.doi.org/10.1016/0031-9422(88)87006-7]

[64] Fraga BM. Natural sesquiterpenoids. Nat Prod Rep 2013; 30(9): 1226-64. Available from: https://cir.nii.ac.jp/crid/1363670320834680320
[http://dx.doi.org/10.1039/c3np70047j] [PMID: 23884176]

[65] Summons RE, Bradley AS, Jahnke LL, Waldbauer JR. Steroids, triterpenoids and molecular oxygen. 2006.
[http://dx.doi.org/10.1098/rstb.2006.1837]

[66] Yao LH, Jiang YM, Shi J, *et al.* Flavonoids in food and their health benefits. Plant Foods Hum Nutr 2004; 59(3): 113-22.
[http://dx.doi.org/10.1007/s11130-004-0049-7] [PMID: 15678717]

[67] Hesse M. Alkaloids: Nature's curse or blessing?. New York: Wiley VCH 2002. Available from: https://books.google.com/books?hl=tr&lr=&id=hLufLzRE0I4C&oi=fnd&pg=PA1&dq=Hesse+M.+Alkaloids:+Nature%E2%80%99s+Curse+or+Blessing%3F+Wiley%3FVCH,+New+York,+2002&ots=1iM-6gNtC1&sig=r7p6wXZHoadWklJ0gwBpCK2Q7-U

[68] Crozier A, Jaganath IB, Clifford MN. Phenols, Polyphenols and Tannins: An Overview. In: Crozier A, Clifford MN, Ashihara H, editors. Plant Secondary Metabolites: Occurrence, Structure and Role in the Human Diet. Wiley-Blackwell; 2006.
[http://dx.doi.org/10.1002/9780470988558.ch1]

[69] Barzegar E, Fouladdel S, Movahhed TK, *et al.* Effects of berberine on proliferation, cell cycle distribution and apoptosis of human breast cancer T47D and MCF7 cell lines. Iran J Basic Med Sci 2015; 18(4): 334-42.
[PMID: 26019795]

[70] Li Q, Zhao H, Chen W, Huang P. Berberine induces apoptosis and arrests the cell cycle in multiple cancer cell lines. Arch Med Sci 2023; 19(5): 1530-7.
[http://dx.doi.org/10.5114/aoms/132969] [PMID: 37732040]

[71] Zhang Q, Lyu Y, Huang J, *et al.* Antibacterial activity and mechanism of sanguinarine against *Providencia rettgeri in vitro.* PeerJ 2020; 8: e9543.
[http://dx.doi.org/10.7717/peerj.9543] [PMID: 32864203]

[72] Zhong H, Hu DD, Hu GH, *et al.* Activity of sanguinarine against *Candida albicans* biofilms. Antimicrob Agents Chemother 2017; 61(5): e02259-16.
[http://dx.doi.org/10.1128/AAC.02259-16] [PMID: 28223387]

[73] Zielińska S, Wójciak-Kosior M, Dziągwa-Becker M, *et al.* The activity of isoquinoline alkaloids and extracts from *Chelidonium majus* against pathogenic bacteria and *Candida* sp. Toxins (Basel) 2019; 11(7): 406.
[http://dx.doi.org/10.3390/toxins11070406] [PMID: 31336994]

[74]    Nigdelioglu Dolanbay S, Kocanci FG, Aslim B. Neuroprotective effects of allocryptopine-rich alkaloid extracts against oxidative stress-induced neuronal damage. Biomed Pharmacother 2021; 140: 111690.
[http://dx.doi.org/10.1016/j.biopha.2021.111690] [PMID: 34004513]

[75]    Wei T, Xiaojun X, Peilong C. Magnoflorine improves sensitivity to doxorubicin (DOX) of breast cancer cells *via* inducing apoptosis and autophagy through AKT/mTOR and p38 signaling pathways. Biomed Pharmacother 2020; 121: 109139.
[http://dx.doi.org/10.1016/j.biopha.2019.109139] [PMID: 31707337]

[76]    Kim J, Ha Quang Bao T, Shin YK, Kim KY. Antifungal activity of magnoflorine against *Candida strains*. World J Microbiol Biotechnol 2018; 34(11): 167.
[http://dx.doi.org/10.1007/s11274-018-2549-x] [PMID: 30382403]

[77]    Guo S, Jiang K, Wu H, *et al.* Magnoflorine ameliorates lipopolysaccharide-induced acute lung injury *via* suppressing NF-κB and MAPK activation. Front Pharmacol 2018; 9: 982.
[http://dx.doi.org/10.3389/fphar.2018.00982] [PMID: 30214410]

[78]    Li J, Qiu DM, Chen SH, Cao SP, Xia XL. Suppression of human breast cancer cell metastasis by coptisine *in vitro*. Asian Pac J Cancer Prev 2014; 15(14): 5747-51.
[http://dx.doi.org/10.7314/APJCP.2014.15.14.5747] [PMID: 25081696]

[79]    Nakonieczna S, Grabarska A, Gawel K, *et al.* Isoquinoline alkaloids from *Coptis chinensis* franch: focus on coptisine as a potential therapeutic candidate against gastric cancer cells. Int J Mol Sci 2022; 23(18): 10330.
[http://dx.doi.org/10.3390/ijms231810330] [PMID: 36142236]

[80]    Li C, Huang P, Wong K, *et al.* Coptisine-induced inhibition of *Helicobacter pylori* : elucidation of specific mechanisms by probing urease active site and its maturation process. J Enzyme Inhib Med Chem 2018; 33(1): 1362-75.
[http://dx.doi.org/10.1080/14756366.2018.1501044] [PMID: 30191728]

[81]    Wang F, Zhang Y, Pang R, Shi S, Wang R. Scoulerine promotes cytotoxicity and attenuates stemness in ovarian cancer by targeting PI3K/AKT/mTOR axis. Acta Pharm 2023; 73(3): 475-88.
[http://dx.doi.org/10.2478/acph-2023-0021] [PMID: 37708956]

[82]    Tian J, Mo J, Xu L, *et al.* Scoulerine promotes cell viability reduction and apoptosis by activating ROS-dependent endoplasmic reticulum stress in colorectal cancer cells. Chem Biol Interact 2020; 327: 109184.
[http://dx.doi.org/10.1016/j.cbi.2020.109184] [PMID: 32590070]

[83]    Ncube B, Van Staden J. Tilting plant metabolism for improved metabolite biosynthesis and enhanced human benefit. Molecules 2015; 20(7): 12698-731.
[http://dx.doi.org/10.3390/molecules200712698] [PMID: 26184148]

[84]    Guerriero G, Berni R, Muñoz-Sanchez JA, *et al.* Production of plant secondary metabolites: Examples, tips and suggestions for biotechnologists. Mdpi. Genes (Basel) 2018; 9(6): 309.
[http://dx.doi.org/10.3390/genes9060309] [PMID: 29925808]

[85]    Akula R, Ravishankar GA. Influence of abiotic stress signals on secondary metabolites in plants. Plant Signal Behav 2011; 6(11): 1720-31.
[http://dx.doi.org/10.4161/psb.6.11.17613] [PMID: 22041989]

[86]    Seigler DS. Plant secondary metabolism 2012. Available from: https://books.google.com/books?hl=tr&lr=&id=uKPwBwAAQBAJ&oi=fnd&pg=PP7&dq=Seigler,+D.S.+1998.+Plant+Secondary+Metabolism.+Springer,+US.&ots=AhPQj6UjmY&sig=vqZWFntJvI4prR_ctYHuefL2cm4

[87]    Zinkel DF, Russell J. Naval Stores: Production, Chemistry, Utilization. New York: Pulp Chemicals Association 1989.
[http://dx.doi.org/10.1007/978-3-642-74075-6_26]

[88]    Dawson GA. The amazing terpenes. Naval Stores Rev 1994; pp. 6-12.

[89]   Britto SJ, Senthilkumar S. Antibacterial activity of *Solanum incanum* L. leaf extracts. Asian J Microbiol Biotechnol Environ Sci 2001; 3(1-2): 65-6.

[90]   Saha S, Pal D. Elicitor Signal Transduction Leading to the Production of Plant Secondary Metabolites. In: Pal, D., Nayak, A.K. (eds) Bioactive Natural Products for Pharmaceutical Applications. Advanced Structured Materials, vol 140. Springer, Cham.
[http://dx.doi.org/10.1007/978-3-030-54027-2_1]

[91]   Krikorian AD, Steward FC. Biochemical differentiation:the biosynthetic potentialities of growing and quiescent tissue. In: Steward FC, Ed. Plant Physiology - A Treatise. New York, London: Academic Press 1969; pp. 227-326.
[http://dx.doi.org/10.1016/B978-0-12-395679-8.50012-1]

[92]   Zenk MH. 6. Chasing the enzymes of secondary metabolism: Plant cell cultures as a pot of gold. Phytochemistry 1991; 30(12): 3861-3.
[http://dx.doi.org/10.1016/0031-9422(91)83424-J]

[93]   Bourgaud F, Gravot A, Milesi S, Gontier E. Production of plant secondary metabolites: a historical perspective. Plant Sci 2001; 161(5): 839-51.
[http://dx.doi.org/10.1016/S0168-9452(01)00490-3]

[94]   Ravishankar GA, Ramachandra Rao S. Biotechnological production of phyto-pharmaceuticals. J Biochem Mol Biol Biophys 2000; 4: 73-102.

[95]   Cheetham PSJ. Biotransformations: new routes to food ingredients. Chem Ind 1995; 7: 265-8.

[96]   Tiwari R, Rana CS. Plant secondary metabolites: a review. Int J Eng Res Gen Sci 2015; 3(5): 661-70.

[97]   Bourgaud F, Gravot A, Milesi S, science, E. G.-P., & undefined. (n.d.). Production of plant secondary metabolites: a historical perspective. Elsevier, 2001.

# CHAPTER 5

# Bioreactor Sytems: Physiology of Cell Cultures

**Ergun Kaya**[1,*] and **Sedat Çiçek**[1]

[1] *Muğla Sıtkı Koçman University, Faculty of Science, Molecular Biology and Genetics Department, 48000, Menteşe, Muğla, Türkiye*

**Abstract:** Cell culture in plants is a technique in which cells of plant tissues are developed *in vitro* in an artificial environment suitable for growth and proliferation. By developing different cell culture environments, it is possible to conduct many experimental studies such as cell proliferation, differentiation, identification of growth factors, understanding the mechanisms underlying the normal functions of various cell types, cell-cell or cell-matrix interactions, and determining the effects of molecules thought to be effective in metabolic pathways. Therefore, cell cultures have become one of the major tools used in cellular and molecular biology. Bioreactor systems, developed as an alternative support to traditional cell culture studies, aim not only for large-scale mass propagation, but also for the application of various physiological approaches, especially in plant protoplast cells, understanding metabolic pathways, and the factors effective in secondary metabolite production, and especially the application of transformation-oriented methods. In this context, this chapter aims to examine the physiology of cell cultures on a bioreactor basis and shed light on physiological processes with current and/or future approaches.

**Keywords:** Bioreactor, Metabolic pathways, Plant growth regulators, Protoplast.

## INTRODUCTION

Bioreactors can be defined as mechanical containers or tanks in which living organisms, cells or tissues are cultured in a liquid nutrient medium and the conditions inside are kept under strict control. The term bioreactor is often used synonymously with fermentor. However, fermenter has a narrow meaning used to obtain alcohol from sugar in an anaerobic environment. The main difference that distinguishes bioreactors from traditional chemical reactors is the control and support of living-biological contents [1, 2]. Bioreactors can also be defined as a tool or device used to carry out one or more biochemical reactions to obtain the desired products from a starting material. Bioreactors represent the latest steps in

---

* **Corresponding author Ergun Kaya:** Muğla Sıtkı Koçman University, Faculty of Science, Molecular Biology and Genetics Department, 48000, Menteşe, Muğla, Türkiye; E-mail: ergunkaya@mu.edu.tr

the development of biologically based processes. In general, the basic function of the bioreactor is to provide optimal conditions for effective cell growth and metabolism by tightly regulating various key environmental (chemical and physical) factors [3, 4].

A bioreactor is a tank with an electronic control panel where probes that detect pH, temperature, and dissolved oxygen in the culture medium are placed, allowing the addition of fresh medium and removal of products (according to the operation mode), pH regulation, air supply, mixing and temperature control without disturbing the aseptic conditions of the culture medium. Thus, the bioreactor is a technological system with mechanical and electronic features that enable close monitoring of culture conditions and carry out the necessary physical and chemical interventions as programmed [5, 6].

The first bioreactors used outside microbial technology in plant cell and tissue cultures were stirred tank reactors. Various types of bioreactors have emerged with their application to plant cell and tissue cultures. Stirred tank reactors provided mixing and aeration by mechanical means. It was also used in bioreactors in gas-blowing units (Fig. **1**) [7]. As we have emphasized several times before, plant secondary metabolites are produced in plant cell and tissue cultures in lower amounts than they are produced in the plant. For this reason, bioreactors of different shapes are trying to solve these problems and produce them at high yields. However, this technology has various problems that need to be solved in the production of secondary metabolites of commercial and medical importance. There are various factors that affect the production of bioreactors such as carbohydrates in the nutrient medium, nutrients, growth regulators, pH, atmospheric gases, oxygen supply, $CO_2$ cycle, viscosity - liquid medium fluidity, cell density, lysis stress of plant cells [8, 9].

Today, bioreactors are seen to be effectively applied in culture types such as cellular biomass production, organ cultures, somatic embryo production, and micropropagation. Liquid culture media increase the uptake of nutrients in micropropagation, and yield by encouraging growth, and provide a larger volumetric area. *In vitro* culture in liquid media also has disadvantages. Effects such as oxygen deficiency (asphyxia), hyperhydricity and shear forces can be considered as disadvantages. Considering these problems, modifications, combinations and the development of new bioreactor systems have been made in bioreactors. In micropropagation and organ cultures, systems (immersion bioreactors SETIS™, RITA®, WE VITRO) that contain two separate tanks and allow plant materials to come into contact with nutrient solutions for certain periods of time have been used in micropropagation and organ cultures in recent years (Fig. **2**) [7]. Although bioreactors serve different application areas, they are important biotechnology applications that will develop in many areas in the next century [11, 12].

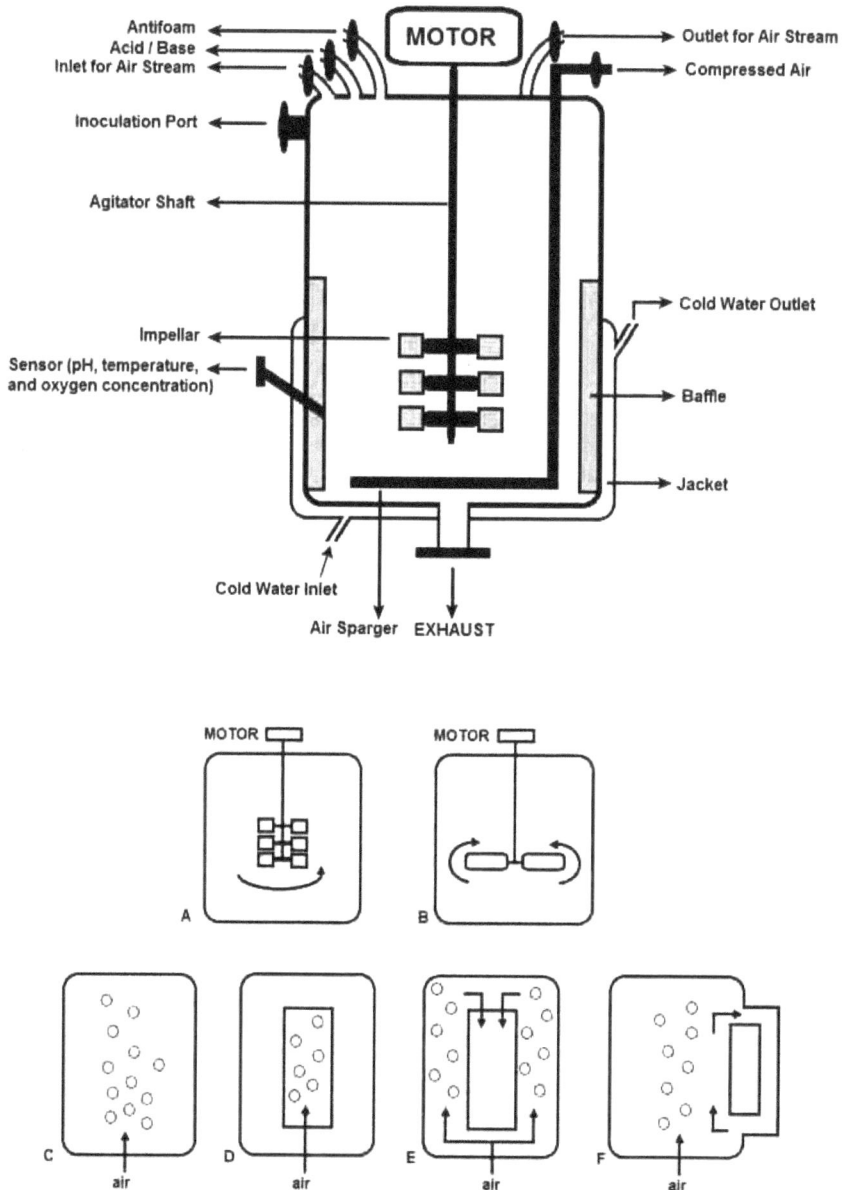

**Fig. (1).** Typical structure of a stirred-tank bioreactor system (upper side). Structure of different types of bioreactor systems. Flat-blade turbine impeller (**A**) and marine propeller (**B**) agitation based bioreactor systems, bubble column (**C, D**), draft tube air lift (**E**) and external loop air-lift (**F**) bioreactor systems (lower side) [7, 10].

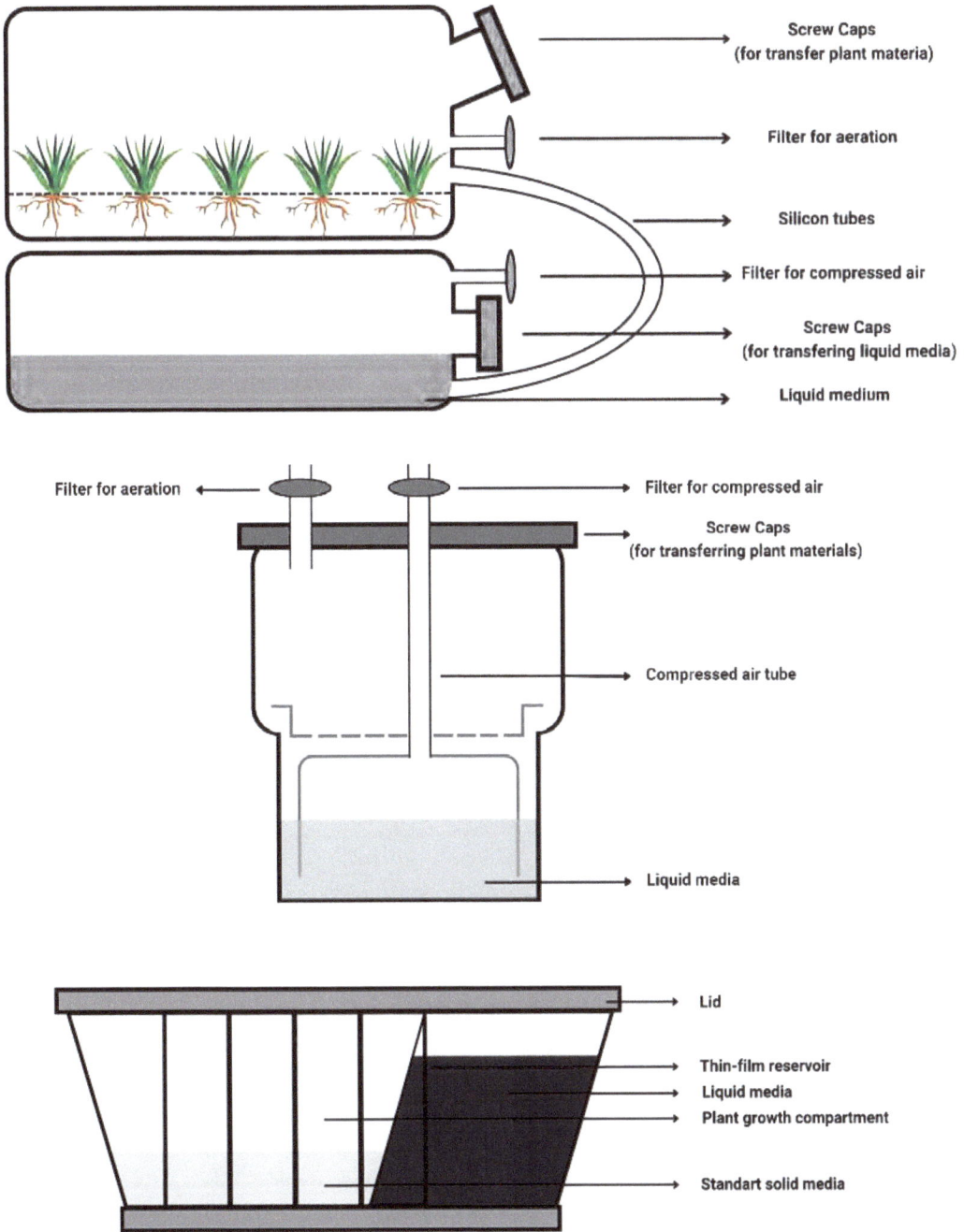

**Fig. (2).** Structure of SETIS™ (**A**); RITA® (**B**); WE VITRO® (**C**) bioreactor systems [7, 10].

# CELL CULTURES

Plant cell cultures exhibit significant differences compared to animal cell cultures. First of all, while cultured plant cells proliferate unlimitedly in vitro, animal cell cultures generally undergo a limited number of cell divisions due to the effect of telomeric shortening. There are also differences between two different cell cultures such as temperature, organic carbon source, pH and light requirement [13].

Cell cultures begin with the transfer of different types of sterile explants to a nutrient medium containing plant growth regulators, followed by the formation of undifferentiated cell clusters (callus) on the culture medium, followed by the development of cells in constantly shaking liquid culture media or bioreactors. The time required to initiate and grow cell culture varies depending on the plant species and growth environment. Callus cultures, which grow slower than cell cultures, allow long-term preservation of plant cell lines and can provide a material source for researchers in case of contamination of cell cultures [14, 15].

The usability of the material in different areas, the high rate of cell growth and the reproducibility of the conditions make cell cultures a simplified model system for cellular and molecular studies. The application areas of plant cell cultures in recent years have been aimed at large-scale production of secondary metabolites as well as in basic research [15, 16].

## Single Cell Cultures

In addition to cell metabolism, the effects of various substances on cellular responses can be better revealed in single-cell cultures compared to organs and/or whole plants. Single cells can be isolated from healthy plant organs or liquid cultures. However, single-cell systems used in basic and applied research can also be obtained from cultured tissues as they offer various advantages [8, 17].

The composition of the culture medium and the initial cell density used are two critical factors for the success of single-cell culture. For high densities of cells, it is recommended to use media with a composition similar to nutrient media used for liquid cultures or callus cultures. The effect of cell density on cell division has been explained on the basis that cells synthesize the compounds necessary for their division. The concentration of these compounds must reach a threshold level before a cell begins to divide. At high cell density, equilibrium is achieved much earlier than at low density and therefore the lag phase is short. Below a critical cell density, equilibrium is never reached and cells cannot divide. In this context, detailed analyses are needed to identify factors responsible for cell division in single-cell cultures [17, 18].

Plant cell cultures allow the complex structure of the vegetal organization to be examined in detail in its basic units, thus enabling easier analysis of various physiological processes. Another common use of plant cell cultures is the production of secondary metabolites [19, 20].

Cell culture studies are applications used in various fields, from scientific research to industry. Cell culture techniques and the number of today's cell lines have come a long way since the beginning of these studies. Their advances are further enhanced by the availability of various molecular biology tools, such as DNA fingerprinting and cytogenetic analysis, to identify and characterize cell lines. With the increase in the number of cell lines, growth, imaging, data collection and analysis methods in cell culture techniques have also improved. Although the use of classical cell culture is widespread, recent research has turned to culture using three-dimensional (3D) structures and more realistic biochemical and biomechanical microenvironments [21, 22].

## Protoplast Cultures

The cells that form tissues in plants are found together through the middle lamella. Plant cells that form tissues can be separated from each other by enzymatic or mechanical means. When the primary cell wall is removed by different applications, the plant cell consisting of the cell membrane and protoplasm is called a protoplast. Protoplasts can maintain their viability under appropriate physical conditions. For example, they can re-form cell walls in isotonic environments. They can divide by mitosis. They even re-form plant cell clumps. Different applications can be made from these cells. This obtained cellular material can be used as an important cellular resource in plants [23, 24].

Obtaining a plant from a single protoplast cell (Fig. **3**), that is, single cell-based plant regeneration, is one of the most important achievements. Plant clones can be produced from a single cell that is diploid or has different levels of polyploidy. At the same time, different plant lines and hybrids can be produced from the obtained protoplasts using a modern approach to plant breeding through fusion, that is, somatic hybridization processes, and these can be propagated by cloning [25, 26].

In order to clone from protoplast, plant regeneration methods must be resolved and applicable. Many different biotechnological and genetic engineering applications can be made from these protoplasts, such as somatic embryo production, cell cultures and cell lines, secondary metabolite production, and molecular agriculture. Genetic transformation, which is considered one of the most important issues in plant molecular studies, provides great convenience for mutant isolation studies. In the fields of plant physiology and cytology, they can be used in cell physiology studies such as the entry and transport of different molecules from gases to nutrients through the cell membrane, isolation of cell organelles, cell cycle, and differentiation [29, 30].

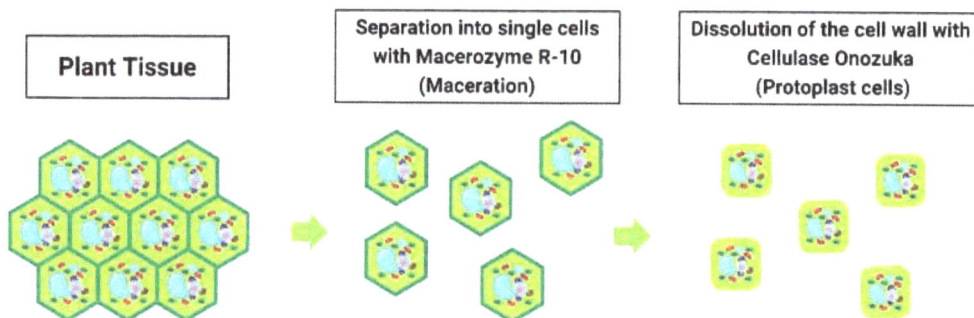

**Fig. (3).** Illustration of protoplast cell preparation [27, 28].

Protoplast culture and somatic hybridization, which are the most modern approaches in plant breeding, are promising in species whose genetic diversity is restricted and decreasing. In the future, protoplast fusion will provide advantages for transferring disease and pest-resistance genes seen in wild species to cultivated plants. Against the negative perspectives created by transgenic studies, this method can be seen as a much more valid and appropriate line of work [31, 32]. Protoplast culture and protoplast fusion are important research topics in overcoming many interspecies' incompatibility problems, in hybridization that cannot be achieved by classical methods, and even in the development of new varieties and new species. Somatic hybridization can be performed under appropriate laboratory and cellular conditions. Thus, hybrid plants can be easily selected, cloned, and propagated. In the literature, you can find information about many varieties and hybrids obtained through somatic crossbreeding. However, there are still problems in practice regarding protoplast culture and somatic hybridization in many important plant groups. Protoplast fusion also enables the transfer of desired qualities such as resistance to diseases (bacterial, fungal, viral), pests, herbicides and other stress factors [33, 34].

**Three-dimensional (3D) Cell Cultures**

Three-dimensional (3D) cell cultures are a model system that allows cell aggregates to form as tissue spheroids or embedded cells in a tissue scaffold or liquid-based methods where structural proteins and other biological molecules found in living tissues mimic the extracellular matrix. Since cells cultured as 3D models show polarization similar to that *in-vivo*, their morphology also approaches *in-vivo*. It mimics *in-vivo* events to a certain extent. For example, by placing cells in an extracellular matrix, conditions related to physiological behaviors such as apical-basal polarization, lumen formation, proliferation changes, and numerous changes in RNA and protein expressions can be more easily monitored [35 - 37].

3D cell culture approaches that aim to model *in-vivo* interactions of tissues and organs also enable the examination of biochemical and biomechanical signals. These models have been proven to produce more realistic implications for study findings in *in-vivo* applications. Although classical 2D cell lines provide homogeneous working material, culturing them as 3D models allows them to be closer to natural conditions. A well-designed microenvironment in 3D culture models can be used to support proliferation, migration, matrix production and stem cell differentiation. To date, the 3D culture approach has been used to study hundreds of cell lines. In addition, drug discovery, cytotoxicity, genotoxicity, cell growth, gene and protein expression studies are important areas where 3D cell culture systems are frequently used. Similarly, co-cultures in 3D systems also provide a better understanding of cell interactions [38 - 40].

## BIOREACTOR APPLICATIONS IN PLANTS

Plant cells can be grown in bioreactors of different shapes, but there are several problems that need to be solved in adapting this technology to the large-scale production of useful secondary metabolites. Among the various factors affecting the culture medium and secondary metabolite production in bioreactors include atmospheric gases, oxygen supply, $CO_2$ cycle, pH, carbohydrates, growth regulators, liquid medium fluidity, and cell density [8, 41].

In somatic embryo culture, large-scale bioreactor application was attempted for the first time for carrot plants, but only a few embryos could be created. Studies on somatic embryo production in this plant continued and it was determined that the optimal amount of dissolved oxygen was 16%, only ammonium was suitable for use as a nitrogen source and successful culture could be achieved with optimal pH [42, 43].

Bioreactors were first used for microbial technology, and these applications were almost limited to stirred tank reactors. When subsequently applied to plant cell cultures, mechanical or gas-blown bioreactors were used to provide mixing and aeration. It has subsequently been effectively applied in micropropagation with biomass production, organogenesis and somatic embryogenesis in bioreactors. Liquid culture medium increased nutrient uptake and promoted more efficient growth in micropropagation. However, in addition to these advantages of *in vitro* culture in liquid media, technical problems such as oxygen deficiency (asphyxia), hyperhydricity and shear forces are also encountered. To overcome these problems, bioreactors have been developed over time and many models have been produced [9, 44].

## Liquid Phase Bioreactors

Liquid phase bioreactors increase the uptake of nutrients in micropropagation, and efficiency by encouraging growth, and provide a larger volumetric area. Despite its widespread use, *in vitro* culture in liquid media also has disadvantages. Effects such as oxygen deficiency (asphyxia), hyperhydricity, and shear forces can be considered as disadvantages. Considering these problems, modifications, combinations or the development of new bioreactor systems have been made in bioreactors [45, 46].

### Stirred Tank Bioreactors

They are mechanically stirred liquid-phase bioreactors. They are preferred in plant cell cultures rather than animal cell cultures. These types of reactors are most commonly used in high-scale biosynthetic processes of plant cell cultures. Stirred tank bioreactors are simple mixers made of glass or stainless steel. The mixing mechanism can be at the top or bottom. The baffle of the bioreactor prevents vortex formation. Most commercial bioreactors are these types of bioreactors. There are also studies in which these bioreactors are used for different plants [47].

For example, a study was conducted comparing Erlenenmayer and stirred bioreactor culture for the production of betalain, an intracellular metabolite obtained from Beta vulgaris, and as a result, it was observed that biomass, betalain production, and growth rate were lower in the bioreactor. However, no difference was observed in cell aggregate size change and cellular viability. In this study, it was observed that cells secrete proteins and polysaccharides against hydrodynamic stress, thus changing the fluidity of the environment has negative effects. On the other hand, it has been reported that somatic embryo cultures of some seedless plants and black spruce have been successfully propagated in stirred bioreactor tanks [48, 49].

### Air-lift Bioreactors

These types of bioreactors can be described as bubble columns in which the environment is mixed and ventilated by the introduction of air or another gas mixture into an airtight tube. Gas circulation is provided through an inner airtight tube or an outer loop. Therefore, the reactor, which has a vertical flow loop, is divided into two parts, the part containing gas and the part not containing gas. They are similar to stirred bioreactors but do not have stirrer systems [50].

Plant cells have large vacuoles and a slow growth rate. Plant roots need lower levels of oxygen. Air-lift bioreactors can provide small amounts of oxygen with a low fragmentation rate effect. Air passes through the glass grille and functions as ventilation. This method has been found to be more successful in stem cell cultures than stirred tank bioreactors [50, 51].

These bioreactors have disadvantages such as foaming caused by excess air and cell growth in the headspace. The reason for foaming and cell growth on the walls of the tank is that the top of the tank has the same radius. To overcome these problems, the upper parts of the bioreactor were modified by expanding them. Using a balloon-shaped tube to lift the cells at the top of the tank greatly reduced foaming. On the other hand, this limitation could be avoided by using antifoam [50, 52].

### Ebb-and-flow Bioreactors

It is a type of tidal bioreactor developed for the propagation of various plants. This type of reactor has a support structure that prevents the plant material from completely submerging in the liquid medium. Nutrient medium is pumped from the supplement tank to the culture tank. To ensure even growth, the nutrient medium is regularly delivered to the plant materials through a series of channels. The nutrient medium remains in the tank for a few minutes and is filtered back into the make-up tank for reuse. The filtration process is controlled by a solenoid valve at periods determined by the plant species and explant type. A programmable pump moves the nutrient medium back and forth, with compressed air delivered through a distributor at the bottom of the medium [9, 53].

### Turbine Blade Bioreactors

In this bioreactor system, the cultivation area and the mixing area are separated by a mesh made of stainless steel. In this way, while the roots do not come into contact with an impeller, the air is given by the paddle mixer at the bottom and the environment is mixed. In a comparative study, it was reported that these bioreactors were more advantageous than other bioreactors in carrot root cells [9, 46].

### Rotating Drum Bioreactors

This bioreactor system has considerably more surface area compared to other bioreactors. As a result, less energy is consumed for material transport. This feature is preferred for tissues sensitive to degradation and photobioreactors. The boiler-shaped carrier tank is mounted on a rotator for support and rotation. The boiler is rotated at a low speed, thus minimizing the stress of disintegration on the roots [9, 46].

## Nutrient Mist Bioreactors (Gas-Phase Bioreactors)

In gas-phase bioreactors, biomass is exposed to air or a gas mixture, and nutrients are delivered *via* droplets. Droplet diameter is achieved in different ways, including mist, spray-spray. When using a gas-phase reactor, difficulties in the transport of materials, especially oxygen, can be significantly reduced or eliminated [9, 55].

Plant cells are extremely sensitive to the gases in their environment, especially $O_2$, $CO_2$ and ethylene. Due to the low solubility of these gases, there are difficulties in delivering them to the roots in liquid media. Mixing, shaking, or bubbling the liquid medium to deliver gases can cause disintegration and damage to plant tissues. Therefore, gas-phase reactors offer many advantages over liquid-phase reactors [56].

## Hybrid Bioreactors

Hybrid bioreactor applications, which are a mixture of gas and liquid bioreactors, are also encountered for the needs of some plant cultures. Due to the disadvantages of gas-phase bioreactors, it has been shown that it is appropriate to first switch to a liquid bioreactor and then to a gas-phase bioreactor compared to plant materials [9, 57].

## Disposible Bioreactors

Single-use bioreactors are generally used for small-scale production. The main advantage of these bioreactors is that they eliminate cleaning and sterilization issues and reduce the investment cost. The development of disposable wave bioreactor systems has provided new advantages. The working principle of this system is based on wave-induced mixing and its most obvious advantage is that it reduces stress sources. In addition, since they are disposable, they save staff and time in terms of cleaning and sterilization. It also facilitates expectations of good manufacturing practices [58, 59].

Disposable reactors are bioreactors in which media are placed in plastic bags, ventilated with plastic valves, and generally mixed by holding them on a shaker. Growth media carriers are typically manufactured from the Food and Drug Administration-approved biocompatible plastics [polyethylene - High-Density Polyethylene (HDPE) or Low-Density Polyethylene (LDPE), polystyrene - PET, polytetrafluoroethylene - PTFE, polypropylene – FEP] [60, 61].

## Microfluidic Bioreactors

Microfluidic technology is one of the highest-pressure techniques ever used to produce products with better performance. Microfluidic technology is a way to process small volumes of liquids between microliters and picoliters. These techniques that assist fluid flow are generally classified as active or passive. Active microfluidics involves the movement, transport, and analysis of biological samples. Small volumes of liquid are used in high-volume scanning, diagnostic, and research applications. Studies on these devices cover a variety of topics, including liquid distribution, system properties, detection techniques, and bioanalytical applications. These devices are preferred due to reasons such as low production costs, affordable disposable chips, and the possibility of mass production [62, 63].

In today's technologies, microfluidic systems are seen to be made of silicon and glass (Fig. **4**). However, since the production of silicone products is expensive, new products are sought instead. Additionally, silicon and glass cannot be combined with optical microscopy due to their low gas permeability. Due to the low gas permeability problem of silicone and glass, they are not suitable for use in the biological field. Researchers are studying organic polymers to develop options that can be optically transparent, flexible, easy to produce and inexpensive. For this reason, microfluidic techniques, related mechanisms and applications are developing day by day [64, 65].

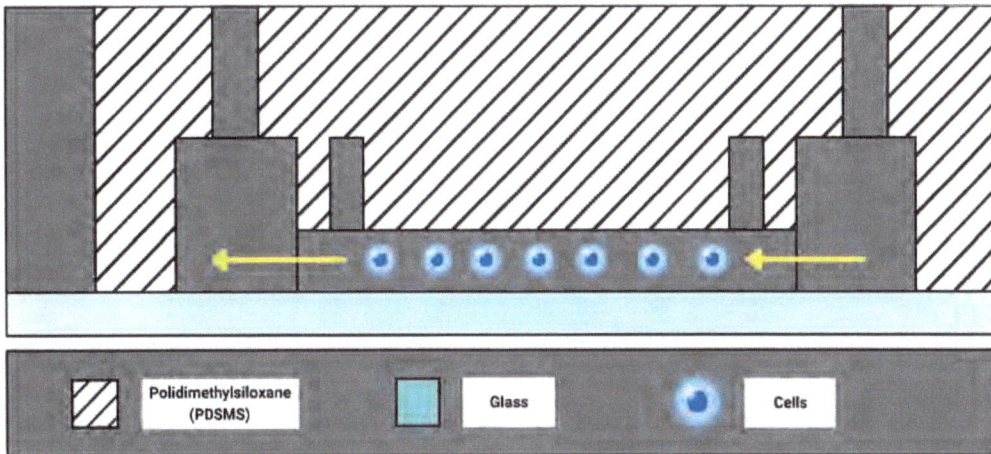

**Fig. (4).** Microfluidic cell culture system [66, 67].

Microfluidic systems have the potential to make a significant impact on cell culture and cell biology research. The cell is the smallest unit of a living thing that can display structural and functional features. It does not circulate in our body but instead binds to an environment called the extracellular matrix for survival. Cell heterogeneity is vital for the correct interpretation of their results. No two cells in a genetically identical group are identical. The orientation of individual cells involved, molecular and cellular stochastic processes, phases of cells, asymmetric division during cell division, and non-homogeneous cell environments are important [68, 69].

Thanks to traditional technologies, the first microfluidic systems were generally made of silicon and glass. Researchers have used silicon and glass with today's technology and developed microfluidic systems. However, these products have some disadvantages. Silicon is expensive and cannot be incorporated into an optical microscope due to its opacity. Additionally, both silicone and glass have low gas permeability. To overcome these drawbacks, researchers have studied organic polymers that are best suited to develop alternatives that are optically transparent, flexible, easy to produce, and inexpensive compared to their predecessors. As a result of these studies, PDMS was developed for use in microfluidic systems in the 1990s [64, 70].

Today, it has become the most used material in microfluidic systems and is widely used. Compared to natural and synthetic materials used in other microelectronic technologies, one of the most important advantages of PDMS is its recognition by cells (biocompatibility) and the ease it provides in growing organisms with simple cell structures. In the late 1990s, microfluidic devices were created for cell biology applications such as cell-based biosensors, cell and protein separation modeling, culture, and research [64].

Towards the 2000s, researchers began to develop and apply microfluidic devices that can be used as tissue and organ samples in new drug research and development. In these studies, an organ-on-chip was developed by examining pathophysiology and biological processes. A well-designed microenvironment in 3D culture models can be used to support proliferation, migration, matrix production, and stem cell differentiation. In addition, drug discovery, cell growth, cytotoxicity, genotoxicity, protein and gene expression studies are among the important areas where 3D cell culture systems are frequently used [38, 71].

**Horticultural plant production with bioreactor systems**

Industrial plant propagation appears to be promising with large-scale plant production by cell tissue and embryo cultures in bioreactors. In the context of biochemistry, bioreactors are typically defined as self-contained, sterile environments that focus on liquid nutrient or liquid/air intake and outflow systems. They are made for intensive culture and provide many opportunities for microenvironmental condition monitoring and control. Clonal propagation through conventional micropropagation is usually a labor-intensive process. Anoectochilus, rice endosperm, carrot root, tomato fruit, potato, apple, Chrysanthemum, garlic, ginseng, grape, Lilium, and Phalaenopsis can all be tested [72 - 76]. The manufacture of important natural goods, including medications, fine chemicals, and flavors and perfumes, is the main goal of using plant cell cultures. Plants may manufacture about 20,000 different compounds, and each year, roughly 1600 new plant chemicals are identified. The business sector has conducted a large portion of the research in this area. Plant cell cultures have not been widely used due to a number of issues including low cell productivity, slow growth, genetic instability, and an inability to sustain photoautotrophic development. Only ginsenosides, berberine, and shikonin have been synthesized on a wide scale in plant cell cultures, despite the potential benefits of doing so. One viable method of lowering micropropagation expenses has been suggested, automating the process in bioreactors. Since then, it has been shown to work with a wide range of plant species and organs, such as somatic embryos, shoots, bulbs, micro tubers, and corms [77 - 79].

## CONCLUSION

The development and availability of *in vitro* plant cell and tissue cultures have increased tremendously over the past hundred years. There are many studies on new plant species in the development stage, and successful examples are seen in the scientific world. As a result of this knowledge in the 21st century, the economically efficient and sustainable production of plants used in agriculture, food, pharmacology, and medicine will expand and spread with applications of biotechnology such as bioreactors. Major changes and advances in the production of biomass and plant secondary metabolites and molecular agriculture products have been experienced so far and will continue to be experienced in the future. Increasing their production, which can be applied on an industrial scale, is seen as the solution to food and medicine problems on a global scale. It is clear that 3D cell culture models have many advantages over classical two-dimensional monolayer culture methods due to improved cell-to-cell and cell-extracellular matrix interactions, as well as *in-vivo*-like cellular structures. *In vitro* studies with 3D cell cultures are in many ways closer to the organism and offer biologically superior structures for studying complex interactions that are not possible with classic cultures. Nowadays, classic cell culture studies are mostly preferred, but it is of great importance to use 3D culture techniques to understand the behavior of some *in-vivo* conditions. There have been great changes and advances in the production of biomass and plant-derived metabolites and therapeutic proteins using *in vitro* cultures. Dozens of these have been implemented in industry and many more are under development. Today, bioreactor designs and equipment have come a long way in meeting the physiological needs of plant cell and tissue cultures, and will continue to progress by constantly adding innovations with new technological developments.

## REFERENCES

[1]  Malhotra N. Bioreactors design, types, influencing factors and potential application in dentistry. A literature review. Curr Stem Cell Res Ther 2019; 14(4): 351-66.
[http://dx.doi.org/10.2174/1574888X14666190111105504] [PMID: 30636614]

[2]  Martin Y, Vermette P. Bioreactors for tissue mass culture: Design, characterization, and recent advances. Biomaterials 2005; 26(35): 7481-503.
[http://dx.doi.org/10.1016/j.biomaterials.2005.05.057] [PMID: 16023202]

[3]  Georgiev MI, Weber J. Bioreactors for plant cells: hardware configuration and internal environment optimization as tools for wider commercialization. Biotechnol Lett 2014; 36(7): 1359-67.
[http://dx.doi.org/10.1007/s10529-014-1498-1] [PMID: 24573443]

[4]  King JA, Miller WM. Bioreactor development for stem cell expansion and controlled differentiation. Curr Opin Chem Biol 2007; 11(4): 394-8.
[http://dx.doi.org/10.1016/j.cbpa.2007.05.034] [PMID: 17656148]

[5]  Wang B, Wang Z, Chen T, Zhao X. Development of novel bioreactor control systems based on smart sensors and actuators. Front Bioeng Biotechnol 2020; 8: 7.
[http://dx.doi.org/10.3389/fbioe.2020.00007] [PMID: 32117906]

[6]  van Kelle MAJ, Oomen PJA, Bulsink JA, *et al.* A bioreactor to identify the driving mechanical stimuli of tissue growth and remodeling. Tissue Eng Part C Methods 2017; 23(6): 377-87.
[http://dx.doi.org/10.1089/ten.tec.2017.0141] [PMID: 28478703]

[7]  Kaya E, Galatalı S, Güldağ S, *et al.* Mass production of medicinal plants for obtaining secondary metabolite using liquid mediums *via* bioreactor systems: SETISTM and RITA®. Turk Bilimsel Derleme Derg 2018; 11(2): 5-10.

[8]  Verdú-Navarro F, Moreno-Cid JA, Weiss J, Egea-Cortines M. The advent of plant cells in bioreactors. Front Plant Sci 2023; 14: 1310405.
[http://dx.doi.org/10.3389/fpls.2023.1310405] [PMID: 38148861]

[9]  Murthy HN, Joseph KS, Paek KY, Park SY. Bioreactor systems for micropropagation of plants: present scenario and future prospects. Front Plant Sci 2023; 14: 1159588.
[http://dx.doi.org/10.3389/fpls.2023.1159588] [PMID: 37152119]

[10]  Spier MR, de Souza Vandenberghe LP, Medeiros ABP, Soccol CR. Application of different types of bioreactors in bioprocesses. In: Antolli PG, Liu Z, Eds. Bioreactors: Design. Properties and Applications Nova Sci Publishers 2011; pp. 55-90.

[11]  García-Ramírez Y. Temporary immersion system for biomass production of Salvia spp.: a mini-review. Methods Mol Biol 2024; 2759: 217-25.
[http://dx.doi.org/10.1007/978-1-0716-3654-1_20] [PMID: 38285153]

[12]  Monja-Mio KM, Olvera-Casanova D, Herrera-Alamillo MÁ, Sánchez-Teyer FL, Robert ML. Comparison of conventional and temporary immersion systems on micropropagation (multiplication phase) of *Agave angustifolia* Haw. 'Bacanora'. 3 Biotech. 2021; 11(2): 77.
[http://dx.doi.org/10.1007/s13205-020-02604-8]

[13]  Hall RD. Plant cell culture initiation. Practical tips. Mol Biotechnol 2000; 16(2): 161-74.
[http://dx.doi.org/10.1385/MB:16:2:161] [PMID: 11131975]

[14]  Evans DE, Coleman JOD, Kearns A. Cell Suspension Cultures. In: Evans DE, Coleman JOD, Kearns A, Eds. Plant Cell Culture. London: Taylor & Francis 2003; pp. 77-91.

[15]  Neumann KH, Kumar A, Imani J. Callus Culture. In: Neumann KH, Kumar A, Imani J, Eds. Plant Cell and Tissue Culture – A Tool in Biotechnology. Cham: Springer 2020; pp. 25-59.

[http://dx.doi.org/10.1007/978-3-030-49098-0_3]

[16]    Raj S, Saudagar P. Plant Cell Culture as Alternatives to Produce Secondary Metabolites. In: Akhtarand MS, Swamy MK, Eds. Natural Bio-active Compounds-Volume 3: Bio-technology, Bioengineering, and Molecular Approaches. Singapore: Springer Nature 2019; pp. 265-86.
[http://dx.doi.org/10.1007/978-981-13-7438-8_11]

[17]    Bhojwani SS, Dantu PK. Plant Tissue Culture: An Introductory Text. 1$^{st}$ ed. India: Springer 2013; p. 309.
[http://dx.doi.org/10.1007/978-81-322-1026-9]

[18]    Dong J, Bowra S, Vincze E. The development and evaluation of single cell suspension from wheat and barley as a model system; a first step towards functional genomics application. BMC Plant Biol 2010; 10(1): 239.
[http://dx.doi.org/10.1186/1471-2229-10-239] [PMID: 21054876]

[19]    McCoy E, O'Connor SE. Natural products from plant cell cultures. Prog Drug Res 2008; 65: 329-370, 331-370.
[http://dx.doi.org/10.1007/978-3-7643-8117-2_9] [PMID: 18084920]

[20]    Houston K, Tucker MR, Chowdhury J, Shirley N, Little A. The plant cell wall: a complex and dynamic structure as revealed by the responses of genes under stress conditions. Front Plant Sci 2016; 7: 984.
[http://dx.doi.org/10.3389/fpls.2016.00984] [PMID: 27559336]

[21]    Lee J, Cuddihy MJ, Kotov NA. Three-dimensional cell culture matrices: state of the art. Tissue Eng Part B Rev 2008; 14(1): 61-86.
[http://dx.doi.org/10.1089/teb.2007.0150] [PMID: 18454635]

[22]    Birgersdotter A, Sandberg R, Ernberg I. Gene expression perturbation *in vitro*—A growing case for three-dimensional (3D) culture systems. Semin Cancer Biol 2005; 15(5): 405-12.
[http://dx.doi.org/10.1016/j.semcancer.2005.06.009] [PMID: 16055341]

[23]    Reed KM, Bargmann BOR. Protoplast regeneration and its use in new plant breeding technologies. 2021.
[http://dx.doi.org/10.3389/fgeed.2021.734951]

[24]    Mathur J, Koncz C. Protoplast isolation, culture, and regeneration. Methods Mol Biol 1998; 82: 35-42.
[http://dx.doi.org/10.1385/0-89603-391-0:35] [PMID: 9664409]

[25]    Hall R, Pedersen C, Krens F. Improvement of protoplast culture protocols for *Beta vulgaris* L. (sugar beet). Plant Cell Rep 1993; 12(6): 339-42.
[http://dx.doi.org/10.1007/BF00237431] [PMID: 24197260]

[26]    Power JB, Davey MR. Protoplasts of higher and lower plants : isolation, culture, and fusion. Methods Mol Biol 1990; 6: 237-60.
[http://dx.doi.org/10.1385/0-89603-161-6:237] [PMID: 21390611]

[27]    Jiang F, Zhu J, Liu HL. Protoplasts: a useful research system for plant cell biology, especially dedifferentiation. Protoplasma 2013; 250(6): 1231-8.
[http://dx.doi.org/10.1007/s00709-013-0513-z] [PMID: 23719716]

[28]    Gilliard G, Huby E, Cordelier S, Ongena M, Dhondt-Cordelier S, Deleu M. Protoplast: a valuable toolbox to investigate plant stress perception and response. Front Plant Sci 2021; 12: 749581.
[http://dx.doi.org/10.3389/fpls.2021.749581] [PMID: 34675954]

[29]    Ozyigit II, Dogan I, Hocaoglu-Ozyigit A, *et al.* Production of secondary metabolites using tissue culture-based biotechnological applications. Front Plant Sci 2023; 14: 1132555.
[http://dx.doi.org/10.3389/fpls.2023.1132555] [PMID: 37457343]

[30]    Murthy HN, Joseph KS, Hahn JE, Lee HS, Paek KY, Park SY. Suspension culture of somatic embryos for the production of high-value secondary metabolites. Physiol Mol Biol Plants 2023; 29(8): 1153-77.
[http://dx.doi.org/10.1007/s12298-023-01365-x] [PMID: 37829704]

[31]  Ranaware AS, Kunchge NS, Lele SS, Ochatt SJ. Protoplast technology and somatic hybridisation in the family apiaceae. plants (Basel). 2023; 12(5): 1060.
[http://dx.doi.org/10.3390/plants12051060]

[32]  Davey MR, Anthony P, Power JB, Lowe KC. Plant protoplasts: status and biotechnological perspectives. Biotechnol Adv 2005; 23(2): 131-71.
[http://dx.doi.org/10.1016/j.biotechadv.2004.09.008] [PMID: 15694124]

[33]  Gamborg OL, Holl FB. Plant protoplast fusion and hybridization. Basic Life Sci 1977; 9: 299-316.
[http://dx.doi.org/10.1007/978-1-4684-0880-5_19] [PMID: 336026]

[34]  Bruznican S, Eeckhaut T, Van Huylenbroeck J, De Keyser E, De Clercq H, Geelen D. An asymmetric protoplast fusion and screening method for generating celeriac cybrids. Sci Rep 2021; 11(1): 4553.
[http://dx.doi.org/10.1038/s41598-021-83970-y] [PMID: 33633203]

[35]  Antoni D, Burckel H, Josset E, Noel G. Three-dimensional cell culture: a breakthrough *in vivo*. Int J Mol Sci 2015; 16(3): 5517-27.
[http://dx.doi.org/10.3390/ijms16035517] [PMID: 25768338]

[36]  Sapudom J, Waschke J, Franke K, Hlawitschka M, Pompe T. Quantitative label-free single cell tracking in 3D biomimetic matrices. Sci Rep 2017; 7(1): 14135.
[http://dx.doi.org/10.1038/s41598-017-14458-x] [PMID: 29075007]

[37]  Tibbitt MW, Anseth KS. Hydrogels as extracellular matrix mimics for 3D cell culture. Biotechnol Bioeng 2009; 103(4): 655-63.
[http://dx.doi.org/10.1002/bit.22361] [PMID: 19472329]

[38]  Duval K, Grover H, Han LH, *et al.* Modeling physiological events in 2D *vs.* 3D cell culture. Physiology (Bethesda) 2017; 32(4): 266-77.
[http://dx.doi.org/10.1152/physiol.00036.2016] [PMID: 28615311]

[39]  Zietarska M, Maugard CM, Filali-Mouhim A, *et al.* Molecular description of a 3D *in vitro* model for the study of epithelial ovarian cancer (EOC). Mol Carcinog 2007; 46(10): 872-85.
[http://dx.doi.org/10.1002/mc.20315] [PMID: 17455221]

[40]  Edmondson R, Broglie JJ, Adcock AF, Yang L. Three-dimensional cell culture systems and their applications in drug discovery and cell-based biosensors. Assay Drug Dev Technol 2014; 12(4): 207-18.
[http://dx.doi.org/10.1089/adt.2014.573] [PMID: 24831787]

[41]  Ruffoni B, Pistelli L, Bertoli A, Pistelli L. Plant cell cultures: bioreactors for industrial production. Adv Exp Med Biol 2010; 698: 203-21.
[http://dx.doi.org/10.1007/978-1-4419-7347-4_15] [PMID: 21520713]

[42]  Scragg AH, Fowler MW. Large-scale culture of plant cells. Methods Mol Biol 1990; 6: 477-94.
[http://dx.doi.org/10.1385/0-89603-161-6:477] [PMID: 21390630]

[43]  Kreis W, Reinhard E. The production of secondary metabolites by plant cells cultivated in bioreactors1. Planta Med 1989; 55(5): 409-16.
[http://dx.doi.org/10.1055/s-2006-962054] [PMID: 17262454]

[44]  Liang H, Xiong Y, Guo B, *et al.* Shoot organogenesis and somatic embryogenesis from leaf and root explants of *Scaevola sericea*. Sci Rep 2020; 10(1): 11343.
[http://dx.doi.org/10.1038/s41598-020-68084-1] [PMID: 32647162]

[45]  Teke GM, Pott RWM. Design and evaluation of a continuous semipartition bioreactor for *in situ* liquid-liquid extractive fermentation. Biotechnol Bioeng 2021; 118(1): 58-71.
[http://dx.doi.org/10.1002/bit.27550] [PMID: 32876954]

[46]  Shahid M, Shahzad A, Malik A, Sahai A. Recent Trends in Biotechnology and Therapeutic Applications of Medicinal Plants. 2013.
[http://dx.doi.org/10.1007/978-94-007-6603-7]

[47]    Reuss M. Stirred tank bioreactors. Bioprocess Technol 1995; 21: 207-55.
        [PMID: 7765640]

[48]    Rodríguez-Monroy M, Galindo E. Broth rheology, growth and metabolite production of *Beta vulgaris*
        suspension culture: a comparative study between cultures grown in shake flasks and in a stirred tank.
        Enzyme Microb Technol 1999; 24(10): 687-93.
        [http://dx.doi.org/10.1016/S0141-0229(99)00002-2]

[49]    Tautorus TE, Lulsdorf MM, Kikcio SI, Dunstan DI. Bioreactor culture of *Picea mariana* Mill. (black
        spruce) and the species complex *Picea glauca-engelmannii* (interior spruce) somatic embryos. Growth
        parameters. Appl Microbiol Biotechnol 1992; 38(1): 46-51.
        [http://dx.doi.org/10.1007/BF00169417]

[50]    Merchuk JC, Gluz M. Bioreactors, Air-lift Reactors. In: Flickinger MC, Drew SW, Eds. Encyclopedia
        of Bioprocess Technology. 2002.
        [http://dx.doi.org/10.1002/0471250589.ebt029]

[51]    Fontana RC, Polidoro TA, Silveira MM. Comparison of stirred tank and airlift bioreactors in the
        production of polygalacturonases by *Aspergillus oryzae*. Bioresour Technol 2009; 100(19): 4493-8.
        [http://dx.doi.org/10.1016/j.biortech.2008.11.062] [PMID: 19467860]

[52]    Feitkenhauer H, Maleski R, Märkl H. Airlift-reactor design and test for aerobic environmental
        bioprocesses with extremely high solid contents at high temperatures. Water Sci Technol 2003; 48(8):
        69-77.
        [http://dx.doi.org/10.2166/wst.2003.0454] [PMID: 14682572]

[53]    Adjout R, Mouget J, Pruvost J, Chentir I, Loiseau C, Baba Hamed MB. Effects of temperature,
        irradiance, and pH on the growth and biochemical composition of *Haslea ostrearia* batch-cultured in
        an airlift plan-photobioreactor. Appl Microbiol Biotechnol 2022; 106(13-16): 5233-47.
        [http://dx.doi.org/10.1007/s00253-022-12055-1] [PMID: 35842874]

[54]    Mancilla-Álvarez E, Pérez-Sato JA, Núñez-Pastrana R, Spinoso-Castillo JL, Bello-Bello JJ.
        Comparison of different semi-automated bioreactors for *in vitro* propagation of taro (*Colocasia
        esculenta* L. Schott). Plants 2021; 10(5): 1010.
        [http://dx.doi.org/10.3390/plants10051010] [PMID: 34069416]

[55]    Towler MJ, Kim Y, Wyslouzil BE, Correll M, Weathers PJ. Design, Development, and Applications
        of Mist Bioreactors for Micropropagation and Hairy Root Culture. In: Gupta SD, Ibaraki Y, Eds. Plant
        Tissue Culture Engineering Focus on Biotechnology. Dordrecht: Springer 2006; Vol. 6: pp. 119-34.
        [http://dx.doi.org/10.1007/978-1-4020-3694-1_7]

[56]    Mirzabe AH, Hajiahmad A, Fadavi A, Rafiee S. Design of nutrient gas-phase bioreactors: a critical
        comprehensive review. Bioprocess Biosyst Eng 2022; 45(8): 1239-65.
        [http://dx.doi.org/10.1007/s00449-022-02728-6] [PMID: 35562481]

[57]    Wang KW, Liu WZ, Kang D, Zhang YX, Cui D. Hybrid bioreactor built-in with fixed bio-carriers for
        denitrification with low C/N ratio: Hydrodynamic optimization and microbial divergence. Environ Res
        2023; 224: 115510.
        [http://dx.doi.org/10.1016/j.envres.2023.115510] [PMID: 36796606]

[58]    Eibl R, Kaiser S, Lombriser R, Eibl D. Disposable bioreactors: the current state-of-the-art and
        recommended applications in biotechnology. Appl Microbiol Biotechnol 2010; 86(1): 41-9.
        [http://dx.doi.org/10.1007/s00253-009-2422-9] [PMID: 20094714]

[59]    Ducos JP, Terrier B, Courtois D. Disposable bioreactors for plant micropropagation and mass plant
        cell culture. Adv Biochem Eng Biotechnol 2009; 115: 89-115.
        [http://dx.doi.org/10.1007/10_2008_28] [PMID: 19475375]

[60]    Wierzchowski K, Pilarek M. Disposable rocking bioreactors: recent applications and progressive
        perspectives. Trends Biotechnol. 2023; S0167-7799(23): 00277-9.
        [http://dx.doi.org/10.1016/j.tibtech.2023.09.003]

[61]  Eibl R, Eibl D. Application of disposable bag bioreactors in tissue engineering and for the production of therapeutic agents. Adv Biochem Eng Biotechnol 2008; 112: 183-207.
[http://dx.doi.org/10.1007/10_2008_3] [PMID: 19290502]

[62]  Saggiomo V, Velders AH. Simple 3D printed scaffold-removal method for the fabrication of intricate microfluidic devices. Adv Sci (Weinh) 2015; 2(9): 1500125.
[http://dx.doi.org/10.1002/advs.201500125] [PMID: 27709002]

[63]  Choi I, Ahn GY, Kim ES, *et al.* Microfluidic bioreactor with fibrous micromixers for *in vitro* mRNA transcription. Nano Lett 2023; 23(17): 7897-905.
[http://dx.doi.org/10.1021/acs.nanolett.3c01699] [PMID: 37435905]

[64]  Folch A, Toner M. Cellular micropatterns on biocompatible materials. Biotechnol Prog 1998; 14(3): 388-92.
[http://dx.doi.org/10.1021/bp980037b] [PMID: 9622519]

[65]  Sackmann EK, Fulton AL, Beebe DJ. The present and future role of microfluidics in biomedical research. Nature 2014; 507(7491): 181-9.
[http://dx.doi.org/10.1038/nature13118] [PMID: 24622198]

[66]  Chokkalingam V, Tel J, Wimmers F, *et al.* Probing cellular heterogeneity in cytokine-secreting immune cells using droplet-based microfluidics. Lab Chip 2013; 13(24): 4740-4.
[http://dx.doi.org/10.1039/c3lc50945a] [PMID: 24185478]

[67]  Volpatti LR, Yetisen AK. Commercialization of microfluidic devices. Trends Biotechnol 2014; 32(7): 347-50.
[http://dx.doi.org/10.1016/j.tibtech.2014.04.010] [PMID: 24954000]

[68]  Ansari MIH, Hassan S, Qurashi A, Khanday FA. Microfluidic-integrated DNA nanobiosensors. Biosens Bioelectron 2016; 85: 247-60.
[http://dx.doi.org/10.1016/j.bios.2016.05.009] [PMID: 27179566]

[69]  Paguirigan AL, Beebe DJ. Microfluidics meet cell biology: bridging the gap by validation and application of microscale techniques for cell biological assays. BioEssays 2008; 30(9): 811-21.
[http://dx.doi.org/10.1002/bies.20804] [PMID: 18693260]

[70]  Chang CW, Cheng YJ, Tu M, *et al.* A polydimethylsiloxane–polycarbonate hybrid microfluidic device capable of generating perpendicular chemical and oxygen gradients for cell culture studies. Lab Chip 2014; 14(19): 3762-72.
[http://dx.doi.org/10.1039/C4LC00732H] [PMID: 25096368]

[71]  Esch EW, Bahinski A, Huh D. Organs-on-chips at the frontiers of drug discovery. Nat Rev Drug Discov 2015; 14(4): 248-60.
[http://dx.doi.org/10.1038/nrd4539] [PMID: 25792263]

[72]  Fausther-Bovendo H, Kobinger G. Plant-made vaccines and therapeutics. Science 2021; 373(6556): 740-1.
[http://dx.doi.org/10.1126/science.abf5375] [PMID: 34385382]

[73]  Ma F, Xu Q, Wang A, *et al.* A universal design of restructured dimer antigens: Development of a superior vaccine against the paramyxovirus in transgenic rice. Proc Natl Acad Sci USA 2024; 121(4): e2305745121.
[http://dx.doi.org/10.1073/pnas.2305745121] [PMID: 38236731]

[74]  He Y, Ning T, Xie T, *et al.* Large-scale production of functional human serum albumin from transgenic rice seeds. Proc Natl Acad Sci USA 2011; 108(47): 19078-83.
[http://dx.doi.org/10.1073/pnas.1109736108] [PMID: 22042856]

[75]  Huebbers JW, Buyel JF. On the verge of the market – Plant factories for the automated and standardized production of biopharmaceuticals. Biotechnol Adv 2021; 46: 107681.
[http://dx.doi.org/10.1016/j.biotechadv.2020.107681] [PMID: 33326816]

[76]    Buyel JF. Plant molecular farming - integration and exploitation of side streams to achieve sustainable biomanufacturing. Front Plant Sci 2019; 9: 1893.
[http://dx.doi.org/10.3389/fpls.2018.01893] [PMID: 30713542]

[77]    Bourgaud F, Gravot A, Milesi S, Gontier E. Production of plant secondary metabolites: a historical perspective. Plant Sci 2001; 161(5): 839-51.
[http://dx.doi.org/10.1016/S0168-9452(01)00490-3]

[78]    Paek KY, Hahn EJ, Son SH. Application of bioreactors for large-scale micropropagation systems of plants. *In Vitro* Cell. Dev Biol Plant 2001; 37(2): 149-57.
[http://dx.doi.org/10.1007/s11627-001-0027-9]

[79]    Sajc L, Grubisic D, Vunjak-Novakovic G. Bioreactors for plant engineering: an outlook for further research. Biochem Eng J 2000; 4(2): 89-99.
[http://dx.doi.org/10.1016/S1369-703X(99)00035-2]

# CHAPTER 6

# Genomics - Proteomics Approaches in Plant Physiology

**Selin Galatalı**[1,*] and **Hacer Ağar**[1]

[1] *Muğla Sıtkı Koçman University, Faculty of Science, Molecular Biology and Genetics Department, 48000, Menteşe, Muğla, Türkiye*

**Abstract:** Today, climate change, the impact of which is felt more and more due to global warming, also affects food security. Increasing population and global warming cause challenges in food demand and medicine supply. New approaches need to be developed to tackle these challenges and maintain the current balance. In particular, plant physiology studies have been carried out in many fields since the beginning of human life due to its vital importance. With modern technologies, plant physiology studies have moved to a higher level. Studies at the molecular level, known as genomics and proteomics, are progressing towards the cultivation of plants with superior properties such as being more resistant and having higher nutrient content. The use of genomic techniques such as whole genome sequencing, comparative genomics, molecular markers, and proteomic approaches such as gel-based, affinity, and reagent-based, mass spectrometry has led to the acquisition of comprehensive data in the field of plant physiology. These data have led to a better understanding of plant morphology and development, the analysis of the complex interactions between genes and proteins, and the collective development of genetic-based breeding efforts. In this chapter, basic genomic and proteomic approaches in the field of plant physiology and past and present studies on economically important plants are presented.

**Keywords:** Comparative genomics, Metabolomics, Stress physiology, Transcriptomics.

## INTRODUCTION

Genomics, which can be summarized as all the studies carried out to explain the organizational-functional information about the genome and genes, mediates the identification of all genes in a living thing separately, the investigation of the interactions of genes with each other and the environment, and the examination of the production and activation of genes in time, place and quantity. It is a branch of

---
* **Corresponding author Selin Galatalı:** Muğla Sıtkı Koçman University, Faculty of Science, Molecular Biology and Genetics Department, 48000, Menteşe, Muğla, Türkiye; E-mail: ergunkaya@mu.edu.tr

**Ergun Kaya (Ed.)**

science that deals with advances in the fields of genomic automation and bioinformatics, which processes, interprets and stores its outputs through information technologies [1, 2]. Genomic studies allow the identification of genes related to disease and physiological processes, revealing structural-functional interactions between genes, and the roles of genes in development and their expression profiles, and comparing different organisms on a genetic basis [3, 4].

Considering that cell functions are regulated by proteins rather than genes and mRNA and the non-linear relationship between mRNA and proteins, it would be useful to supplement transcriptomics with information about proteins. Indeed, genomic data need to be supported by post-genomic studies to understand the extent to which the organism uses genes, which ones it responds to the situations it encounters, which genes are expressed and converted into proteins, their locations in the cell and their relative amounts (Fig. **1**). This discipline, which aims to provide collective information about proteins, is called proteomics. Genomic studies are carried out through traditional and high-throughput methods [5, 6].

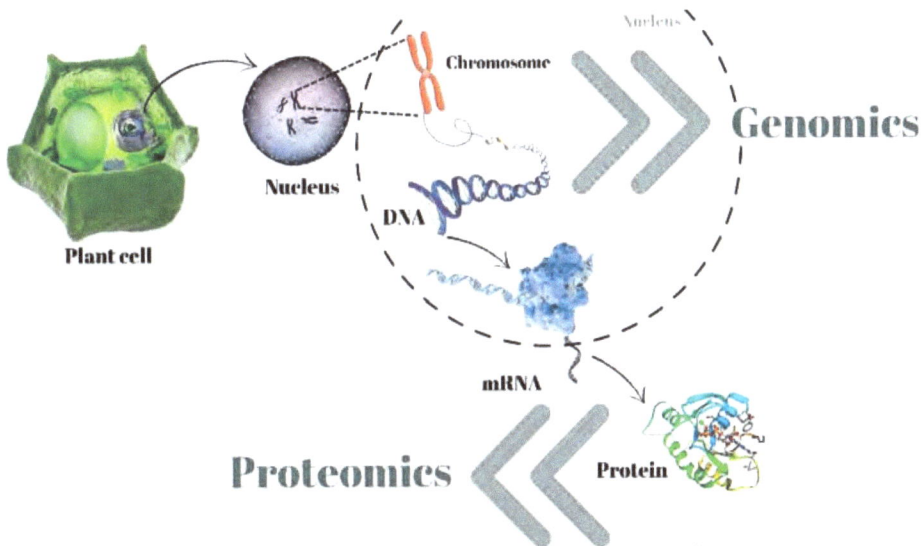

**Fig. (1).** Genomics and proteomics: Two complementary perspectives on life.

The proteome describes all proteins encoded by the genome. It is the entire set of proteins possessed and expressed by a given organism and includes not only the polypeptide structures encoded by genes but also post-translational modifications. The term proteomics refers to processes such as the expression of different proteins in different cell types and different cell parts and different developmental stages, environmental conditions, various diseases, and aging. In other words, the proteome is a dynamic structure that varies according to tissues and cells, phases of the cell cycle, internal and external stimuli, environmental conditions, and similar situations. Proteomics aims to define the proteome. It refers to the examination of the proteome in the context of structure, location, quantity, post-translational modifications, function in tissues/cells, and interaction with other proteins and macromolecules. Unlike genomics, proteomics, which is a dynamic concept, can also be defined as the quantitative analysis technology of proteins in cells, tissues, or body fluids under different conditions [7 - 9].

Proteomic studies, which enable comparative studies, as in genomic studies, provide supporting information for genomics in situations that limit the effectiveness of genomics. The difference of proteomic studies from traditional biochemical methods is that they enable the study of a large number of proteins (proteome) simultaneously. Various methods can be used for proteomic studies that need strong bioinformatics support to interpret and make sense of the data it provides. However, studying proteins is not as easy as studying genes due to various reasons such as they derive certain characters from their three-dimensional structures. They are sometimes found in very low amounts, there is no clear relationship between their behavior and quantities, there are reversible post-translational modifications, and they can form splice variants due to RNA splicing [10, 11].

## GENOMIC APPROACHES IN PLANT PHYSIOLOGY

Genomics is an interdisciplinary field that studies genome structure, function, mapping, evolution, and regulation. Genomics also includes the sequencing and analysis of genomes through the use of high-throughput DNA sequencing and bioinformatics to assemble and analyse the function and structure of whole genomes.

With modern technologies, the study of plant physiology in the field of genomics has progressed to a much higher level. Especially recently, the genome sequences of many plant species have become the main theme in plant physiology research. Open access and continuous updates create a fertile environment for the development of economically important plants. Recent technological advances and especially agricultural challenges have led to the emergence of high-throughput tools for research and utilisation of plant genomes for crop improvement. These genomics-based approaches aim to decrypt the whole genome, including gene and intergenic regions, to understand plant molecular responses and to develop strategies that enable improvement in many areas under plant physiology.

## Sequence-Based Approaches

The history of plant genomics began with expressed sequence tagging (EST), a high-throughput gene discovery method [12]. EST data of economically important crops such as maize, soybean, wheat, and rice are stored in the database of the National Centre for Biotechnological Information (NCBI, https://www.ncbi.nlm.nih.gov/). For EST, which is one of the most pioneering methods of gene discovery and sequencing, cDNA libraries obtained from various plant tissues and regions are used as sources, but their lengths are generally shorter than the source cDNAs and cannot provide the opportunity to completely analyse a transcriptome [13 - 17]. The Serial Analysis of Gene Expression (SAGE) method was developed to analyse the simultaneous or parallel expression of multiple genes [18]. This method uses short sequence tags (10-14 bp) specific to each transcript to identify the transcript. It has also been used to identify novel genes expressed in a tissue or under special conditions [13, 19]. The sequence tag-based Massively Parallel Signature Sequencing (MPSS) method, similar to the SAGE method, was developed to detect and analyse large numbers of genes at a lower cost [20]. Using longer tags that can detect long transcripts, the analysis of genes has been achieved with high efficiency and sensitivity [13, 16]. MPSS has also been used to detect small RNA sequences as well as transcripts in plant physiology studies, especially in stress defence mechanisms [21, 22]. The data obtained since the development of the method are stored in MPSS databases [23].

## Molecular Markers

Molecular markers are short DNA sequences that allow the detection of variation or polymorphism among individuals in a population-specific to certain regions of DNA. The identification of these sequences laid the foundations of modern genomics. Since it is stable and detectable in every cell regardless of the morphological or physiological state of a living organism, it has a great advantage over traditional methods and molecular marker-based techniques have developed rapidly [24 - 26]. Emerging methods have greatly improved and facilitated research in many fields such as plant taxonomy, physiology, genetics, and breeding [24, 26 - 28]. There are many molecular markers but the most common ones are Restriction Fragment Length Polymorphisms (RFLPs), Amplified Fragment Length Polymorphisms (AFLPs), Randomly Amplified Polymorphic DNA (RAPD), Simple Sequence Repeats (SSRs), and Inter Simple Sequence Repeats (ISSR).

### *Restriction Fragment Length Polymorphism (RFLP)*

RFLP, the first molecular marker used in human genome mapping, was soon used in plant genomic studies [24, 29]. These marker methods, which are usually polymorphic, are not based on the PCR method (Fig. **2**). Hybridisation and DNA fragment profiles are obtained by Southern blotting using chemically or radioactively labelled probes [25].

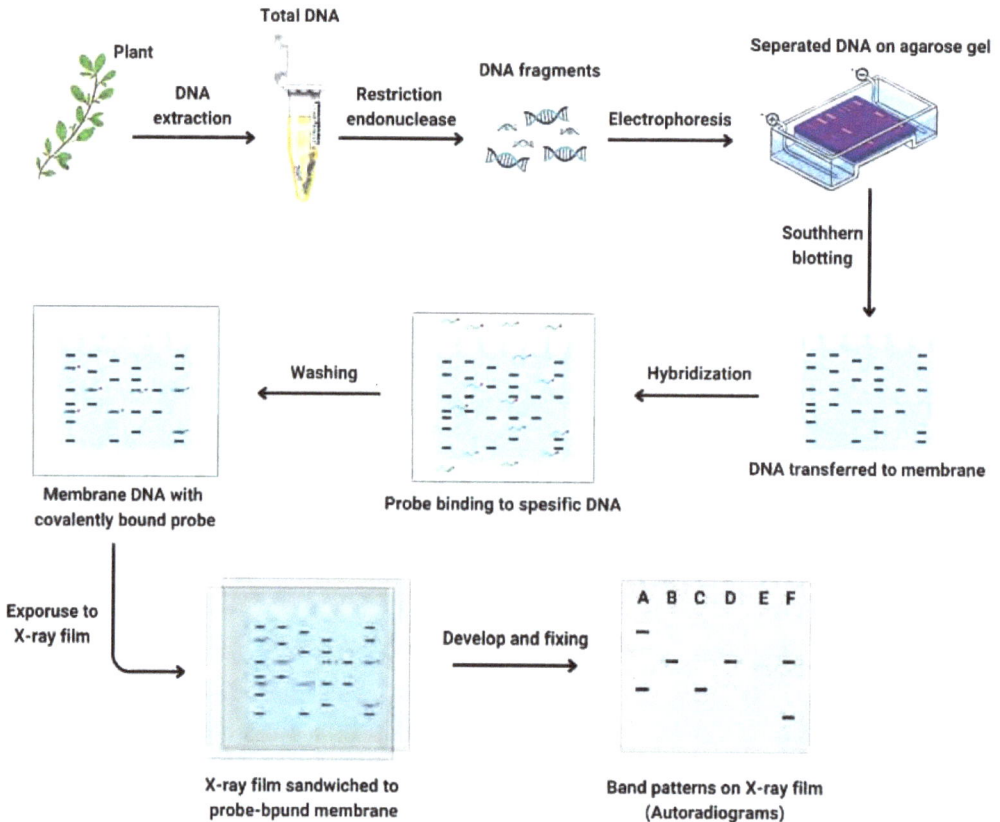

**Fig. (2).** Schematic representation of RFLP technique (Figure Made in https://www.biorender.com/).

## Amplified Fragment Length Polymorphism (AFLP)

In the AFLP method, genomic DNA is digested with restriction enzymes, and adaptor sequence ligation is performed on the cut sites and amplified by PCR. DNA fragments are usually visualized on polyacrylamide gel electrophoresis (PAGE) and DNA fragments are detected by autoradiographic methods (Fig. **3**). Due to its easier applicability and high-quality results compared to other molecular marker methods, it has been used alone, especially in plant genetics studies [30 - 33].

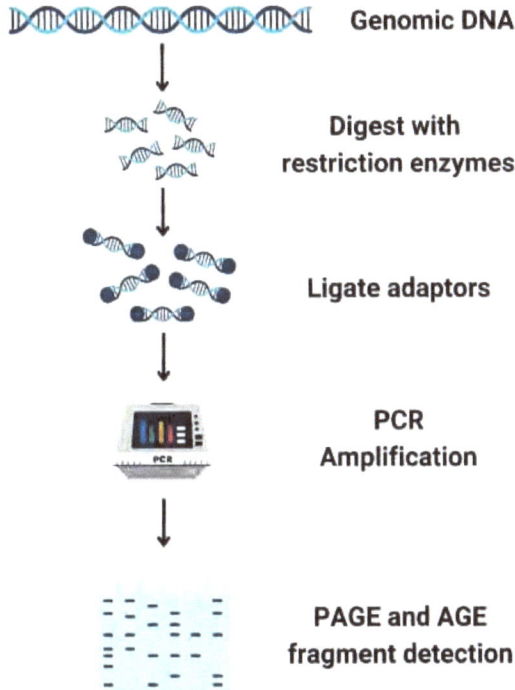

**Fig. (3).** Schematic representation of AFLP technique (Figure Made in https://www.biorender.com/).

## Simple Sequence Repeats (SSR)

Simple sequence repeats are 2-5 bp units repeated many times at a gene locus. These microsatellite-specific SSR-based analyses are widely applied due to the abundance, common dominant inheritance, and diversity of microsatellites. In plant genetics studies, they have been used primarily for the discovery of new species and interspecific discrimination. SSR markers are among the current popular methods used in plant physiology in many fields such as germplasm conservation, genetic diversity, stress physiology, and plant metabolism [34 - 37].

## Inter Simple Sequence Repeat (ISSR)

They are sequence-specific markers between two identical but oppositely oriented microsatellite repeat regions. They are advantageous in terms of ease of analysis, not requiring knowledge of the genome sequence and providing highly accurate results. In recent years, it has been widely used especially in plant genetic variation, genetic accuracy and interpopulation polymorphism studies [33, 38, 39].

## Random Amplified Polymorphic DNA (RAPD)

It is the first PCR-based DNA marker method. This single primer, which has a high G-C ratio and is approximately 10 bp long, binds randomly to DNA. After PCR amplification, the DNA fragments are visualized and analysed in the gel (Fig. **4**). It gives high-efficiency results even in low amounts of DNA samples [33, 40]. The RAPD method is also used in many genetic studies alone or in combination with other marker methods [41, 42].

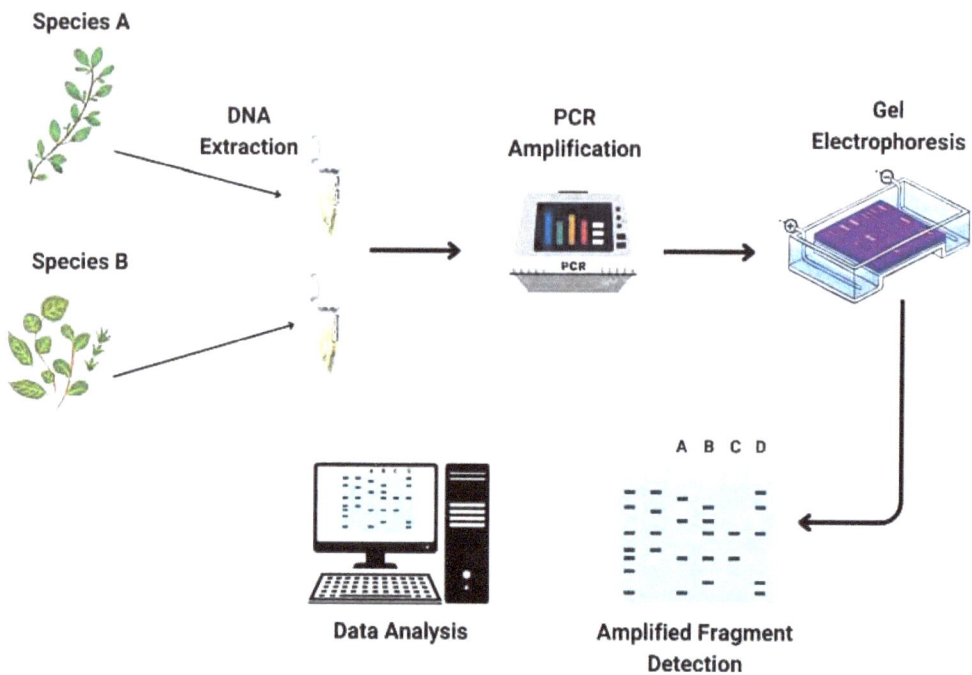

**Fig. (4).** Schematic representation of RAPD technique (Figure Made in https://www.biorender.com/).

## Genome Sequencing Technologies

In 1997, the genomics era started with the method known as Sanger sequencing developed by Sanger *et al.* using the dideoxy chain termination methodology of fluorescent molecules [43 - 45]. Plant genome sequencing analyses started in 2000 with the *Arabidopsis thaliana* genome obtained using the Sanger Sequencing Method, and then the genomes of important plants such as rice, maize and soya were sequenced with this method [46 - 49].

The Sanger sequencing method has significant disadvantages such as high cost, high labour, and low efficiency. With the developing technology, Sanger sequencing technology has been largely replaced by next-generation sequencing

(NGS) methods. These methods are based on the reverse chain termination reaction and allow the sequencing of high-quality and large numbers of fragments [50 - 52]. Instrument platforms such as Illumina, Ion Torrent/Proton, SOLID, and BGI, which are leading NGS methods, perform billions of short reads at a time. All of these platforms have unique enzymes, sequencing, hardware, and software systems [35, 53 - 55]. They are used to define the molecular basis of plant physiological functions in all omic domains. By utilising plant genome data with next-generation sequencing technologies, plant development studies have accelerated significantly to develop plants tolerant to biotic/abiotic stresses and high agricultural yields [44].

## Comparative Genomic Approaches

Sequence data obtained from genomic studies are of great importance in the identification and analysis of all coded and non-coded DNA sequences. These processes, which fall under comparative genomic analysis approaches, are one of the current areas of modern biotechnology. This field mainly reveals important similarities and differences between species and enables the discovery of the functions of many genes [56].

Comparative genomics requires new approaches for processing and analysing big data. These approaches, which continue to develop with modern technologies in plant physiology, facilitate the understanding of the correlation between various genes in many metabolic pathways, as well as increasing gene discovery in plant species [56, 57]. Online platforms have been developed to store and manage the collectively increasing amount of genomic data and to perform comparative genomics studies. Phytozome, PLAZA, GreenPhylDB, and PlantsDB are the most recent and widely used plant comparative genomics databases [58 - 62].

## PROTEOMIC APPROACHES IN PLANT PHYSIOLOGY

Although the functions of many genes have been identified by genomic and functional genomics methods in recent years, researchers have focused on the proteins that are directly involved in these processes for cellular metabolic and/or physiological processes that cannot be fully elucidated. These developments have given rise to the systems biology approach in the post-genomic era, including proteomics, transcriptomics, metabolomics and artificial intelligence-based mathematical modeling systems [63, 64].

Although many genes with different expressions associated with a certain biological process have been identified through genomic, functional genomic, and transcriptomic approaches, it is still not clear whether they are functional or not. Transcriptional changes often cannot fully reflect the proteins that govern biological processes or changes at the protein level [65]. One of the important reasons for this is that a gene can encode more than one protein, and the mRNA product can be transformed into different proteins with different biological functions through post-transcriptional modifications. Another important reason is post-translational modifications and differences in the stability of proteins. In most cases, post-translational modifications are directly related to the function of the protein. Therefore, proteomic approaches provide results that are very close to reality in elucidating the molecular mechanism of a biological process [66, 67].

Proteomic studies are more difficult than genomic studies due to the complex and dynamic structures of proteins, and they only allow the analysis of proteins present in a biological system under a certain condition and in a certain time period. In addition, it enables the detailed examination of the complex metabolic and regulatory reaction pathways developed by the organism against various situations and stresses it encounters throughout its life [68, 69]. In the early years of proteomic approaches, proteomic studies of plants were limited because of difficulties such as hard cell walls in plant tissues, the presence of complex and diverse secondary metabolites, large amounts of pigments, proteases, polyphenols, polysaccharides, starch, and lipids which caused many problems during the preparation of total protein samples and the separation of proteins. However, thanks to the ongoing efforts to overcome each problem, the use of proteomic studies in the plant world has become widespread with the rapid development of technology [70, 71].

In the early years of studies on plant proteomics, proteome profiling of various plant tissues and organs was generally carried out in studies carried out on model plants such as *Arabidopsis thaliana* and *Oryza sativa*. These studies also enabled the development of tissue-specific proteomic methods. The next stage is subcellular proteomics studies. Due to methodological limitations and the low amount of protein obtained, the number of subcellular proteomics studies in all organisms is extremely low, and this is quite limited in plants. In addition, mitochondrial proteomics [72], chloroplast proteomics [73, 74], and proteomics studies cover cell wall soluble proteins in *Zea mays* [75], and endoplasmic reticulum, plasma membrane, vacuole membrane and Golgi apparatus are studied using isotope labeling method in *Arabidopsis thaliana*. Studies on proteome profiling [74] and comparative examination of endoplasmic reticulum proteomes during the germination and development periods of *Ricinus communis* seeds [76] are important examples in this field.

In recent years, the most common application area of proteomics in plants is biotic and abiotic stress tolerance and its molecular mechanisms [77, 78]. The secretory proteins (secretome) of the plant, which are particularly effective in biotic stress tolerance, have been an area of great interest in the last decade. Secretome profiles, especially in the interaction of pathogens, which are obligate biotrophs, with the plant they infect, are carried out by isolating the extracellular fluid between the plant cell wall and the plasma membrane, extracting the proteins in the isolated fluid and identifying them with proteomic techniques. The number of studies published in this very new field is quite low [79, 80].

## PROTEOMIC RESPONSES OF PLANTS UNDER BIOTIC/ABIOTIC STRESS

Despite being exposed to stress conditions, many plant species can survive and continue their life cycles in their environment. They can complete it. As a result of research conducted in species known to be resistant to stress or in various model organisms, physiological, biochemical and molecular responses to stress in plants have been elucidated [81]. In the light of the data obtained, genomic and/or proteomic molecular response mechanisms in plants against stress conditions, homeostasis of macromolecules and ions, synthesis of protective molecules, and reactive oxygen are examined. It can be expressed as the formation and detoxification of reactive oxygen species (ROS) [82].

Functional genomic and transcriptomic analyses are widely used to understand the molecular basis of mechanisms sensitive to biotic and/or abiotic stresses in plants. For example, genes associated with drought stress are genes that encode osmotic pressure regulating enzymes, aquaporins, detoxifying enzymes, late embryogenesis abundant (LEA) proteins, enzymes that clear reactive oxygen species, and chaperones that protect the integrity of cell membranes and ensure ion transport/balance [69, 83]. Additionally, various transcription factors and protein kinases that regulate gene expression and signal transduction under stress conditions are also important in the stress response [84, 85].

Plant proteomics is a dynamic discipline aimed at investigating the biological functions of plant proteome and proteins in plants exposed to stress. In recent years, the number of studies on changes in the proteome of plants under stress has been increasing. Molecular information obtained from plants exposed to stress allows the identification of possible candidate genes for the genetic development of stress-tolerant plants [86, 87]. Proteomic studies conducted in different plant species have provided important information in understanding the molecular basis of responses to drought stress [88, 89]. Although the effect of stress conditions on proteome changes varies depending on plant species, genotypes or the severity of stress, proteomic analyses have revealed levels of proteins related to carbohydrates and energy metabolism, as well as signal transduction, scavenging of reactive oxygen species, osmotic regulation, protein synthesis and processing, and regulation of cell structure, which have shown significant changes [90, 91].

## Salt Stress and Proteome Changes in Plants

Proteomics is based on comparing the compositions of different proteomes. In the research field of plant abiotic stress, the most common situation is the comparison of proteomes isolated from treatment groups under control and stress conditions [92]. Proteomic studies on salt stress-induced changes are numerous; mainly focused on quantitative changes in protein abundance between control and stressed plants [93 - 95]. Specifically, it occurs as a general regulation of proteins involving carbohydrates, nitrogen and energy metabolism, especially glycolytic and trichloroacetic acid enzymes. Additionally, salt stress can cause metabolic imbalances that promote the formation and accumulation of reactive oxygen species (ROS) [96]. It is therefore not surprising that ROS-fighting proteins have been commonly identified, including superoxide dismutase (SOD), ascorbate peroxidase (APX) and glutathione reductase (GR), which function to reduce oxidative damage [97]. In many studies, proteins that provide cytoskeletal stability as well as other proteins involved in protein synthesis, processing, activity and degradation have been identified. Regarding photosynthetic functions, a general decrease in the level of proteins associated with chlorophyll biosynthesis was determined, while an increase in proteins involved in light-dependent reactions was observed. Some identified proteins are determinants of the general stress response pathway in plants. Proteins identified in the signaling, trafficking, transport, and cell structure categories are less common [98].

Salt stress signaling includes signals to regulate ionic, osmotic, detoxification, cell division, and expansion [99]. Transient salt signaling receptors localized in the plasma membrane or cytoplasm have been identified in proteomic studies in plants under salt stress [100, 101]. Signaling of salt tolerance is an important issue, and many salt response signaling pathways have been identified, such as the salt hypersensitive signaling pathway, ABA signaling pathway, $Ca^{2+}$ signal transduction pathway, protein kinase pathway, phospholipid pathway, ethylene signaling pathway, and jasmonic acid-promoted signaling pathway [102, 103].

The heterotrimeric G-protein complex and related G-protein coupling receptors play an important role in response to abiotic stress conditions [104]. Heterotrimeric GTP-binding proteins (G protein) are known as specific signal cascade regulators to convert a signal into a specific cellular response [105], and the subunit of this heterotrimeric protein is involved in salt tolerance in plants [106]. Salt-induced G protein and several small G proteins have been identified in some plant species by proteomic studies [107, 108]. Additionally, genes encoding two subunits of the G protein were upregulated by NaCl treatment [99, 107], and a G protein transcript was also increased in Mesembryanthemum crystallinum under salt stress [108]. Proteomic results provide new information on changes in the amount of G proteins in response to salt. Proteomic analyses of G protein mutants and protein-protein associations have increased knowledge about signaling networks and G protein function in salinity response [109].

Salt stress-promoted $Ca^{2+}$ dependent signaling network mediates $Na^+$ homeostasis and salt resistance [103, 104]. In plant cells, $Ca^{2+}$ is a unique secondary messenger in numerous signaling pathways. Different $Ca^{2+}$ binding proteins such as calmodulin (CaM), calreticulin, and developmentally regulated plasma membrane polypeptides (DREPP)-like proteins (with glutamate-rich regions for $Ca^{2+}$ binding) involved in the regulation of cytosolic $Ca^{2+}$ homeostasis are promoted in salt-stressed plants. It has been revealed by proteomic studies [100, 110 - 112].

The salt hypersensitive (SOS) stress signaling pathway has been identified as an essential regulator of plant ion homeostasis and salt tolerance [113, 114]. In the SOS signal transduction pathway, the myristoylated calcium-binding protein SOS3 is sensitive to salt stress-induced cytosolic calcium changes. SOS3 physically associates with and activates SOS2 (serine/threonine kinase). The SOS3/SOS2 kinase complex phosphorylates and promotes the plasma membrane $Na^+/H^+$ antiporter encoded by the SOS1 gene. It has been suggested that SOS1 (plasma membrane $Na^+/H^+$ antiporter) may have a regulatory role in addition to its transport function and may even be a novel sensor for $Na^+$ [115]. Comparative proteomics studies have revealed more NaCl-responsive proteins in protein kinase cascades [116 - 118]. However, a phosphoproteomic study revealed dynamic changes in phosphoproteins and salt stress signal transduction [109].

## Drought Stress and Proteome Changes in Plants

During drought stress, various defense mechanisms are regulated at physiological, biochemical and molecular levels. Drought stress affects plant growth by affecting various physiological and biochemical processes such as photosynthesis, respiration, translocation, ion uptake, water potential, stomatal closure, sugar and nutrient metabolism, antioxidant system, as well as phytohormones [119]. The expression products of drought-responsive genes mainly include proteins involved in signal transduction pathways and transcriptional regulation, functional proteins that protect cellular membranes, abscisic acid (ABA) biosynthesis-related proteins, and other proteins such as LEA [120]. Many proteomic studies have shown that some classes of proteins are promoted in response to drought stress and that these proteins may play important roles in defense and adaptation processes [88, 121, 122].

Environmental stresses such as drought cause damage to processes such as photosynthesis, mitochondrial respiration and photorespiration and increase the production of reactive oxygen species [123]. Under normal conditions, plants use reactive oxygen species as signaling molecules in various cellular events such as programmed cell death, abiotic stress responses and defense against pathogens. Excessive reactive oxygen species produced in plants exposed to stress cause oxidative damage to lipids, nucleic acids, and proteins, leading to cell death [124]. Complex arrays of enzymatic and non-enzymatic mechanisms have evolved to maintain levels of reactive oxygen species. It has been reported that the abundance of protein in enzymes such as superoxide dismutase (SOD), ascorbate peroxidase (APX), catalase (CAT), peroxidase (POD), thioredoxin peroxidase, and glutathione S-transferase (GST) is affected and changed in many plant species exposed to drought stress [125, 126]. SOD is known as the first line of defense against oxidative damage by converting the highly toxic superoxide anion into less toxic hydrogen peroxide. The increase in the expression level of SOD protein was determined in plant species such as rice and wheat under drought stress [89, 127].

The importance of the SOD enzyme in conferring tolerance to abiotic stresses was revealed using transgenic plants that overexpress the genes encoding this enzyme [128, 129]. Increased APX expression level has been determined in many plant species under drought stress, and it has been reported that increased ascorbate-glutathione cycle activity under drought stress conditions is more prominent in tolerant genotypes [130, 131]. GST is an important enzyme that plays a role in the detoxification of tripeptide glutathione and xenobiotics and alleviates oxidative stress [132]. It has been reported that GST expression is downregulated in drought-sensitive wheat genotypes and this genotype fails to detoxify toxic molecules [88]. Similarly, the increase in GST abundance was reported to occur only in drought-tolerant barley genotypes [133, 134]. However, increased drought tolerance has been reported in transgenic Arabidopsis plants overexpressing the tomato GST gene [134].

When plants are exposed to stress, stress signals are generally detected through special receptors and then these signals are sent to the signal transduction mechanism to regulate gene expression [135]. Protein phosphatase 2C functions as a general regulator of signaling pathways activated by plant hormones such as abscisic acid and gibberellic acid, as well as by various stresses such as drought, salt, wounding, and cold [136]. Many environmental stresses, including drought, cause an increase in free amino acids and amines in plant cells, possibly due to a decrease in the electron transport chain [137]. These amino acids and amines can function as osmoprotectants, osmotic regulators, or scavengers of reactive oxygen species. Regulation of proteins involved in the synthesis of osmoprotectants such as proline and glycine betaine (GB) plays an important role in the stress tolerance of plants. Glutamine synthetase (GS) plays an important role in nitrogen metabolism and has been stated to function in regulating proline levels in plants. It has been suggested that the increased abundance of GS protein in soybean under drought stress is associated with high proline content [138].

LEA (Late-embryogenesis abundant) proteins are water-soluble proteins synthesized at high concentrations in desiccation-tolerant plants. It has been reported that dehydrin and ferritin proteins are increased in soybean plants under drought stress [138]. Dehydrins are LEA proteins and can effectively increase plant growth under stress by reducing the harmful effects of reactive oxygen species [139]. It has been reported that many proteins detected in Medicago truncatula plants are LEA proteins and these proteins are associated with drought tolerance [140].

## Heavy Metal Toxicity and Proteome Changes in Plants

Since some heavy metals resemble elements necessary for plant development, they can directly enter the roots and be taken up in the same way as essential elements. These heavy metals are then transported to above-ground organs *via* the same pathway or other transport mechanisms for essential elements with which they are structurally similar [141]. In addition to crop loss, the accumulation of heavy metals in agricultural plants threatens human health and the entire ecosystem [142]. Therefore, removing heavy metals from contaminated areas is very important. Unlike organic pollutants, heavy metals cannot be degraded by any known biological process and therefore it is necessary to remediate contaminated areas with an effective and environmentally friendly technology such as phytoremediation, which uses plants to accumulate and detoxify heavy metals [143]. Phytoremediation is a promising approach using hyperaccumulator plants to remove heavy metals from contaminated areas. Therefore, unlike many other plants, hyperaccumulator plants must be able to accumulate high levels of heavy metals in their above-ground organs [144]. However, the application of this approach is limited by the lack of hyperaccumulator plants with sufficient biomass and fast growth rate [145]. To overcome these limitations, it is necessary to develop genetically modified plant species with desired characteristics. Therefore, the physiological and molecular regulation of the heavy metal accumulation mechanism in hyperaccumulator plants needs to be well understood.

The majority of proteins differentially expressed in plants under heavy metal stress are proteins related to photosynthesis [146, 147]. RuBisCO (ribulose 1,5-bisphosphate carboxylase/oxygenase) constitutes 30-70% of soluble leaf protein and plays an important role in photosynthesis. RuBisCO catalyzes $CO_2$ fixation, which is the first step of carbon metabolism. However, the RuBisCO-binding subunit, RuBisCO activase and phosphoribulokinase are essential components for $CO_2$ fixation. The RuBisCO binding subunit is a protein required for the folding of the RuBisCO large subunit and RuBisCO activase provides the catalytic activity of RuBisCO [148 - 150] Phosphoribulokinase catalyzes the formation of ribulose-1,5-bisphosphate, a receptor molecule for $CO_2$ fixation of the Calvin cycle [151]. Heavy metal stress also affects many catabolic pathways and reduces ATP production. In plants exposed to heavy metal stress, the amount of many enzymes related to energy metabolism (carbohydrate metabolism, pentose phosphate pathway and tricarboxylic acid cycle) changes in response to the increased energy need [146, 152, 153].

In addition to carbon metabolism, sulfur metabolism is also affected by heavy metal stress [154, 155]. In plants, S-adenosyl-L-methionine (SAM) is an important methyl donor in several transmethylation reactions [156]. SAM is formed from L-methionine and ATP by SAM synthetase. Cobalamin-independent methionine synthase (MS) catalyzes the transfer of a methyl group to L-homocysteine [157]. In proteomic studies conducted in many plant species exposed to metal stress, an increase in the amount of SAMS and MS enzymes was determined [147, 158]. Additionally, SAM applied to cells under Cd stress has been shown to have a protective effect [159]. Due to the function of these enzymes in catalyzing intermediate reactions in the formation of methionine and the methyl cycle, it has been reported that an increase in the amount of these proteins can cause an increase in the level of methionine, which may result in the provision of methyl groups necessary for methylation reactions in many biosynthetic pathways to overcome Cd toxicity [147, 159]. However, it has been reported that Cu stress in rice plants increases the amount of SAM synthetase, whereas it decreases MS expression, and this may lead to a decrease in methionine level [155].

## APPROACHES TO PLANT PROTEOMICS STUDIES

Two approaches are basically used in proteomics studies: gel-based and gel-free proteomics approaches. In gel-based approaches, mass spectrometry (MS) analyses combined two-dimensional polyacrylamide gel electrophoresis (2D-PAGE) and 2D-PAGE method in the same gel with the control sample of the test subjected to a certain condition with a single 2D gel electrophoresis. MS analysis allows separation using the DIGE technique. After the control and test samples are labeled with Cyanine dyes with different fluorescent properties, both samples are mixed and subsequent processes are continued as in classical 2D-PAGE. Another technique is the 1D-gel/LCMS/MS approach, in which separation is achieved by one-dimensional gel electrophoresis and MS is combined with liquid chromatography (LC-MS/MS) analysis [160, 161].

## Gel-Based Approaches

There are denaturing and reducing agents used in the sodium dodecyl sulfate-polyacrylamide gel electrophoresis (SDS-PAGE) technique to separate proteins. Thanks to these substances, the subunits of proteins are separated from each other and examined. In SDS-PAGE, which is used for the one-dimensional separation of proteins, the purity and structure of a particular protein are examined, while at the same time, mixtures consisting of a small number of proteins are separated. However, since proteomic studies involve a mixture of all the proteins that make up a cell, methods with higher discrimination power are needed [162, 163].

2D PAGE, a combination of the SDS-PAGE method and isoelectric focusing (IEF), is an important technique used in the separation of complex protein mixtures. In this method, molecules are separated according to their isoelectric points (pI) in the first dimension and their molecular weight in the second dimension. Unlike classical electrophoresis, IEF is based on the principle of separating proteins according to their pI values in a pH gradient. In this method, the pH gradient is created with the help of low molecular weight amphoteric substances. Because these molecules have a pI value like proteins, their net electrical charge is zero at appropriate pH points [164, 165].

## Gel-Free Approaches

Limited resolution, loss of hydrophobic proteins, and low reproducibility are the most important problems in using the 2D-PAGE method to separate a proteome. However, a significant portion of the proteins found in a proteome are hydrophobic. Therefore, the analysis of these proteins is needed to accurately reveal the expression of genes in a cell, and it has become clear that much more powerful analytical techniques are needed. Among gel-free approaches, the use of Shotgun proteomic approaches has increased in recent years for all sample types. Two-dimensional liquid chromatography (2D-LC, two-dimensional liquid chromatography) techniques were developed as an alternative to the 2D-PAGE technique in order to expand the protein profile. The aim of liquid chromatography's approach to proteomics is to separate protein mixtures into fractions at high resolution, display them in virtual gel maps, and reveal protein quantity differences [166, 167].

Today, High-performance liquid chromatography (HPLC) technique is one of the most used analytical techniques. The 2D-LC system, which has taken its place among the new generation technologies in recent years and was created by combining the HPLC technique with the fractionation system, was developed to provide high-resolution protein separation for proteomic studies. In the 2D-LC separation technique, proteins are separated and fractionated according to their pI points in the first dimension by chromatofocusing (CF), and in the second dimension by high-performance reversed phase chromatography (HPRP) according to their hydrophobicity properties. Another important feature of the system is the inclusion of advanced computer programs for the detection of proteins separated in 2D-LC fractions and the determination of differences between samples. Another advantage of this system designed for proteome analysis is that crude protein extracts are analyzed after purification in several steps. Thus, it ensures that the data obtained as a result of comparing protein profiles is more reproducible and reliable compared to the 2D-PAGE method. In other words, the use of 2D-LC techniques instead of two-dimensional gel systems allows faster, reliable, and powerful separation of large numbers of total protein samples. Another important feature of the system is that low-expression proteins, which have a hydrophobic structure and many of which are regulatory proteins, which constitute a significant part of the proteome of a cell/organism, can be separated by the 2D-LC method, although they are lost in 2D-PAGE because they are insoluble in water [168 - 170].

## CONCLUSION

The aim of genomics and proteomics studies is not only to determine the quality of the identities of the expressed genes and/or proteins but also to know when and where these genes and proteins are expressed, and with what modifications. In particular, some modifications are both very unstable and very few. Being able to identify these modifications allows an understanding of the role of the relevant stress condition in the hypothesis under investigation or the molecular mechanisms underlying the events in question. One of the areas that need to be investigated more deeply in the future in plant genomic and proteomic research is subcellular genomic and proteomic studies, the number of which is actually limited for all eukaryotic organisms. On the one hand, the fact that these structures are membrane-based structures and on the other hand, the fact that they require intensive labor are the factors that limit the studies on these organelles. The most studied subcellular organelles in plants are mitochondria and chloroplasts because these organelles are the energy centers of the plant and are also active organelles in response to various biotic/abiotic stresses. The necessity of approaching omics approaches as a whole gave birth to systems biology. Thus, it enabled the processing of genomic transcriptomic, proteomic and metabolomic findings through various bioinformatic tools. These bioinformatics tools are informatics programs such as understanding the subcellular localization of molecules, determining whether they are secretory proteins, and determining

DNA-protein and protein-protein interactions. Most of these programs are web-based programs and are freely accessible.

## REFERENCES

[1]     Stencel A, Crespi B. What is a genome? Mol Ecol 2013; 22(13): 3437-43.
        [http://dx.doi.org/10.1111/mec.12355] [PMID: 23967454]

[2]     Del Giacco L, Cattaneo C. Introduction to Genomics. Methods Mol Biol 2012; 823: 79-88.
        [http://dx.doi.org/10.1007/978-1-60327-216-2_6] [PMID: 22081340]

[3]     Martin DB, Nelson PS. From genomics to proteomics: techniques and applications in cancer research.
        Trends Cell Biol 2001; 11(11): S60-5.
        [http://dx.doi.org/10.1016/S0962-8924(01)02123-7] [PMID: 11684444]

[4]     Del Boccio P, Urbani A. Homo sapiens proteomics: clinical perspectives. Ann Ist Super Sanita 2005;
        41(4): 479-82.
        [PMID: 16569916]

[5]     Tefferi A. Genomics basics: DNA structure, gene expression, cloning, genetic mapping, and molecular
        tests. Semin Cardiothorac Vasc Anesth 2006; 10(4): 282-90.
        [http://dx.doi.org/10.1177/1089253206294343] [PMID: 17200086]

[6]     Williams GA, Liede S, Fahy N, *et al.* Regulating the unknown: a guide to regulating genomics for
        health policy-makers. Copenhagen, Denmark: European Observatory on Health Systems and Policies
        2020. Available from: https://www.ncbi.nlm.nih.gov/books/NBK569502/

[7]     Aslam B, Basit M, Nisar MA, Khurshid M, Rasool MH. Proteomics: technologies and their
        applications. J Chromatogr Sci 2017; 55(2): 182-96.
        [http://dx.doi.org/10.1093/chromsci/bmw167] [PMID: 28087761]

[8]     Aizat WM, Hassan M. Proteomics in systems biology. Adv Exp Med Biol 2018; 1102: 31-49.
        [http://dx.doi.org/10.1007/978-3-319-98758-3_3] [PMID: 30382567]

[9]     Anderson NL, Anderson NG. Proteome and proteomics: New technologies, new concepts, and new
        words. Electrophoresis 1998; 19(11): 1853-61.
        [http://dx.doi.org/10.1002/elps.1150191103] [PMID: 9740045]

[10]    Zhu H, Bilgin M, Snyder M. Proteomics. Annu Rev Biochem 2003; 72(1): 783-812.
        [http://dx.doi.org/10.1146/annurev.biochem.72.121801.161511] [PMID: 14527327]

[11]    Kwon SJ, Choi EY, Choi YJ, Ahn JH, Park OK. Proteomics studies of post-translational modifications
        in plants. J Exp Bot 2006; 57(7): 1547-51.
        [http://dx.doi.org/10.1093/jxb/erj137] [PMID: 16551683]

[12]    Adams MD, Kelley JM, Gocayne JD, *et al.* Complementary DNA sequencing: expressed sequence
        tags and human genome project. Science 1991; 252(5013): 1651-6.
        [http://dx.doi.org/10.1126/science.2047873] [PMID: 2047873]

[13]    Campos-De Quiroz H. Plant genomics: an overview. Biol Res 2002; 35(3-4): 385-99.
        [http://dx.doi.org/10.4067/S0716-97602002000300013] [PMID: 12462991]

[14]    Tabata S. Impact of genomics approaches on plant genetics and physiology. J Plant Res 2002; 115(4):
        271-5.
        [http://dx.doi.org/10.1007/s10265-002-0036-8] [PMID: 12582730]

[15]    Rudd S. Expressed sequence tags: alternative or complement to whole genome sequences? Trends
        Plant Sci 2003; 8(7): 321-9.
        [http://dx.doi.org/10.1016/S1360-1385(03)00131-6] [PMID: 12878016]

[16]    Akpınar BA, Lucas SJ, Budak H. Genomics approaches for crop improvement against abiotic stress.
        ScientificWorldJournal 2013; 2013(1): 361921.

[http://dx.doi.org/10.1155/2013/361921] [PMID: 23844392]

[17]   Rashid B, Husnain T, Riazuddin S. Genomic Approaches and Abiotic Stress Tolerance in Plants. In: Ahmad P, Rasool S, Eds. Emerging Technologies and Management of Crop Stress Tolerance. Academic Press 2014; pp. 1-37.
[http://dx.doi.org/10.1016/B978-0-12-800876-8.00001-1]

[18]   Velculescu VE, Zhang L, Vogelstein B, Kinzler KW. Serial analysis of gene expression. Science 1995; 270(5235): 484-7.
[http://dx.doi.org/10.1126/science.270.5235.484] [PMID: 7570003]

[19]   Kaur B, Sandhu KS, Kamal R, *et al.* Omics for the improvement of abiotic, biotic, and agronomic traits in major cereal crops: applications, challenges, and prospects. Plants 2021; 10(10): 1989.
[http://dx.doi.org/10.3390/plants10101989] [PMID: 34685799]

[20]   Brenner S, Johnson M, Bridgham J, *et al.* Gene expression analysis by massively parallel signature sequencing (MPSS) on microbead arrays. Nat Biotechnol 2000; 18(6): 630-4.
[http://dx.doi.org/10.1038/76469] [PMID: 10835600]

[21]   Meyers BC, Souret FF, Lu C, Green PJ. Sweating the small stuff: microRNA discovery in plants. Curr Opin Biotechnol 2006; 17(2): 139-46.
[http://dx.doi.org/10.1016/j.copbio.2006.01.008] [PMID: 16460926]

[22]   Nobuta K, Venu RC, Lu C, *et al.* An expression atlas of rice mRNAs and small RNAs. Nat Biotechnol 2007; 25(4): 473-7.
[http://dx.doi.org/10.1038/nbt1291] [PMID: 17351617]

[23]   Meyerslab. Available from: https://mpss.meyerslab.org/

[24]   Agarwal M, Shrivastava N, Padh H. Advances in molecular marker techniques and their applications in plant sciences. Plant Cell Rep 2008; 27(4): 617-31.
[http://dx.doi.org/10.1007/s00299-008-0507-z] [PMID: 18246355]

[25]   Adhikari S, Saha S, Biswas A, Rana TS, Bandyopadhyay TK, Ghosh P. Application of molecular markers in plant genome analysis: a review. Nucleus 2017; 60(3): 283-97.
[http://dx.doi.org/10.1007/s13237-017-0214-7]

[26]   Ahmad R, Akbar Anjum M, Naz S, Mukhtar Balal R. Applications of molecular markers in fruit crops for breeding programs-a review. Phyton (B Aires) 2021; 90(1): 17-34.
[http://dx.doi.org/10.32604/phyton.2020.011680]

[27]   Tanksley SD. Molecular markers in plant breeding. Plant Mol Biol Report 1983; 1(1): 3-8.
[http://dx.doi.org/10.1007/BF02680255]

[28]   Lidder P, Sonnino A. Biotechnologies for the management of genetic resources for food and agriculture. Adv Genet 2012; 78: 1-167.
[http://dx.doi.org/10.1016/B978-0-12-394394-1.00001-8] [PMID: 22980921]

[29]   Botstein D, White RL, Skolnick M, Davis RW. Construction of a genetic linkage map in man using restriction fragment length polymorphisms. Am J Hum Genet 1980; 32(3): 314-31.
[PMID: 6247908]

[30]   Vos P, Hogers R, Bleeker M, *et al.* AFLP: a new technique for DNA fingerprinting. Nucleic Acids Res 1995; 23(21): 4407-14.
[http://dx.doi.org/10.1093/nar/23.21.4407] [PMID: 7501463]

[31]   Bobo-Pinilla J, Salmerón-Sánchez E, Mota JF, Peñas J. Genetic conservation strategies of endemic plants from edaphic habitat islands: The case of *Jacobaea auricula* (Asteraceae). J Nat Conserv 2021; 61: 126004.
[http://dx.doi.org/10.1016/j.jnc.2021.126004]

[32]   Adhikari S, Biswas A, Saha S, Bandyopadhyay TK, Ghosh P. AFLP-based assessment of genetic variation in certain Indian elite cultivars of Cymbopogon species. J Appl Res Med Aromat Plants

2022; 29: 100372.
[http://dx.doi.org/10.1016/j.jarmap.2022.100372]

[33]    Choudhury A, Deb S, Kharbyngar B, Rajpal VR, Rao SR. Dissecting the plant genome: through new generation molecular markers. Genet Resour Crop Evol 2022; 69(8): 2661-98.
[http://dx.doi.org/10.1007/s10722-022-01441-3]

[34]    Hon CC, Chow YC, Zeng FY, Leung FCC. Genetic authentication of ginseng and other traditional Chinese medicine. Acta Pharmacol Sin 2003; 24(9): 841-6.
[PMID: 12956929]

[35]    Ganie SH, Upadhyay P, Das S, Prasad Sharma M. Authentication of medicinal plants by DNA markers. Plant Gene 2015; 4: 83-99.
[http://dx.doi.org/10.1016/j.plgene.2015.10.002] [PMID: 32289060]

[36]    Leela M, Kavitha C, Soorianathasundaram K. Genetic diversity and population structure analysis of papaya (*Carica papaya* L.) germplasm using simple sequence repeat (SSR) markers. Genet Resour Crop Evol 2024; 1-15.
[http://dx.doi.org/10.1007/s10722-024-01883-x]

[37]    Verma S, Chaudhary HK, Singh K, *et al.* Genetic diversity dissection and population structure analysis for augmentation of bread wheat (*Triticum aestivum* L.) germplasm using morpho-molecular markers. Genet Resour Crop Evol 2024; 1-22.
[http://dx.doi.org/10.1007/s10722-023-01851-x]

[38]    Mir MA, Mansoor S, Sugapriya M, Alyemeni MN, Wijaya L, Ahmad P. Deciphering genetic diversity analysis of saffron (*Crocus sativus* L.) using RAPD and ISSR markers. Saudi J Biol Sci 2021; 28(2): 1308-17.
[http://dx.doi.org/10.1016/j.sjbs.2020.11.063] [PMID: 33613060]

[39]    Chirumamilla P, Gopu C, Jogam P, Taduri S. Highly efficient rapid micropropagation and assessment of genetic fidelity of regenerants by ISSR and SCoT markers of *Solanum khasianum* Clarke. Plant Cell Tissue Organ Cult 2021; 144(2): 397-407.
[http://dx.doi.org/10.1007/s11240-020-01964-6]

[40]    Welsh J, McClelland M. Fingerprinting genomes using PCR with arbitrary primers. Nucleic Acids Res 1990; 18(24): 7213-8.
[http://dx.doi.org/10.1093/nar/18.24.7213] [PMID: 2259619]

[41]    Bhandare PP, Waghmare TE, Naik GR. Analysis of genetic diversity of neem using RAPD markers. Indian Journal of Pure & Applied Biosciences 2021; 9(2): 66-76.
[http://dx.doi.org/10.18782/2582-2845.7658]

[42]    Oliya BK, Chand K, Thakuri LS, Baniya MK, Sah AK, Pant B. Assessment of genetic stability of micropropagated plants of *Rhynchostylis retusa* (L.) using RAPD markers. Sci Hortic (Amsterdam) 2021; 281: 110008.
[http://dx.doi.org/10.1016/j.scienta.2021.110008]

[43]    Sanger F, Nicklen S, Coulson AR. Proceedings of the national academy of sciences PNAS. 5463-7.

[44]    Varshney RK, Nayak SN, May GD, Jackson SA. Next-generation sequencing technologies and their implications for crop genetics and breeding. Trends Biotechnol 2009; 27(9): 522-30.
[http://dx.doi.org/10.1016/j.tibtech.2009.05.006] [PMID: 19679362]

[45]    Edwards D, Batley J. Plant genome sequencing: applications for crop improvement. Plant Biotechnol J 2010; 8(1): 2-9.
[http://dx.doi.org/10.1111/j.1467-7652.2009.00459.x] [PMID: 19906089]

[46]    Schnable PS, Ware D, Fulton RS, *et al.* The B73 maize genome: complexity, diversity, and dynamics. Science 2009; 326(5956): 1112-5.
[http://dx.doi.org/10.1126/science.1178534] [PMID: 19965430]

[47]    Schmutz J, Cannon SB, Schlueter J, *et al.* Genome sequence of the palaeopolyploid soybean. Nature

2010; 463(7278): 178-83.
[http://dx.doi.org/10.1038/nature08670] [PMID: 20075913]

[48]   Analysis of the genome sequence of the flowering plant *Arabidopsis thaliana*. Nature 2000; 408(6814): 796-815.
[http://dx.doi.org/10.1038/35048692] [PMID: 11130711]

[49]   Sasaki T. The map-based sequence of the rice genome. Nature 2005; 436(7052): 793-800.
[http://dx.doi.org/10.1038/nature03895] [PMID: 16100779]

[50]   Shendure J, Mitra RD, Varma C, Church GM. Advanced sequencing technologies: methods and goals. Nat Rev Genet 2004; 5(5): 335-44.
[http://dx.doi.org/10.1038/nrg1325] [PMID: 15143316]

[51]   Hamilton JP, Robin Buell C. Advances in plant genome sequencing. Plant J 2012; 70(1): 177-90.
[http://dx.doi.org/10.1111/j.1365-313X.2012.04894.x] [PMID: 22449051]

[52]   Reuter JA, Spacek DV, Snyder MP. High-throughput sequencing technologies. Mol Cell 2015; 58(4): 586-97.
[http://dx.doi.org/10.1016/j.molcel.2015.05.004] [PMID: 26000844]

[53]   Mardis ER. Next-generation DNA sequencing methods. Annu Rev Genomics Hum Genet 2008; 9(1): 387-402.
[http://dx.doi.org/10.1146/annurev.genom.9.081307.164359] [PMID: 18576944]

[54]   Shendure J, Ji H. Next-generation DNA sequencing. Nat Biotechnol 2008; 26(10): 1135-45.
[http://dx.doi.org/10.1038/nbt1486] [PMID: 18846087]

[55]   Dmitriev AA, Pushkova EN, Melnikova NV. Plant genome sequencing: modern technologies and novel opportunities for breeding. Mol Biol 2022; 56(4): 495-507.
[http://dx.doi.org/10.1134/S0026893322040045] [PMID: 35964310]

[56]   Ong Q, Nguyen P, Phuong Thao N, Le L. Bioinformatics approach in plant genomic research. Curr Genomics 2016; 17(4): 368-78.
[http://dx.doi.org/10.2174/1389202917666160331202956] [PMID: 27499685]

[57]   Bradbury LMT, Niehaus TD, Hanson AD. Comparative genomics approaches to understanding and manipulating plant metabolism. Curr Opin Biotechnol 2013; 24(2): 278-84.
[http://dx.doi.org/10.1016/j.copbio.2012.07.005] [PMID: 22898705]

[58]   Duvick J, Fu A, Muppirala U, *et al.* PlantGDB: a resource for comparative plant genomics. Nucleic Acids Res 2007; 36(Database) (Suppl. 1): D959-65.
[http://dx.doi.org/10.1093/nar/gkm1041] [PMID: 18063570]

[59]   Rouard M, Guignon V, Aluome C, *et al.* GreenPhylDB v2.0: comparative and functional genomics in plants. Nucleic Acids Res 2011; 39(Database issue) (Suppl. 1): D1095-102.
[http://dx.doi.org/10.1093/nar/gkq811] [PMID: 20864446]

[60]   Van Bel M, Proost S, Wischnitzki E, *et al.* Dissecting plant genomes with the PLAZA comparative genomics platform. Plant Physiol 2012; 158(2): 590-600.
[http://dx.doi.org/10.1104/pp.111.189514] [PMID: 22198273]

[61]   Goodstein DM, Shu S, Howson R, *et al.* Phytozome: a comparative platform for green plant genomics. Nucleic Acids Res 2012; 40(D1): D1178-86.
[http://dx.doi.org/10.1093/nar/gkr944] [PMID: 22110026]

[62]   Hawkins RD, Hon GC, Ren B. Next-generation genomics: an integrative approach. Nat Rev Genet 2010; 11(7): 476-86.
[http://dx.doi.org/10.1038/nrg2795] [PMID: 20531367]

[63]   Thiellement H, Bahrman N, Damerval C, *et al.* Proteomics for genetic and physiological studies in plants. Electrophoresis 1999; 20(10): 2013-26.
[http://dx.doi.org/10.1002/(SICI)1522-2683(19990701)20:10<2013::AID-ELPS2013>3.0.CO;2-#]

[PMID: 10451110]

[64]   Ytterberg AJ, Jensen ON. Modification-specific proteomics in plant biology. J Proteomics 2010; 73(11): 2249-66.
       [http://dx.doi.org/10.1016/j.jprot.2010.06.002] [PMID: 20541636]

[65]   Zivy M, de Vienne D. Proteomics: a link between genomics, genetics and physiology. Plant Mol Biol 2000; 44(5): 575-80.
       [http://dx.doi.org/10.1023/A:1026525406953] [PMID: 11198419]

[66]   Barbier-Brygoo H, Joyard J. Focus on plant proteomics. Plant Physiol Biochem 2004; 42(12): 913-7.
       [http://dx.doi.org/10.1016/j.plaphy.2004.10.012] [PMID: 15707829]

[67]   Ruan SL, Ma HS, Wang SH, *et al.* Advances in plant proteomics. II. Application of proteome techniques to plant biology research. Yi Chuan 2006; 28(12): 1633-48.
       [http://dx.doi.org/10.1360/yc-006-1633] [PMID: 17138554]

[68]   Sahithi BM, Razi K, Al Murad M, *et al.* Comparative physiological and proteomic analysis deciphering tolerance and homeostatic signaling pathways in chrysanthemum under drought stress. Physiol Plant 2021; 172(2): 289-303.
       [http://dx.doi.org/10.1111/ppl.13142] [PMID: 32459861]

[69]   Hakeem KR, Chandna R, Ahmad P, Iqbal M, Ozturk M. Relevance of proteomic investigations in plant abiotic stress physiology. OMICS 2012; 16(11): 621-35.
       [http://dx.doi.org/10.1089/omi.2012.0041] [PMID: 23046473]

[70]   Yin X, Komatsu S. Plant nuclear proteomics for unraveling physiological function. N Biotechnol 2016; 33(5) (5 Pt B): 644-54.
       [http://dx.doi.org/10.1016/j.nbt.2016.03.001] [PMID: 27004615]

[71]   Garibay-Hernández A, Barkla BJ, Vera-Estrella R, Martinez A, Pantoja O. Membrane proteomic insights into the physiology and taxonomy of an oleaginous green microalga. Plant Physiol 2017; 173(1): 390-416.
       [http://dx.doi.org/10.1104/pp.16.01240] [PMID: 27837088]

[72]   Hochholdinger F, Guo L, Schnable PS. Cytoplasmic regulation of the accumulation of nuclear-encoded proteins in the mitochondrial proteome of maize. Plant J 2004; 37(2): 199-208.
       [http://dx.doi.org/10.1046/j.1365-313X.2003.01955.x] [PMID: 14690504]

[73]   Lonosky PM, Zhang X, Honavar VG, Dobbs DL, Fu A, Rodermel SR. A proteomic analysis of maize chloroplast biogenesis. Plant Physiol 2004; 134(2): 560-74.
       [http://dx.doi.org/10.1104/pp.103.032003] [PMID: 14966246]

[74]   Majeran W, Cai Y, Sun Q, van Wijk KJ. Functional differentiation of bundle sheath and mesophyll maize chloroplasts determined by comparative proteomics. Plant Cell 2005; 17(11): 3111-40.
       [http://dx.doi.org/10.1105/tpc.105.035519] [PMID: 16243905]

[75]   Zhu J, Alvarez S, Marsh EL, *et al.* Cell wall proteome in the maize primary root elongation zone. II. Region-specific changes in water soluble and lightly ionically bound proteins under water deficit. Plant Physiol 2007; 145(4): 1533-48.
       [http://dx.doi.org/10.1104/pp.107.107250] [PMID: 17951457]

[76]   Dunkley TPJ, Hester S, Shadforth IP, *et al.* Mapping the *Arabidopsis* organelle proteome. Proc Natl Acad Sci USA 2006; 103(17): 6518-23.
       [http://dx.doi.org/10.1073/pnas.0506958103] [PMID: 16618929]

[77]   Maltman DJ, Gadd SM, Simon WJ, Slabas AR. Differential proteomic analysis of the endoplasmic reticulum from developing and germinating seeds of castor (*Ricinus communis*) identifies seed protein precursors as significant components of the endoplasmic reticulum. Proteomics 2007; 7(9): 1513-28.
       [http://dx.doi.org/10.1002/pmic.200600694] [PMID: 17407185]

[78]   Fernando U, Chatur S, Joshi M, *et al.* Redox signalling from NADPH oxidase targets metabolic enzymes and developmental proteins in *Fusarium graminearum*. Mol Plant Pathol 2019; 20(1): 92-

106.
[http://dx.doi.org/10.1111/mpp.12742] [PMID: 30113774]

[79]    Rampitsch C, Huang M, Djuric-Cignaovic S, Wang X, Fernando U. Temporal quantitative changes in the resistant and susceptible wheat leaf apoplastic proteome during infection by wheat leaf rust (*Puccinia triticina*). Front Plant Sci 2019; 10: 1291.
[http://dx.doi.org/10.3389/fpls.2019.01291] [PMID: 31708941]

[80]    Fang X, Chen J, Dai L, *et al.* Proteomic dissection of plant responses to various pathogens. Proteomics 2015; 15(9): 1525-43.
[http://dx.doi.org/10.1002/pmic.201400384] [PMID: 25641875]

[81]    Boscaiu M, Lull C, Lidon A, *et al.* Plant responses to abiotic stress in their natural habitats. Bulletin UASVM Hortic 2008; 65(1): 53-8.

[82]    Huang H, Ullah F, Zhou DX, Yi M, Zhao Y. Mechanisms of ROS regulation of plant development and stress responses. Front Plant Sci 2019; 10: 800.
[http://dx.doi.org/10.3389/fpls.2019.00800] [PMID: 31293607]

[83]    Magwanga RO, Lu P, Kirungu JN, *et al.* Characterization of the late embryogenesis abundant (LEA) proteins family and their role in drought stress tolerance in upland cotton. BMC Genet 2018; 19(1): 6.
[http://dx.doi.org/10.1186/s12863-017-0596-1] [PMID: 29334890]

[84]    Pautasso C, Rossi S. Transcriptional regulation of the protein kinase A subunits in *Saccharomyces cerevisiae*: Autoregulatory role of the kinase A activity. Biochim Biophys Acta Gene Regul Mech 2014; 1839(4): 275-87.
[http://dx.doi.org/10.1016/j.bbagrm.2014.02.005] [PMID: 24530423]

[85]    Hoang XLT, Nhi DNH, Thu NBA, Thao NP, Tran LSP. Transcription factors and their roles in signal transduction in plants under abiotic stresses. Curr Genomics 2017; 18(6): 483-97.
[http://dx.doi.org/10.2174/1389202918666170227150057] [PMID: 29204078]

[86]    Kosová K, Vítámvás P, Urban MO, Prášil IT, Renaut J. Plant abiotic stress proteomics: the major factors determining alterations in cellular proteome. Front Plant Sci 2018; 9: 122.
[http://dx.doi.org/10.3389/fpls.2018.00122] [PMID: 29472941]

[87]    Barkla BJ, Vera-Estrella R, Raymond C. Single-cell-type quantitative proteomic and ionomic analysis of epidermal bladder cells from the halophyte model plant *Mesembryanthemum crystallinum* to identify salt-responsive proteins. BMC Plant Biol 2016; 16(1): 110.
[http://dx.doi.org/10.1186/s12870-016-0797-1] [PMID: 27160145]

[88]    Michaletti A, Naghavi MR, Toorchi M, Zolla L, Rinalducci S. Metabolomics and proteomics reveal drought-stress responses of leaf tissues from spring-wheat. Sci Rep 2018; 8(1): 5710.
[http://dx.doi.org/10.1038/s41598-018-24012-y] [PMID: 29632386]

[89]    Xin L, Zheng H, Yang Z, *et al.* Physiological and proteomic analysis of maize seedling response to water deficiency stress. J Plant Physiol 2018; 228: 29-38.
[http://dx.doi.org/10.1016/j.jplph.2018.05.005] [PMID: 29852332]

[90]    Miller G, Shulaev V, Mittler R. Reactive oxygen signaling and abiotic stress. Physiol Plant 2008; 133(3): 481-9.
[http://dx.doi.org/10.1111/j.1399-3054.2008.01090.x] [PMID: 18346071]

[91]    Sachdev S, Ansari SA, Ansari MI, Fujita M, Hasanuzzaman M. Abiotic stress and reactive oxygen species: generation, signaling, and defense mechanisms. Antioxidants 2021; 10(2): 277.
[http://dx.doi.org/10.3390/antiox10020277] [PMID: 33670123]

[92]    Kosová K, Vítámvás P, Prášil IT, Renaut J. Plant proteome changes under abiotic stress — Contribution of proteomics studies to understanding plant stress response. J Proteomics 2011; 74(8): 1301-22.
[http://dx.doi.org/10.1016/j.jprot.2011.02.006] [PMID: 21329772]

[93]    Yang L, Ma C, Wang L, Chen S, Li H. Salt stress induced proteome and transcriptome changes in

sugar beet monosomic addition line M14. J Plant Physiol 2012; 169(9): 839-50.
[http://dx.doi.org/10.1016/j.jplph.2012.01.023] [PMID: 22498239]

[94]   Li B, He L, Guo S, *et al.* Proteomics reveal cucumber Spd-responses under normal condition and salt stress. Plant Physiol Biochem 2013; 67: 7-14.
[http://dx.doi.org/10.1016/j.plaphy.2013.02.016] [PMID: 23524299]

[95]   Liu Y, Du H, He X, Huang B, Wang Z. Identification of differentially expressed salt-responsive proteins in roots of two perennial grass species contrasting in salinity tolerance. J Plant Physiol 2012; 169(2): 117-26.
[http://dx.doi.org/10.1016/j.jplph.2011.08.019] [PMID: 22070977]

[96]   Suzuki N, Koussevitzky S, Mittler R, Miller G. ROS and redox signalling in the response of plants to abiotic stress. Plant Cell Environ 2012; 35(2): 259-70.
[http://dx.doi.org/10.1111/j.1365-3040.2011.02336.x] [PMID: 21486305]

[97]   Gill SS, Tuteja N. Reactive oxygen species and antioxidant machinery in abiotic stress tolerance in crop plants. Plant Physiol Biochem 2010; 48(12): 909-30.
[http://dx.doi.org/10.1016/j.plaphy.2010.08.016] [PMID: 20870416]

[98]   Barkla BJ, Vera-Estrella R, Pantoja O. Progress and challenges for abiotic stress proteomics of crop plants. Proteomics 2013; 13(12-13): 1801-15.
[http://dx.doi.org/10.1002/pmic.201200401] [PMID: 23512887]

[99]   Zhu JK. Salt and drought stress signal transduction in plants. Annu Rev Plant Biol 2002; 53(1): 247-73.
[http://dx.doi.org/10.1146/annurev.arplant.53.091401.143329] [PMID: 12221975]

[100]  Cheng Y, Qi Y, Zhu Q, *et al.* New changes in the plasma membrane-associated proteome of rice roots under salt stress. Proteomics 2009; 9(11): 3100-14.
[http://dx.doi.org/10.1002/pmic.200800340] [PMID: 19526560]

[101]  Zhang L, Tian LH, Zhao JF, Song Y, Zhang CJ, Guo Y. Identification of an apoplastic protein involved in the initial phase of salt stress response in rice root by two-dimensional electrophoresis. Plant Physiol 2009; 149(2): 916-28.
[http://dx.doi.org/10.1104/pp.108.131144] [PMID: 19036832]

[102]  Cao YR, Chen SY, Zhang JS. Ethylene signaling regulates salt stress response. Plant Signal Behav 2008; 3(10): 761-3.
[http://dx.doi.org/10.4161/psb.3.10.5934] [PMID: 19513226]

[103]  Mahajan S, Pandey GK, Tuteja N. Calcium- and salt-stress signaling in plants: Shedding light on SOS pathway. Arch Biochem Biophys 2008; 471(2): 146-58.
[http://dx.doi.org/10.1016/j.abb.2008.01.010] [PMID: 18241665]

[104]  Misra S, Wu Y, Venkataraman G, Sopory SK, Tuteja N. Heterotrimeric G-protein complex and G-protein-coupled receptor from a legume ( *Pisum sativum* ): role in salinity and heat stress and cross-talk with phospholipase C. Plant J 2007; 51(4): 656-69.
[http://dx.doi.org/10.1111/j.1365-313X.2007.03169.x] [PMID: 17587233]

[105]  Perfus-Barbeoch L, Jones AM, Assmann SM. Plant heterotrimeric G protein function: insights from *Arabidopsis* and rice mutants. Curr Opin Plant Biol 2004; 7(6): 719-31.
[http://dx.doi.org/10.1016/j.pbi.2004.09.013] [PMID: 15491922]

[106]  Tuteja N. How pea phospholipase c functions in salinity stress tolerance. ISB News Report. 2007: 4-7.

[107]  Wang MC, Peng ZY, Li CL, Li F, Liu C, Xia GM. Proteomic analysis on a high salt tolerance introgression strain of *Triticum aestivum* / *Thinopyrum ponticum*. Proteomics 2008; 8(7): 1470-89.
[http://dx.doi.org/10.1002/pmic.200700569] [PMID: 18383010]

[108]  Bolte S, Schiene K, Dietz KJ. Characterization of a small GTP-binding protein of the rab 5 family in *Mesembryanthemum crystallinum* with increased level of expression during early salt stress. Plant Mol Biol 2000; 42(6): 923-35.

[http://dx.doi.org/10.1023/A:1006449715236] [PMID: 10890538]

[109]   Zhao Q, Zhang H, Wang T, Chen S, Dai S. Proteomics-based investigation of salt-responsive mechanisms in plant roots. J Proteomics 2013; 82: 230-53.
[http://dx.doi.org/10.1016/j.jprot.2013.01.024] [PMID: 23385356]

[110]   Jiang Y, Yang B, Harris NS, Deyholos MK. Comparative proteomic analysis of NaCl stress-responsive proteins in *Arabidopsis* roots. J Exp Bot 2007; 58(13): 3591-607.
[http://dx.doi.org/10.1093/jxb/erm207] [PMID: 17916636]

[111]   Li XJ, Yang MF, Chen H, Qu LQ, Chen F, Shen SH. Abscisic acid pretreatment enhances salt tolerance of rice seedlings: Proteomic evidence. Biochim Biophys Acta Proteins Proteomics 2010; 1804(4): 929-40.
[http://dx.doi.org/10.1016/j.bbapap.2010.01.004] [PMID: 20079886]

[112]   Zörb C, Schmitt S, Mühling KH. Proteomic changes in maize roots after short-term adjustment to saline growth conditions. Proteomics 2010; 10(24): 4441-9.
[http://dx.doi.org/10.1002/pmic.201000231] [PMID: 21136597]

[113]   Hasegawa PM, Bressan RA, Zhu JK, Bohnert HJ. Plant cellular and molecular responses to high salinity. Annu Rev Plant Physiol Plant Mol Biol 2000; 51(1): 463-99.
[http://dx.doi.org/10.1146/annurev.arplant.51.1.463] [PMID: 15012199]

[114]   Sanders D, Pelloux J, Brownlee C, Harper JF. Calcium at the crossroads of signaling. Plant Cell 2002; 14(Suppl) (Suppl. 1): S401-17.
[http://dx.doi.org/10.1105/tpc.002899] [PMID: 12045291]

[115]   Shi H, Ishitani M, Kim C, Zhu JK. The *Arabidopsis thaliana* salt tolerance gene SOS1 encodes a putative $Na^+/H^+$ antiporter. Proc Natl Acad Sci USA 2000; 97(12): 6896-901.
[http://dx.doi.org/10.1073/pnas.120170197] [PMID: 10823923]

[116]   Peng Z, Wang M, Li F, Lv H, Li C, Xia G. A proteomic study of the response to salinity and drought stress in an introgression strain of bread wheat. Mol Cell Proteomics 2009; 8(12): 2676-86.
[http://dx.doi.org/10.1074/mcp.M900052-MCP200] [PMID: 19734139]

[117]   Zhou S, Sauvé RJ, Liu Z, *et al.* Identification of salt-induced changes in leaf and root proteomes of the wild tomato, *Solanum chilense.* J Am Soc Hortic Sci 2011; 136(4): 288-302.
[http://dx.doi.org/10.21273/JASHS.136.4.288]

[118]   Chitteti BR, Peng Z. Proteome and phosphoproteome differential expression under salinity stress in rice (*Oryza sativa*) roots. J Proteome Res 2007; 6(5): 1718-27.
[http://dx.doi.org/10.1021/pr060678z] [PMID: 17385905]

[119]   Prasad PVV, Pisipati SR, Momčilović I, Ristic Z. Independent and combined effects of high temperature and drought stress during grain filling on plant yield and chloroplast EF-Tu expression in spring wheat. J Agron Crop Sci 2011; 197(6): 430-41.
[http://dx.doi.org/10.1111/j.1439-037X.2011.00477.x]

[120]   Nakashima K, Yamaguchi-Shinozaki K, Shinozaki K. The transcriptional regulatory network in the drought response and its crosstalk in abiotic stress responses including drought, cold, and heat. Front Plant Sci 2014; 5: 170.
[http://dx.doi.org/10.3389/fpls.2014.00170] [PMID: 24904597]

[121]   Khodadadi E, Fakheri BA, Aharizad S, Emamjomeh A, Norouzi M, Komatsu S. Leaf proteomics of drought-sensitive and -tolerant genotypes of fennel. Biochim Biophys Acta Proteins Proteomics 2017; 1865(11): 1433-44.
[http://dx.doi.org/10.1016/j.bbapap.2017.08.012] [PMID: 28887228]

[122]   Nemati M, Piro A, Norouzi M, Moghaddam Vahed M, Nisticò DM, Mazzuca S. Comparative physiological and leaf proteomic analyses revealed the tolerant and sensitive traits to drought stress in two wheat parental lines and their F6 progenies. Environ Exp Bot 2019; 158: 223-37.
[http://dx.doi.org/10.1016/j.envexpbot.2018.10.024]

[123]  Mittler R. Oxidative stress, antioxidants and stress tolerance. Trends Plant Sci 2002; 7(9): 405-10.
[http://dx.doi.org/10.1016/S1360-1385(02)02312-9] [PMID: 12234732]

[124]  Pitzschke A, Forzani C, Hirt H. Reactive oxygen species signaling in plants. Antioxid Redox Signal 2006; 8(9-10): 1757-64.
[http://dx.doi.org/10.1089/ars.2006.8.1757] [PMID: 16987029]

[125]  Faghani E, Gharechahi J, Komatsu S, *et al.* Comparative physiology and proteomic analysis of two wheat genotypes contrasting in drought tolerance. J Proteomics 2015; 114: 1-15.
[http://dx.doi.org/10.1016/j.jprot.2014.10.018] [PMID: 25449836]

[126]  Li JW, Chen XD, Hu XY, Ma L, Zhang SB. Comparative physiological and proteomic analyses reveal different adaptive strategies by *Cymbidium sinense* and *C. tracyanum* to drought. Planta 2018; 247(1): 69-97.
[http://dx.doi.org/10.1007/s00425-017-2768-7] [PMID: 28871432]

[127]  Ji K, Wang Y, Sun W, *et al.* Drought-responsive mechanisms in rice genotypes with contrasting drought tolerance during reproductive stage. J Plant Physiol 2012; 169(4): 336-44.
[http://dx.doi.org/10.1016/j.jplph.2011.10.010] [PMID: 22137606]

[128]  Faize M, Burgos L, Faize L, *et al.* Involvement of cytosolic ascorbate peroxidase and Cu/Zn-superoxide dismutase for improved tolerance against drought stress. J Exp Bot 2011; 62(8): 2599-613.
[http://dx.doi.org/10.1093/jxb/erq432] [PMID: 21239380]

[129]  Negi NP, Shrivastava DC, Sharma V, Sarin NB. Overexpression of CuZnSOD from *Arachis hypogaea* alleviates salinity and drought stress in tobacco. Plant Cell Rep 2015; 34(7): 1109-26.
[http://dx.doi.org/10.1007/s00299-015-1770-4] [PMID: 25712013]

[130]  Chmielewska K, Rodziewicz P, Swarcewicz B, *et al.* Analysis of drought-induced proteomic and metabolomic changes in barley (*Hordeum vulgare* L.) leaves and roots unravels some aspects of biochemical mechanisms involved in drought tolerance. Front Plant Sci 2016; 7: 1108.
[http://dx.doi.org/10.3389/fpls.2016.01108] [PMID: 27512399]

[131]  Wang Y, Fan K, Wang J, *et al.* Proteomic analysis of *Camellia sinensis* (L.) reveals a synergistic network in the response to drought stress and recovery. J Plant Physiol 2017; 219: 91-9.
[http://dx.doi.org/10.1016/j.jplph.2017.10.001] [PMID: 29096085]

[132]  Marrs KA. The functions and regulation of glutathione S-transferases in plants. Annu Rev Plant Physiol Plant Mol Biol 1996; 47(1): 127-58.
[http://dx.doi.org/10.1146/annurev.arplant.47.1.127] [PMID: 15012285]

[133]  Kausar R, Arshad M, Shahzad A, Komatsu S. Proteomics analysis of sensitive and tolerant barley genotypes under drought stress. Amino Acids 2013; 44(2): 345-59.
[http://dx.doi.org/10.1007/s00726-012-1338-3] [PMID: 22707152]

[134]  Xu J, Xing XJ, Tian YS, *et al.* Transgenic *Arabidopsis* plants expressing tomato glutathione S-transferase showed enhanced. PLoS One 2015; 10(9): e0136960.
[http://dx.doi.org/10.1371/journal.pone.0136960] [PMID: 26327625]

[135]  Yan SP, Zhang QY, Tang ZC, Su WA, Sun WN. Comparative proteomic analysis provides new insights into chilling stress responses in rice. Mol Cell Proteomics 2006; 5(3): 484-96.
[http://dx.doi.org/10.1074/mcp.M500251-MCP200] [PMID: 16316980]

[136]  Liu L, Hu X, Song J, Zong X, Li D, Li D. Over-expression of a Zea mays L. *protein phosphatase 2C* gene (ZmPP2C) in *Arabidopsis thaliana* decreases tolerance to salt and drought. J Plant Physiol 2009; 166(5): 531-42.
[http://dx.doi.org/10.1016/j.jplph.2008.07.008] [PMID: 18930563]

[137]  Reggiani R, Nebuloni M, Mattana M, Brambilla I. Anaerobic accumulation of amino acids in rice roots: role of the glutamine synthetase/glutamate synthase cycle. Amino Acids 2000; 18(3): 207-17.
[http://dx.doi.org/10.1007/s007260050018] [PMID: 10901618]

[138] Alam I, Sharmin SA, Kim KH, Yang JK, Choi MS, Lee BH. Proteome analysis of soybean roots subjected to short-term drought stress. Plant Soil 2010; 333(1-2): 491-505.
[http://dx.doi.org/10.1007/s11104-010-0365-7]

[139] Hossain Z, Khatoon A, Komatsu S. Soybean proteomics for unraveling abiotic stress response mechanism. J Proteome Res 2013; 12(11): 4670-84.
[http://dx.doi.org/10.1021/pr400604b] [PMID: 24016329]

[140] Boudet J, Buitink J, Hoekstra FA, *et al.* Comparative analysis of the heat stable proteome of radicles of *Medicago truncatula* seeds during germination identifies late embryogenesis abundant proteins associated with desiccation tolerance. Plant Physiol 2006; 140(4): 1418-36.
[http://dx.doi.org/10.1104/pp.105.074039] [PMID: 16461389]

[141] Ma JF, Yamaji N, Mitani N, *et al.* Transporters of arsenite in rice and their role in arsenic accumulation in rice grain. Proc Natl Acad Sci USA 2008; 105(29): 9931-5.
[http://dx.doi.org/10.1073/pnas.0802361105] [PMID: 18626020]

[142] Satarug S, Baker JR, Urbenjapol S, *et al.* A global perspective on cadmium pollution and toxicity in non-occupationally exposed population. Toxicol Lett 2003; 137(1-2): 65-83.
[http://dx.doi.org/10.1016/S0378-4274(02)00381-8] [PMID: 12505433]

[143] Pilon-Smits E. Phytoremediation. Annu Rev Plant Biol 2005; 56(1): 15-39.
[http://dx.doi.org/10.1146/annurev.arplant.56.032604.144214] [PMID: 15862088]

[144] Baker AJM, Brooks RR. Terrestrial higher plants which hyperacummulate metallic elements – a review of their distribution, ecology and phytochemistry. Biorecovery 1989; 1: 81-126.

[145] Eapen S, D'Souza SF. Prospects of genetic engineering of plants for phytoremediation of toxic metals. Biotechnol Adv 2005; 23(2): 97-114.
[http://dx.doi.org/10.1016/j.biotechadv.2004.10.001] [PMID: 15694122]

[146] Ahsan N, Lee DG, Kim KH, *et al.* Analysis of arsenic stress-induced differentially expressed proteins in rice leaves by two-dimensional gel electrophoresis coupled with mass spectrometry. Chemosphere 2010; 78(3): 224-31.
[http://dx.doi.org/10.1016/j.chemosphere.2009.11.004] [PMID: 19948354]

[147] Zhao L, Sun YL, Cui SX, *et al.* Cd-induced changes in leaf proteome of the hyperaccumulator plant *Phytolacca americana.* Chemosphere 2011; 85(1): 56-66.
[http://dx.doi.org/10.1016/j.chemosphere.2011.06.029] [PMID: 21723586]

[148] Douillard R, de Mathan O. Leaf Proteins for Food Use: Potential of RUBISCO. In: Hudson BJ, Ed. New and Developing Sources of Food Proteins. London: Chapman & Hall 1994; pp. 307-42.
[http://dx.doi.org/10.1007/978-1-4615-2652-0_10]

[149] Boston RS, Viitanen PV, Vierling E. Molecular chaperones and protein folding in plants. Plant Mol Biol 1996; 32(1-2): 191-222.
[http://dx.doi.org/10.1007/BF00039383] [PMID: 8980480]

[150] Portis AR, Jr . Rubisco activase - Rubisco's catalytic chaperone. Photosynth Res 2003; 75(1): 11-27.
[http://dx.doi.org/10.1023/A:1022458108678] [PMID: 16245090]

[151] Miziorko HM. Phosphoribulokinase: current perspectives on the structure/function basis for regulation and catalysis. Adv Enzymol Relat Areas Mol Biol 2000; 74: 95-127.
[http://dx.doi.org/10.1002/9780470123201.ch3] [PMID: 10800594]

[152] Sharmin SA, Alam I, Kim KH, *et al.* Chromium-induced physiological and proteomic alterations in roots of *Miscanthus sinensis.* Plant Sci 2012; 187: 113-26.
[http://dx.doi.org/10.1016/j.plantsci.2012.02.002] [PMID: 22404839]

[153] Visioli G, Vincenzi S, Marmiroli M, Marmiroli N. Correlation between phenotype and proteome in the Ni hyperaccumulator *Noccaea caerulescens* subsp. caerulescens. Environ Exp Bot 2012; 77: 156-64.
[http://dx.doi.org/10.1016/j.envexpbot.2011.11.016]

[154]  Chen YA, Chi WC, Huang TL, *et al.* Mercury-induced biochemical and proteomic changes in rice roots. Plant Physiol Biochem 2012; 55: 23-32.
[http://dx.doi.org/10.1016/j.plaphy.2012.03.008] [PMID: 22522577]

[155]  Song Y, Cui J, Zhang H, Wang G, Zhao FJ, Shen Z. Proteomic analysis of copper stress responses in the roots of two rice (*Oryza sativa* L.) varieties differing in Cu tolerance. Plant Soil 2013; 366(1-2): 647-58.
[http://dx.doi.org/10.1007/s11104-012-1458-2]

[156]  Van Breusegem F, Dekeyser R, Gielen J, Van Montagu M, Caplan A. Characterization of a S-adenosylmethionine synthetase gene in rice. Plant Physiol 1994; 105(4): 1463-4.
[http://dx.doi.org/10.1104/pp.105.4.1463] [PMID: 7972513]

[157]  Pejchal R, Ludwig ML. Cobalamin-independent methionine synthase (MetE): a face-to-face double barrel that evolved by gene duplication. PLoS Biol 2004; 3(2): e31.
[http://dx.doi.org/10.1371/journal.pbio.0030031] [PMID: 15630480]

[158]  Ge C, Ding Y, Wang Z, *et al.* Responses of wheat seedlings to cadmium, mercury and trichlorobenzene stresses. J Environ Sci (China) 2009; 21(6): 806-13.
[http://dx.doi.org/10.1016/S1001-0742(08)62345-1] [PMID: 19803087]

[159]  Noriega GO, Balestrasse KB, Batlle A, Tomaro ML. Cadmium induced oxidative stress in soybean plants also by the accumulation of δ-aminolevulinic acid. Biometals 2007; 20(6): 841-51.
[http://dx.doi.org/10.1007/s10534-006-9077-0] [PMID: 17216352]

[160]  Ünlü M, Morgan ME, Minden JS. Difference gel electrophoresis. A single gel method for detecting changes in protein extracts. Electrophoresis 1997; 18(11): 2071-7.
[http://dx.doi.org/10.1002/elps.1150181133] [PMID: 9420172]

[161]  Wittig I, Braun HP, Schägger H. Blue native PAGE. Nat Protoc 2006; 1(1): 418-28.
[http://dx.doi.org/10.1038/nprot.2006.62] [PMID: 17406264]

[162]  Smith BJ. SDS polyacrylamide gel electrophoresis of proteins. Methods Mol Biol 1984; 1: 41-56.
[http://dx.doi.org/10.1385/0-89603-062-8:41] [PMID: 20512673]

[163]  Pavlova AS, Dyudeeva ES, Kupryushkin MS, Amirkhanov NV, Pyshnyi DV, Pyshnaya IA. SDS-PAGE procedure: Application for characterization of new entirely uncharged nucleic acids analogs. Electrophoresis 2018; 39(4): 670-4.
[http://dx.doi.org/10.1002/elps.201700415] [PMID: 29112277]

[164]  Lognonné JL. 2D-page analysis: a practical guide to principle critical parameters. Cell Mol Biol 1994; 40(1): 41-55.
[PMID: 8003935]

[165]  Meleady P. Two-dimensional gel electrophoresis and 2D-DIGE. Methods Mol Biol 2018; 1664: 3-14.
[http://dx.doi.org/10.1007/978-1-4939-7268-5_1] [PMID: 29019120]

[166]  Van Damme P, Impens F, Vandekerckhove J, Gevaert K. Protein processing characterized by a gel-free proteomics approach. Methods Mol Biol 2008; 484: 245-62.
[http://dx.doi.org/10.1007/978-1-59745-398-1_16] [PMID: 18592184]

[167]  Blankley RT, Gaskell SJ, Whetton AD, Dive C, Baker PN, Myers JE. A proof-of-principle gel-free proteomics strategy for the identification of predictive biomarkers for the onset of pre-eclampsia. BJOG 2009; 116(11): 1473-80.
[http://dx.doi.org/10.1111/j.1471-0528.2009.02283.x] [PMID: 19663911]

[168]  Gaspard SJ, Brodkorb A. The use of high performance liquid chromatography for the characterization of the unfolding and aggregation of dairy proteins. Methods Mol Biol 2019; 2039: 103-15.
[http://dx.doi.org/10.1007/978-1-4939-9678-0_8] [PMID: 31342422]

[169]  Papatheocharidou C, Samanidou V. Two-dimensional high-performance liquid chromatography as a powerful tool for bioanalysis: the paradigm of antibiotics. Molecules 2023; 28(13): 5056.

[http://dx.doi.org/10.3390/molecules28135056] [PMID: 37446719]

[170] Morisaka H, Kirino A, Kobayashi K, Ueda M. Two-dimensional protein separation by the HPLC system with a monolithic column. Biosci Biotechnol Biochem 2012; 76(3): 585-8.
[http://dx.doi.org/10.1271/bbb.110770] [PMID: 22451405]

CHAPTER 7

# Seed Physiology: Future Applications and Current Limitations

## Ergun Kaya[1,*]

[1] *Muğla Sıtkı Koçman University, Faculty of Science, Molecular Biology and Genetics Department, 48000, Menteşe, Muğla, Türkiye*

**Abstract:** The basis of the evolutionary development of a plant is reproduction. Because reproduction ensures the continuity of species, therefore, immediately after fertilization, the seed becomes the recipient of substances assimilated in the plant. As growth continues, the seed basically goes through three main physiological stages. The period in which 80% of the growth occurs is the period of cell division and elongation, the supply of nutrients from the mother plant through the funiculus, and the intense increase in seed weight. The funiculus degenerates and the seed breaks its connection with the parent plant. The seed has reached the maximum dry matter. This period is called mass maturity. Although physiological maturity has been considered for many years, it has been determined that the seed is not physiologically mature during this period. In the drying period, the seed begins to dry. During this period, environmental conditions, rain, temperature, diseases and pests to which the seed is exposed reduce the quality. It is imperative to carefully combine advances in seed management, vegetative propagation, biotechnology, and molecular genetics to support forestry practices that strive to balance socioeconomic requirements, biodiversity, and climate change with sustainable production. The production, preservation, and repair of seeds all start with seed physiology and technology. This chapter aims to explain the physiological processes that affect the formation, growth, and development of seeds, which are the basis of the continuity of plant generations, by evaluating the perspectives of future applications and current limitations.

**Keywords:** Abscisic acid auxin, Cytokinin, Ethylene, Gibberellic acid.

## INTRODUCTION

A seed is morphologically defined as a fertilized egg. This occurs as a result of double fertilization. Firstly, the endosperm is formed by the fertilization of one of the polar nuclei, and the embryo is formed by the fertilization of the egg cell [1, 2]. After double fertilization, endosperm development begins before embryo development. First of all, with the supply of energy, the absorption of nutrients

---

* **Corresponding author Ergun Kaya:** Muğla Sıtkı Koçman University, Faculty of Science, Molecular Biology and Genetics Department, 48000, Menteşe, Muğla, Türkiye; E-mail: ergunkaya@mu.edu.tr

from the main plant cells begins. The outer surface of the endosperm is covered with the aleurone layer, which is a layer containing high levels of protein. This layer plays an important role in enzyme synthesis during the germination period. The aleurone layer is a well-developed layer in grasses and lettuce but is less present in other species [3, 4]. In some species, the endosperm completes its development early and the nucellus (2n) is filled with nutrient tissue. With this, it meets the energy needs of the seed during the germination period. Beetroot and spinach can be given as examples of these species. During seed maturation, this tissue is called perisperm [5, 6].

Some seeds do not contain endosperm or have no perisperm. These seeds are called nonendospermic seeds. These types of seeds contain large embryos. These families include Fabaceae, Cucurbitaceae, and Asteraceae. The endosperm typically contains large amounts of starch, an energy-supplying compound. They contain lower amounts of storage proteins or some types of fat [7, 8].

With the fertilization of the egg, a zygote is formed, which ultimately gives rise to the embryo. Cell division does not occur without the formation of some endosperm. In the first stage, two cells are formed, the one closest to the micropile is elongated and more dominant. The cell in the upper layer ultimately gives rise to the embryo. The circuit of several cells is called proembryo. During this period, the embryo goes through four separate stages. The cotyledons of dicotyledon seeds are quite large. While endospermic seeds have thin, delicate and leaf-like cotyledons, non-endospermic seeds (peas, cowpeas and beans) have very large cotyledons and constitute 90% of the dry weight of the seed [9, 10].

As a general rule, the cotyledons that remain in the soil (hypogeal) when germinated are larger, and the cotyledons that rise to the soil surface (epigeal) are smaller. It can shed the seeds of some species before they are fully mature in terms of development. For example, carrot seeds leave the plant with an immature embryo and continue their development later [11, 12].

Under normal conditions, the seed moisture content is 80% at fertilization. Depending on the characteristics of the seed during the harvest period, it may drop to 50-60% in recalcitrant species, 20-25% in sub orthodox species (intermediate species, when kept at -20°C, the intermediate seeds from this plant may only last five years. They also have a tendency to deteriorate more quickly than traditional seeds. When they are dried between 45 and 65% RH, they last the longest), and up to 10% in orthodox species (This plant's seeds can withstand freezing temperatures and drying to an interior seed moisture content of less than 10%). The source of assimilates necessary for seed development is the mother plant. Starch formation occurs from the use of sucrose produced during photosynthesis. In oilseeds, sucrose changes through some chemical reactions. Protein synthesis occurs by transporting amino acids such as asparagine and glutamine from the roots and leaves [13, 14].

Sugar, fat, and protein accumulation occurs equally throughout the seed development period. During the development period, the seed goes through two important periods in terms of dry matter accumulation: Physiological mass maturity: This is the period when the seed has the maximum dry matter. This period has been determined as the period when seed quality reaches its maximum level in some species [15, 16]. However, in some species, this period occurred one or two weeks later. Harvest maturity is the period when the seed is suitable for machine harvesting. It is also known as the period when seed moisture is 12-20%. It is especially important for dried fruit species. Looking at the changes in seed moisture during the development period, we can say that seed moisture decreases from 55-80% to 10-12% or 35-45%. It decreases to 10-12% in dried fruits and to 35-45% in juicy fruits [17, 18].

## STRUCTURE AND DEVELOPMENT OF SEEDS

In general, the seed is the formation of the embryo through the division of the zygote in plants. Although the embryos look different in monocots and dicots, their developmental structures are the same, but the development of endosperm and other tissues is different. There are structural differences in the mature seed. The developing seed receives water and nutrients to accelerate cell division and elongation. Initially, the amount of sucrose and other sugars in the shell and endosperm is high, which are then converted to starch. Protein accumulation in the endosperm increases proportionally with time [19, 20].

Plants produce starch, also known as amylum, which is a polysaccharide that is produced and used as a renewable source of industrial raw materials. Starch is found in the center of the food and feed chains. Two different types of glucose polymers make up starch: amylopectin, which is highly branched and contains shorter $\alpha$-1,4-linked glucan chains joined by $\alpha$-1,6-glycosidic bonds, and amylose, which is linear and moderately branched and comprises glucose moieties linked together by $\alpha$-1,4-glycosidic connections. Higher plants create two types of starch: transient starch made in chloroplasts in photosynthetic tissues and long-term storage starch made in amyloplasts in non-photosynthetic tissues including seeds, storage roots, and tubers. During the day, transient starch produced from photosynthates is broken down at night to support metabolism and supply energy. Long-term storage ensures that storage starch is ready for remobilization during germination, sprouting, or regeneration [21 - 23].

## Oily Seeds

Oil crops consist of plants belonging to various families. For this reason, seeds may show different properties from each other depending on the characteristic structures of these families. For example, sunflower has an achene-type fruit structure, consisting of seeds and attached fruit shells. The size of the achenes is small in the middle part of the tables and large in the edges. All achenes form fruit shells, whether they have seeds or not. The outermost epidermal cells of these seeds are thick and their outer walls are covered with a thin cuticle (Fig. **1**). When the fruit peel develops, the hypodermis, which consists of a single row of cells, can be seen under the epidermis. The cells in this layer have thin walls and large nuclei [24, 25].

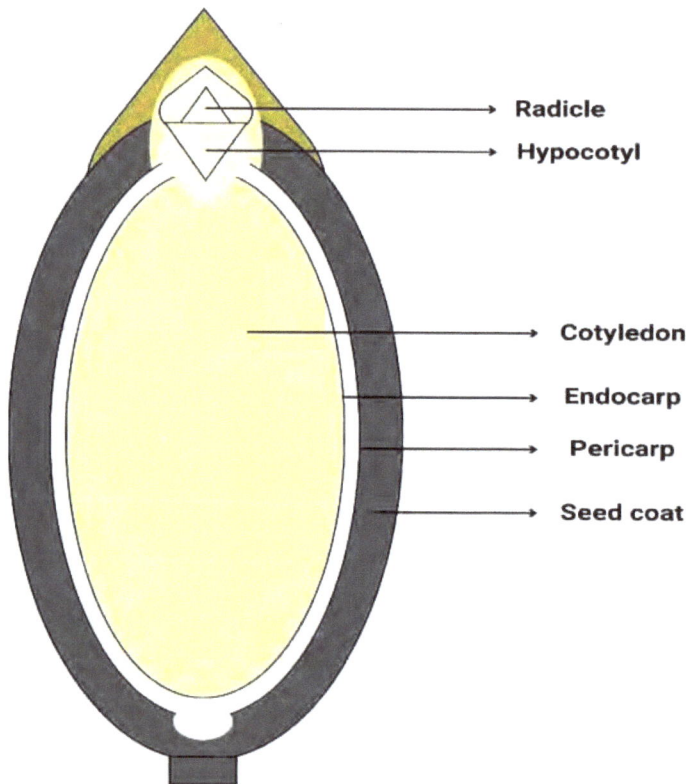

**Fig. (1).** Morphological structure of the achene-type fruit of sunflower [25, 26].

## Legume Seeds (Dicotyledon seeds)

Legume seeds, which are attached to the fruit shell by an umbilical cord within the fruit, may vary in color, size, and shape depending on the genus, species, variety, and environmental conditions. Seed colors of legumes range from white to black. The grain shape can vary from flat disk to round and rectangular prism. In legumes, the seed consists of three parts; (*i*) seed coat, (*ii*) cotyledons, and (*iii*) Embryo (Fig. **2**) [27, 28].

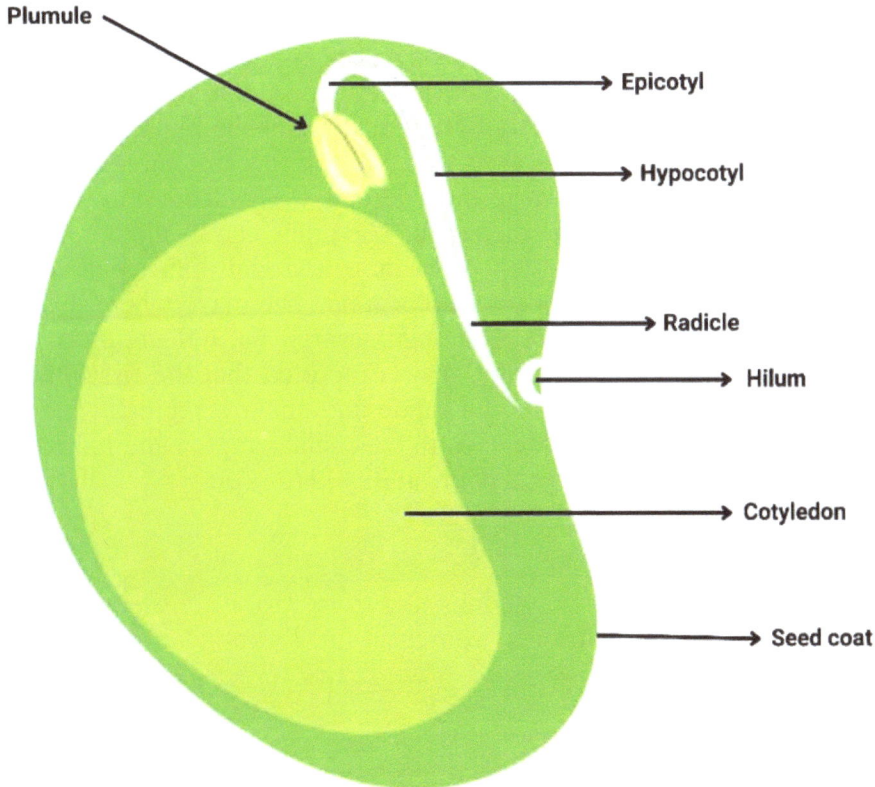

**Fig. (2).** Morphological structure of the Legume Seeds [28].

## Seed Coat (Testa)

In all legumes, the shell layer surrounding the cotyledon and embryo consists only of the seed coat (testa). The seed coat consists of three different cell layers. There are outermost longitudinal palisade cells. The color pigments that generally give the seed its color are collected in this layer. Below this is the layer consisting of T-shaped transverse cells. Under these two layers, there is one or more rows of parenchyma tissue. The thickness of the seed coat varies depending on the genus, species, their varieties and environmental conditions. The thinness of the testa also makes germination easy and fast. The hilum, the umbilical cord residue seen on the seed coat, is the residue on the grain of the part where the grain connects to the fruit shell. The colour, shape and size of the umbilical shell vary depending on the genus, species and varieties [29, 30].

This feature is used as a criterion to distinguish species and varieties from each other. Depending on the shape of the seed, there is a micropyle, which is the residue on the grain shell of the part where the pollen tube enters the ovary, in the form of a bubble at the bottom or top of the umbilical cord. Below this point is the tip of the grass rootlet and it comes out from this point. Micropyle is generally darker and more transparent than the normal seed coat color and is in the form of a slight blister. Micropyle and hilum have been reported to assist in water uptake during germination. There is microphyll at one end of the hilum and strophyll at the other end. There is a raphe on the seed coat, and this structure is located at the other end of the hilum compared to the microphyll. Transmission bundles from the umbilical cord extend within the raphe. The raphe can be distinguished by its externally narrow, elongated, hardened structure, and shell-like appearance, which also attracts attention with its distinct color tone. The place where the seed line ends is called chalaza. It appears as a small, warty, hard elevation on the seed coat. In a study conducted on beans, it was reported that the first place where water is absorbed into the seed is the raphe and chalaza regions. The fact that there is more pectic in the palisade cells in the chalaza region and more phenol in the osteoscleroid cells also prevents permeability [31, 32].

## *Cotyledons*

All legume species have two cotyledons that store various nutrients that serve as the endosperm in grains. Nutrients are rich in fat, carbohydrates, and especially protein [31].

## *Embryo*

It is the living part of the seed. It is located between two cotyledon leaves as a small sample of the plant that will form the new plant. The embryo consists of the plumula, which will form the above-ground organs after germination, the hypocotyl, and the radicle (radicula), which will form the underground organs [31].

## Gramine Seeds (Monocotyle seeds)

This group of seeds is very tolerant to dehydration and the fruit shell (pericarp) and the seed shell are adjacent, this structure is called caryopsis (Fig. **3**) [33, 34].

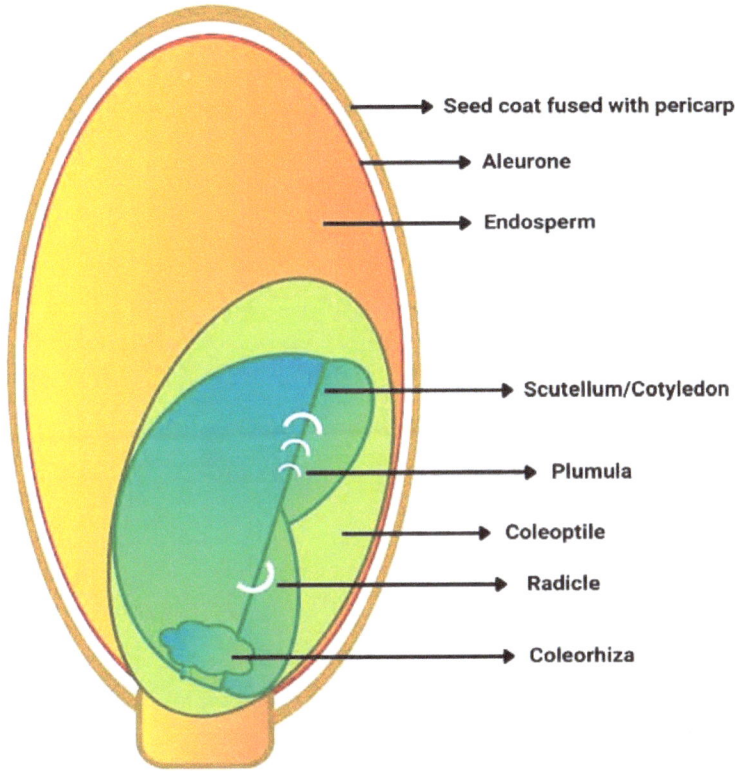

**Fig. (3).** Morphological structure of the *Zea mays* L. (corn) Seeds [31].

## Seed coat

It surrounds the embryo with the cotyledon located in the outermost part of the seed. It occurs by the differentiation of integuments in the ovule during seed development. During seed development, the cell layers in the inner integuments are lost. The outer integuments, consisting of five rows of cells, differentiate and form the seed coat layers. The outermost epidermis of the outer integuments forms the palisade layer, which consists of a single row of long and narrow cells arranged side by side. This layer, which contains color pigments, carries scleroid cells with thick membranes, more or less hexagonal in cross-section [34, 35]. With the differentiation of the second and third cell layers of the outer integument, the subepidermal layer and parenchyma are formed from tangentially arranged large cells. The cells of the subepidermal layer are different from others. These are also called watchglass cells. There are large intercellular spaces between these cells. During the ripening of the grain, the hardening of the seed coat occurs as a result of the formation of a selerol area in the layer of palisade cells and watch glass cells. The thickest part of the seed coat is parenchyma.

Conduction channels are located in this layer. Transmission channels are very well developed. Conduction channels branch and spread from the core of the grain towards the interior. The innermost cell layer of the outer integuments forms the inner parenchyma part, which consists of long-armed mushroom-like cells that appear very compacted in the mature grain [34, 36].

## Cotyledons

The number of these structures is between 1-3 in each seed. The first leaflet is on the opposite side of the shield, and the 2nd and 3rd leaves are located opposite each other, making a 180° angle with each other. These leaflets appear first on the soil surface. Under and in the middle of the first leaflets is the meristematic tissue that will form in the first stem of the plant [33, 35].

## Stalk

The region of the embryo between the radicle and the bud grows and develops and forms the node and internodes of the plant. In addition, there is a small epiblast (embryo flake) structure on the stalk, on the opposite side of the shield, extending towards the embryonic axis [33, 37].

## Radicula

This structure is surrounded by a radicle sheath. There are meristem cells at the tip of the radicula. Radicula is one of the cool climate grains. As germination begins, the radicle sheath stops growing after piercing the shell of the grain, and the actual radicle emerges from the hole at its tip. At the root tip, cells show rapid mitosis division. At the tip, many absorbent hairs form perpendicular to the rootlet [33, 37].

## Endosperm

Aleurone surrounds the outer wall of the endosperm. It is examined separately from the actual endosperm in terms of the shape and size of the cells and the nutrients they store. The cells are large rectangular prism shaped or cubic shaped. They form a layer consisting of one row of cells on the dorsal part of the seed and 2-3 rows of cells on the ventral part. The aleurone layer covers the grain only up to the embryo. Cells in the aleuron layer contain primitive proteins and oils in amide form that have enzymatic activities during germination. There is little or no starch in these cells. Endosperm constitutes 80-85% of the seed weight. The cells in the middle part of this layer are larger, while those closer to the aleurone layer are smaller [37, 38].

## *Shield*

The largest part of the embryo is the shield, and this name was given because it is shield-shaped. It is a structure adjacent to the endosperm, and a row of thin epithelial cells at the end, located on the face facing the endosperm, secretes plant growth regulators. These plant growth regulators stimulate the formation of enzymes in the aleurone layer, and these enzymes break down the large-molecule nutrients that will pass into the actual endosperm and make them able to pass through the cell membrane. Although the embryo removed from the endosperm along with the shield can germinate under appropriate humidity, temperature and weather conditions, the fact that the embryo with the shield removed cannot germinate shows that the shield has a function that initiates and continues germination. Germination of the seed depends on the state of the growth regulators in the shield [33, 37].

## *Embryo*

The embryo is a very small and primitive example of a new plantlet. It divides very quickly during germination and extends into the radicle in the ventral direction of the seed and into the seedling in the dorsal direction. The bud consists of apical meristem tissue along with several embryonic leaves. It grows and forms the above-ground organs and the actual body of the plant. The coleoptile is located in its uppermost and outermost part and has a protective sheath covering the plumula. Coleoptile pierces the seed coat layer during germination, allowing the first leaves to emerge from the hole at the tip to the soil surface without being damaged [37, 38].

These cells, which multiply rapidly after fertilization, store carbohydrates, proteins, minerals and a small amount of fat in a mature seed. First, the zygote is formed, undergoes successive mitosis and reproduces. It produces a new plant under suitable conditions and ensures the continuity of the germplasm. The structure of the embryo contains 40% protein, 15% fat, 6% pentosan and also contains color pigments [10, 39].

## SEED GERMINATION

Seed Germination can be defined as the activation of the metabolic mechanism of the embryo to produce a new plant. In order for a seed to germinate; (*i*) the seed must be alive and capable of germination, (*ii*) the seed must be placed in suitable environmental conditions (available water, appropriate temperature regime, oxygen and sometimes light supply), (*iii*) the seed must have emerged from internal dormancy [40, 41].

During the ripening phase of the seed, most seeds undergo dehydration, lose water and dry out (orthodox seeds). These seeds may or may not be dormant when separated from the plant. However, some seeds do not enter the maturation phase of seed development and germinate before leaving the plant. Some other seeds (recalcitrant seeds) are very sensitive to dehydration and can tolerate only a small amount of drying. Viviparous and recalcitrant seeds can germinate before the maturation phase of seed development is completed [42, 43].

Germination of a seed is the emergence of a new plantlet from the embryo of the seed and the full emergence of the organs that will form a normal plant under favorable soil conditions. In general, a seed consists of three parts: shell, endosperm and embryo, and these parts may vary depending on plant groups. Germination of a seed generally occurs in three stages; *(i) First stage – Absorption of water:* After the completion of seed development (especially orthodox seeds), it is in a dry state (moisture content below 15%). During this phase of germination, water uptake occurs through the absorption of water from the seed coat. This process occurs in two stages. First of all, water intake occurs rapidly in the first 10-30 minutes. The next stage is the slow intake of water. This stage is 1 hour for small seeds and 5-10 hours for large seeds. Water uptake through absorption continues until the lag phase, which is the second stage of seed germination. Seed volume increases during the first stage, which is water intake through absorption [42, 43].

Another characteristic event in the process of water absorption by the seed is the leakage of compounds such as amino acids, organic acids, inorganic substances, sugars, phenolic substances and proteins from the seed. This is due to the fact that cell membranes lose their selective permeability due to the oxidation of unsaturated fatty acids in their lipids. This situation occurs as a result of seed deterioration. In fact, leakage occurs during the absorption of water in all seeds. However, as the seed deteriorates, its amount increases. Seeds may be physically damaged during the absorption of water. This is due to the excessive intake of water by the dry seed during absorption. This situation can be prevented by slightly increasing the moisture content of the seed before planting [40, 44]. *(ii) second phase – Lag phase:* This phase is a period in which water intake is absent or decreased, but it is a phase in which physiological events are quite active. At this stage, new proteins needed for germination are synthesized. The transformation of storage materials for germination begins. *(iii) third stage – The stage in which the radicle appears:* The first visible sign of germination is the appearance of the radicle. This is the result of cell growth rather than cell division. Immediately afterward, cell division occurs at the tip of the rootlet and the rootlet begins to lengthen [42, 43].

## Shoot Formation

The embryo consists of an axis that produces one or two cotyledons. The radicle, which extends from the growth point at the bottom of the embryo axis, is the development point of the root. The plumula, located on the cotyledons at the top of the embryo axis, is the growth point of the shoot. The body of the plant, which consists of seeds, is divided into two parts: hypocotyl and epicotyl. The hypocotyl is the part below the cotyledons and the epicotyl is above it [42, 44].

## ENVIRONMENTAL FACTORS AFFECTING ON GERMINATION

### Water/Moisture

In the absence of dormancy, the availability of available water for many seeds is the most important factor for germination at a suitable temperature. Water stress can reduce the germination rate. Many seeds can germinate at levels of water in the soil ranging from field capacity to the permanent wilting point. Germination of some seeds with dormancy problems (such as beet, lettuce, endive, and celery) is prevented when the moisture level decreases. In species such as spinach, the seeds intensively produce a substance that limits the embryo's oxygen intake when exposed to excessive water. Moisture stress also significantly reduces plant emergence rate. This decrease in output rate occurs as water drops from field capacity to approximately half the level towards the permanent wilting point [45, 46].

Priming treatment is defined as a pre-sowing application that allows the seeds to absorb water until the first stage of germination in an osmotic solution or water, but does not allow the radicle to emerge from the seed coat. Germination remains in the lag phase. The seeds to which this application (priming) is applied are dried again to their original water content before root emergence. Generally used priming materials are PEG 6000 (polyethylene glycol), which is a high molecular weight compound with no toxic effect, PEG 8000, inorganic salts such as potassium, sodium, and magnesium, and low molecular weight organic compounds such as mannitol, glycerol and sucrose. In addition, growth regulators such as gibberellic acid ($GA_3$) and ethylene are also used alone or together [45, 47].

### Heat

It is an effective factor in germination rate and speed. The germination rate decreases at low temperatures. The germination rate also decreases at high temperatures where the seed is damaged. There are three temperature levels in seed germination: minimum, optimum and maximum. These temperature levels vary depending on plant species. The minimum temperature is the lowest and the maximum temperature is the highest for germination to occur. In many plants, the optimum temperature for non-dormant seeds is 25-30 °C. In some species, this value is 15°C [45, 46].

## Gases

The exchange of gases between the germination medium and the embryo is important for rapid and uniform germination. Oxygen is essential for respiration in the germinating seed. Carbon dioxide is produced as a result of respiration and accumulates in the soil under poor ventilation conditions. Excessive irrigation and water accumulation and filling of soil pores with water have a negative effect on aeration and therefore germination. However, in aquatic plants, seeds germinate under these conditions [48, 49].

## Light

It is an effective factor on germination. While the seeds of some plants germinate only in light, some only in the dark, some have no response to light for germination. Light also plays a role in the onset and termination of dormancy. It is effective in terms of both quality (wavelength) and photoperiod (duration). Light is also effective on seedling development. Relatively high intensity light is suitable for obtaining mature and strong plants, especially if the plants are to be transplanted to another environment. Low light intensity causes etiolation and reduces photosynthesis. This reduces the seedling quality. However, high light intensity causes the temperature to rise and high temperature damage to young plants. Shadowing should be done against this negativity [50, 51].

## Dormancy

In some species, the seeds are dormant and inactive only when they leave the plant. Germination in these seeds begins with the absorption of water under appropriate temperature conditions. However, the seeds of some other species have primary dormancy. In this dormancy, seeds cannot germinate even if environmental conditions (water, temperature, and ventilation) are suitable. Secondary dormancy is a situation triggered by unsuitable environmental conditions. Dormancy prevents immediate germination of seeds, that is, it controls germination in terms of time, conditions, and place, and facilitates seed preparation, transportation and storage. Researchers have classified the dormancy states that occur in seeds in different ways [40, 52].

## *Primary Dormancy*

In general, it includes external dormancy, internal dormancy, and bilateral dormancy. *(i) External dormancy (physical, mechanical, and chemical dormancy):* It is the type of dormancy in the seed caused by factors outside the embryo (tissues surrounding the embryo). These factors include the seed and fruit tissues (such as seed coat, and perisperm) or the mechanical effect of the endosperm on the radicle. Tissues surrounding the embryo (seed coat, endosperm, and fruit tissues) prevent germination, and water uptake, thereby mechanically preventing the growth of the embryo and the development of the radicle, and negatively affecting gas exchange (such as limited oxygen reaching the embryo). This prevents the removal of inhibitory substances (inhibitors) from the embryo, and limits it in ways such as providing inhibitor support [53, 54].

External dormancy can be broken by abrasion, separation of seed layers (seed coat, endocarp) (such as breaking the seed coat), removal of fruit tissues surrounding the seed, and folding in hot or cold weather. *(ii) Intrinsic dormancy (morphological and physiological dormancy):* It is the dormancy arising from the embryo itself and can be seen in various forms. In morphological dormancy in some species, the embryo is not fully developed when the seed leaves the plant, and the development of the embryo occurs after the water is absorbed before germination begins. It is appropriate to keep it at a normal temperature for a certain period of time for embryo development (to break dormancy). Germination in these seeds can be cured by exposure to temperatures of 15°C or lower, exposure to changing temperatures, and application of substances such as potassium nitrate or gibberellic acid [40, 55].

Physiological dormancy, non-severe physiological dormancy, or post-harvest maturity period is a period required by seeds in dry storage conditions for the loss of dormancy. This type of internal dormancy is not permanent and is lost during dry storage before the seed is planted by the grower. This type of dormancy is a problem for laboratories performing germination tests. This problem is solved by short-term cooling, changing temperatures, and applications of potassium nitrate and gibberellic acid [40, 52].

Photodormancy, seeds that require light or dark conditions for germination are called photodormants. The main mechanism of photosensitivity in seeds is a photochemically reactive pigment called phytochrome. Exposure of seeds to red light (660-760 nm) causes the conversion of phytochrome to far-red phytochrome (Pfr), which stimulates germination. Exposing the seeds to far-red light (760-800 nm) or keeping them in the dark causes the phytochrome to transform into the alternative red form (Pr), which prevents germination. This dormancy is eliminated by exposure to red light [41, 56].

In moderately severe physiological dormancy, the embryo can germinate when separated from the seed coat. This dormancy can be broken by cooling the seeds of medium length (up to 8 weeks) in moist and airy conditions (at 2-7 °C) and by applying cold stratification. In severe physiological dormancy, when the embryo leaves the seed coat, it cannot germinate or becomes physiologically stunted. This dormancy is eliminated by long-term (more than 8 weeks) cold stratification. Moderate and severe dormancy states are very common in the seeds of trees, shrubs and some annual plants growing in the temperate zone. The duration of the layering application varies depending on the plant species [40, 41].

In epicotyl dormancy, the radicle is not dormant and develops at the appropriate temperature, but the epicotyl remains dormant and does not develop unless it is placed in warm conditions after cold stratification. To break this dormancy, normal temperature is required following cold stratification. *(iii) Double dormancy:* It is the combination of external and internal dormancy. In morphophysiological dormancy, there is an underdeveloped embryo and physiological dormancy. This dormancy can be broken by folding transformation in hot and cold conditions. In another dual situation, a hard seed coat and a moderate physiological dormancy may together cause dormancy. In this case, dormancy can be broken by abrasion following cold stratification [56, 57].

### Secondary Dormancy

This type of dormancy occurs in some seeds when environmental conditions do not allow germination and stimulates dormancy in seeds that were not previously dormant. High temperatures, very low temperatures, long periods of darkness, white light or far-red light, water stress, and lack of oxygen can cause this dormancy. This dormancy can be broken by chilling or hot stratification, growth regulator applications or cold stratification [58, 59].

Thermodormancy is a type of dormancy induced by high temperatures; for example, dormancy may occur in species such as lettuce, celery, and pansies at temperatures above 25 °C and the seeds cannot germinate [54, 60].

### Dormancy Caused by Chemical Substances that Prevent Germination

Chemical substances in the seed that prevent germination can also cause dormancy. These endogenous substances can be isolated from many species. These inhibitors can be found in the cotyledons, radicula, embryo, as well as in the endosperm. These substances are simple organic molecules with low molecular weight. Hydrogen cyanide, ammonia and ethylene are the simplest. Natural inhibitors found in seeds include cyanide and ammonia compounds, mustard oils, alkaloids, organic acids, essential oils and phenolic compounds. The last group of substances that have a germination inhibitory effect are synthetic substances. These are substances such as CCC-AM-01618, phosphane D and fleuorene-9-carboxylic acid [41, 61].

The degradation of starch during germination occurs with the catalysis of the amylase enzyme. Stopping the activity of this enzyme causes germination to stop. This is where the function of blockers comes into play. And they delay germination by preventing amylase enzyme activity. In addition, coumarin, which plays a role in the degradation of proteins stored in seeds and prevents protease enzyme activity, also plays an important role in delaying germination [40, 41].

## *GA Synthesis and Transport*

Gibberellic acid is synthesized in many plant tissues, but especially in actively growing tissues such as embryos or meristematic or developing tissues. A significant proportion of plant gibberels are bound and inactive at any given time. GA synthesis starts from mevalonic acid. GA usually contains twenty carbon atoms. Gibberellins are passively transported throughout the plant in both the phloem and xylem [62, 63].

## *Effect of GA on Germination and Dormancy*

Germination first begins with imbibition. Imbibition stimulates GA secretion from the embryo. GA stimulates the secretion of hydrolytic enzymes in the aleurone layer. The synthesized hydrolytic enzymes are transported to the endosperm. Enzymes break down food and create the energy required for respiration. In some plants, GA replaces chilling. Oats and corn do not respond to GA. Some dicotyledons do not respond to GA. Here cytokinins have a stimulating effect [64, 65].

The activity of amylase in the endosperm is controlled by GAs. Studies have proven that the embryo releases GA in the early stages of germination and is transmitted to the aleurone cells in the endosperm. Indole Acetic Acid (IAA) is also required. It increases protein synthesis, loosens the cell wall, increases the cell's water intake and accelerates respiration. Abscisic acid increases (prolongs) dormancy and delays germination. Ethylene promotes GA synthesis in rice and inhibits root and stem elongation in peas. Cytokinin is effective in the disintegration of the endosperm and the germination of seeds of plants that do not respond to GA [66, 67].

## CONCLUSION

In plant production, the first stage of cultivation is planting seeds and germinating them under suitable conditions. However, negative ecological conditions at this stage, technical errors (such as low soil temperature, formation of a creamy layer in the soil) and negativities arising from the structure of the seed affect germination and seedling emergence. Many studies have been conducted to determine the relationship between seed dormancy and germination. In general, mature and healthy seeds of many fruit species do not germinate even if environmental conditions such as temperature, humidity, oxygen and light are suitable. Seed germination, which is the first stage of plant development, decreases or does not occur at all as a result of the inhibition of substances in the seed itself, the hard and impermeable structure of the seed, various technical errors made during seed planting, and negative environmental factors. For this reason, taking into account the seed characteristics and environmental conditions that vary according to plant species and varieties, making some preliminary applications in the future that will optimize seed germination may directly affect seed germination and indirectly plant development positively.

Today, agriculture and seed growing, which forms the basis of agriculture, is a strategic sector. Seed farming can now be seen as the key to countries' self-sufficiency and even freedom. The importance of agricultural products and water is increasing due to reasons such as the rapid increase in the world population, the effects of the problems that climate change will cause in the medium term, the continuous increase in the need for nutrition and food, and the necessity of a balanced diet. This sensitive situation adds new dimensions to production relations and can reshape the balance between countries. Food security, seed diversity, and productivity are at the center of countries' agricultural policies. It is very difficult for countries that do not have a strong seed policy to progress in the agricultural field. Seed farming, which requires advanced technology and intensive scientific studies, is now one of the shaping sectors of future scenarios.

## REFERENCES

[1]    Bleckmann A, Alter S, Dresselhaus T. The beginning of a seed: regulatory mechanisms of double fertilization. Front Plant Sci 2014; 5: 452.
[http://dx.doi.org/10.3389/fpls.2014.00452] [PMID: 25309552]

[2]    Dresselhaus T, Sprunck S, Wessel GM. Fertilization mechanisms in flowering plants. Curr Biol 2016; 26(3): R125-39.
[http://dx.doi.org/10.1016/j.cub.2015.12.032] [PMID: 26859271]

[3]    Miray R, Kazaz S, To A, Baud S. Molecular control of oil metabolism in the endosperm of seeds. Int J Mol Sci 2021; 22(4): 1621.
[http://dx.doi.org/10.3390/ijms22041621] [PMID: 33562710]

[4]    Sabelli PA, Larkins BA. The development of endosperm in grasses. Plant Physiol 2009; 149(1): 14-26.

[http://dx.doi.org/10.1104/pp.108.129437] [PMID: 19126691]

[5]    Lu J, Magnani E. Seed tissue and nutrient partitioning, a case for the nucellus. Plant Reprod 2018; 31(3): 309-17.
[http://dx.doi.org/10.1007/s00497-018-0338-1] [PMID: 29869727]

[6]    Xu W, Fiume E, Coen O, Pechoux C, Lepiniec L, Magnani E. Endosperm and nucellus develop antagonistically in Arabidopsis seeds. Plant Cell 2016; 28(6): 1343-60.
[http://dx.doi.org/10.1105/tpc.16.00041] [PMID: 27233529]

[7]    Farhana N, Singh RM, Gulzar Ahmed M, *et al.* Seed biology and phytochemistry for sustainable future. IntechOpen 2022.
[http://dx.doi.org/10.5772/intechopen.106208]

[8]    Sliwinska E, Bewley JD. Overview of seed development, anatomy and morphology Seeds: the Ecology of Regeneration in Plant Communities. Wallingford, UK: CABI 2014; pp. 1-17.
[http://dx.doi.org/10.1079/9781780641836.0001]

[9]    Kruglova NN, Titova GE, Seldimirova OA, Zinatullina AE, Veselov DS. Embryo of flowering plants at the critical stage of embryogenesis relative autonomy (by example of Cereals). Russ J Dev Biol 2020; 51(1): 1-15.
[http://dx.doi.org/10.1134/S1062360420010026]

[10]   Radchuk V, Borisjuk L. Physical, metabolic and developmental functions of the seed coat. Front Plant Sci 2014; 5: 510.
[http://dx.doi.org/10.3389/fpls.2014.00510] [PMID: 25346737]

[11]   Chandler JW. Cotyledon organogenesis. J Exp Bot 2008; 59(11): 2917-31.
[http://dx.doi.org/10.1093/jxb/ern167] [PMID: 18612170]

[12]   Kabeya D, Sakai S. The role of roots and cotyledons as storage organs in early stages of establishment in *Quercus crispula*: a quantitative analysis of the nonstructural carbohydrate in cotyledons and roots. Ann Bot (Lond) 2003; 92(4): 537-45.
[http://dx.doi.org/10.1093/aob/mcg165] [PMID: 12907467]

[13]   Matilla AJ. The orthodox dry seeds are alive: a clear example of desiccation tolerance. Plants 2021; 11(1): 20.
[http://dx.doi.org/10.3390/plants11010020] [PMID: 35009023]

[14]   Lan QY, Luo YL, Ma SM, *et al.* Development and storage of recalcitrant seeds of *Hopea hainanensis*. Seed Sci Technol 2012; 40(2): 200-8.
[http://dx.doi.org/10.15258/sst.2012.40.2.05]

[15]   Wang S, Shen Y, Bao H. Morphological, physiological and biochemical changes in *Magnolia zenii* Cheng seed during development. Physiol Plant 2021; 172(4): 2129-41.
[http://dx.doi.org/10.1111/ppl.13445] [PMID: 33937990]

[16]   Sghaier-Hammami B, B M Hammami S, Baazaoui N, Gómez-Díaz C, Jorrín-Novo JV. Dissecting the seed maturation and germination processes in the non-orthodox *Quercus ilex* species based on protein signatures as revealed by 2-DE coupled to MALDI-TOF/TOF proteomics strategy. Int J Mol Sci 2020; 21(14): 4870.
[http://dx.doi.org/10.3390/ijms21144870] [PMID: 32660160]

[17]   Corbineau F. The effects of storage conditions on seed deterioration and ageing: How to improve seed longevity. Seeds 2024; 3(1): 56-75.
[http://dx.doi.org/10.3390/seeds3010005]

[18]   Baskin CC, Baskin JM. Seeds Ecology, Biogeography, and Evolution of Dormancy and Germination. London, UK: Academic Press 1998; p. 666.

[19]   Moïse JA, Han S, Gudynaitę-Savitch L, Johnson DA, Miki BLA. Seed coats: Structure, development, composition, and biotechnology. *In Vitro* Cell. Dev Biol Plant 2005; 41(5): 620-44. Available from: http://www.jstor.org/stable/4293908

[http://dx.doi.org/10.1079/IVP2005686]

[20]   Pires ND. Seed evolution: parental conflicts in a multi-generational household. Biomol Concepts 2014; 5(1): 71-86.
[http://dx.doi.org/10.1515/bmc-2013-0034] [PMID: 25372743]

[21]   Pfister B, Zeeman SC. Formation of starch in plant cells. Cell Mol Life Sci 2016; 73(14): 2781-807.
[http://dx.doi.org/10.1007/s00018-016-2250-x] [PMID: 27166931]

[22]   Tetlow IJ, Morell MK, Emes MJ. Recent developments in understanding the regulation of starch metabolism in higher plants. J Exp Bot 2004; 55(406): 2131-45.
[http://dx.doi.org/10.1093/jxb/erh248] [PMID: 15361536]

[23]   Tetlow IJ. Understanding storage starch biosynthesis in plants: a means to quality improvement. Can J Bot 2006; 84(8): 1167-85.
[http://dx.doi.org/10.1139/b06-089]

[24]   Hu ZY, Hua W, Zhang L, *et al.* Seed structure characteristics to form ultrahigh oil content in rapeseed. PLoS One 2013; 8(4): e62099.
[http://dx.doi.org/10.1371/journal.pone.0062099] [PMID: 23637973]

[25]   Mantese A, Medan D, Hall AJ. Achene structure, development and lipid accumulation in sunflower cultivars differing in oil content at maturity. Ann Bot (Lond) 2006; 97(6): 999-1010.
[http://dx.doi.org/10.1093/aob/mcl046] [PMID: 16675608]

[26]   Biriş SŞ, Ionescu M, Gheorghiţă NE, Ungureanu N, Vlăduţ NV. Study of the compression behavior of sunflower seeds using the finite element method. AGROFOR 2019; 4(1): 128-36.
[http://dx.doi.org/10.7251/AGRENG1901128B]

[27]   Zablatzká L, Balarynová J, Klčová B, Kopecký P, Smýkal P. Anatomy and histochemistry of seed coat development of wild (*Pisum sativum* subsp. elatius (M. Bieb.) Asch. et Graebn. and domesticated pea (*Pisum sativum* subsp. sativum L.). Int J Mol Sci 2021; 22(9): 4602.
[http://dx.doi.org/10.3390/ijms22094602] [PMID: 33925728]

[28]   Van Dongen JT, Ammerlaan AM, Wouterlood M, Van Aelst AC, Borstlap AC. Structure of the developing pea seed coat and the post-phloem transport pathway of nutrients. Ann Bot (Lond) 2003; 91(6): 729-37.
[http://dx.doi.org/10.1093/aob/mcg066] [PMID: 12714370]

[29]   Ramtekey V, Cherukuri S, Kumar S, *et al.* Seed longevity in legumes: deeper insights into mechanisms and molecular perspectives. Front Plant Sci 2022; 13: 918206.
[http://dx.doi.org/10.3389/fpls.2022.918206] [PMID: 35968115]

[30]   Smýkal P, Vernoud V, Blair MW, Soukup A, Thompson RD. The role of the testa during development and in establishment of dormancy of the legume seed. Front Plant Sci 2014; 5: 351.
[http://dx.doi.org/10.3389/fpls.2014.00351] [PMID: 25101104]

[31]   Kigel J, Rosental L, Fait A. Seed Physiology and Germination of Grain Legumes. In: De Ron AM, Ed. Grain Legumes Handbook of Plant Breeding. New York: Springer 2015; pp. 327-63.
[http://dx.doi.org/10.1007/978-1-4939-2797-5_11]

[32]   Nierle W, Wahab El Bayd A. Examination and composition of some legume seeds. Z Lebensm Unters Forsch 1977; 164(1): 23-7.
[http://dx.doi.org/10.1007/BF01135419] [PMID: 560092]

[33]   Ahmad S, Zafar M, Ahmad M, *et al.* Seed morphology using SEM techniques for identification of useful grasses in Dera Ghazi Khan, Pakistan. Microsc Res Tech 2020; 83(3): 249-58.
[http://dx.doi.org/10.1002/jemt.23408] [PMID: 31738478]

[34]   Yang T, Wu X, Wang W, Wu Y. Regulation of seed storage protein synthesis in monocot and dicot plants: A comparative review. Mol Plant 2023; 16(1): 145-67.
[http://dx.doi.org/10.1016/j.molp.2022.12.004] [PMID: 36495013]

[35] Anjum F, Mir A, Shakir Y, *et al.* Seed coat morphology and sculpturing of selected invasive alien plants from lesser Himalaya Pakistan and their systematic implications. BioMed Res Int 2022; 1-11.
[http://dx.doi.org/10.1155/2022/8225494] [PMID: 35924271]

[36] Rodrigues DB, Cavalcante JA, Almeida AS, Nunes CA. SerrÃo AFA, Konzen LH, SuÑÉ AS, Tunes LVM. Seed morphobiometry, morphology of germination and emergence of quinoa seeds 'BRS Piabiru'. An Acad Bras Cienc. 2020.

[37] Thadeo M, Hampilos KE, Stevenson DW. Anatomy of fleshy fruits in the monocots. Am J Bot 2015; 102(11): 1757-79.
[http://dx.doi.org/10.3732/ajb.1500204] [PMID: 26507114]

[38] Berger F. Endosperm: the crossroad of seed development. Curr Opin Plant Biol 2003; 6(1): 42-50.
[http://dx.doi.org/10.1016/S1369526602000043] [PMID: 12495750]

[39] Yan D, Duermeyer L, Leoveanu C, Nambara E. The functions of the endosperm during seed germination. Plant Cell Physiol 2014; 55(9): 1521-33.
[http://dx.doi.org/10.1093/pcp/pcu089] [PMID: 24964910]

[40] Penfield S. Seed dormancy and germination. Curr Biol 2017; 27(17): R874-8.
[http://dx.doi.org/10.1016/j.cub.2017.05.050] [PMID: 28898656]

[41] Bentsink L, Koornneef M. Seed dormancy and germination. Arabidopsis Book 2008; 6: e0119.
[http://dx.doi.org/10.1199/tab.0119] [PMID: 22303244]

[42] Nonogaki H. Seed germination and dormancy: The classic story, new puzzles, and evolution. J Integr Plant Biol 2019; 61(5): 541-63.
[http://dx.doi.org/10.1111/jipb.12762] [PMID: 30565406]

[43] Bewley JD. Seed germination and dormancy. Plant Cell 1997; 9(7): 1055-66.
[http://dx.doi.org/10.1105/tpc.9.7.1055] [PMID: 12237375]

[44] Carrera-Castaño G, Calleja-Cabrera J, Pernas M, Gómez L, Oñate-Sánchez L. An updated overview on the regulation of seed germination. Plants 2020; 9(6): 703.
[http://dx.doi.org/10.3390/plants9060703] [PMID: 32492790]

[45] Javaid MM, Mahmood A, Alshaya DS, *et al.* Influence of environmental factors on seed germination and seedling characteristics of perennial ryegrass (*Lolium perenne* L.). Sci Rep 2022; 12(1): 9522.
[http://dx.doi.org/10.1038/s41598-022-13416-6] [PMID: 35681016]

[46] Luna B, Chamorro D. Germination sensitivity to water stress of eight Cistaceae species from the Western Mediterranean. Seed Sci Res 2016; 26(2): 101-10.
[http://dx.doi.org/10.1017/S096025851600009X]

[47] Ramírez-Tobías HM, Peña-Valdivia CB, Trejo C, Aguirre R JR, Vaquera H H. Seed germination of Agave species as influenced by substrate water potential. Biol Res 2014; 47(1): 11.
[http://dx.doi.org/10.1186/0717-6287-47-11] [PMID: 25027050]

[48] Ahmed AKA, Shi X, Hua L, *et al.* Influences of air, oxygen, nitrogen, and carbon dioxide nanobubbles on seed germination and plant growth. J Agric Food Chem 2018; 66(20): 5117-24.
[http://dx.doi.org/10.1021/acs.jafc.8b00333] [PMID: 29722967]

[49] Negm FB, Smith OE. Effects of ethylene and carbon dioxide on the germination of osmotically inhibited lettuce seed. Plant Physiol 1978; 62(4): 473-6.
[http://dx.doi.org/10.1104/pp.62.4.473] [PMID: 16660541]

[50] Cavallaro V, Muleo R. The effects of LED light spectra and intensities on plant growth. plants. 2022; 11: 1911.
[http://dx.doi.org/10.3390/plants11151911]

[51] Xu Y, Yang M, Cheng F, Liu S, Liang Y. Effects of LED photoperiods and light qualities on *in vitro* growth and chlorophyll fluorescence of *Cunninghamia lanceolata*. BMC Plant Biol 2020; 20(1): 269.
[http://dx.doi.org/10.1186/s12870-020-02480-7] [PMID: 32517650]

[52]    Finch-Savage WE, Leubner-Metzger G. Seed dormancy and the control of germination. New Phytol 2006; 171(3): 501-23.
[http://dx.doi.org/10.1111/j.1469-8137.2006.01787.x] [PMID: 16866955]

[53]    Chahtane H, Kim W, Lopez-Molina L. Primary seed dormancy: a temporally multilayered riddle waiting to be unlocked. J Exp Bot 2016; 68(4): erw377.
[http://dx.doi.org/10.1093/jxb/erw377] [PMID: 27729475]

[54]    Pawłowski TA, Bujarska-Borkowska B, Suszka J, *et al.* Temperature regulation of primary and secondary seed dormancy in *Rosa canina* L.: findings from proteomic analysis. Int J Mol Sci 2020; 21(19): 7008.
[http://dx.doi.org/10.3390/ijms21197008] [PMID: 32977616]

[55]    Nabors MW, Kugrens P, Ross C. Photodormant lettuce seeds: Phytochrome-induced protein and lipid degradation. Planta 1974; 117(4): 361-5.
[http://dx.doi.org/10.1007/BF00388031] [PMID: 24458467]

[56]    Dhyani A, Phartyal SS, Nautiyal BP, Nautiyal MC. Epicotyl morphophysiological dormancy in seeds of *Lilium polyphyllum* (Liliaceae). J Biosci 2013; 38(1): 13-9.
[http://dx.doi.org/10.1007/s12038-012-9284-5] [PMID: 23385808]

[57]    Jayasuriya KMGG, Wijetunga ASTB, Baskin JM, Baskin CC. Physiological epicotyl dormancy and recalcitrant storage behaviour in seeds of two tropical Fabaceae (subfamily Caesalpinioideae) species. AoB Plants 2012; 2012(0): pls044.
[http://dx.doi.org/10.1093/aobpla/pls044] [PMID: 23264873]

[58]    Buijs G. A Perspective on secondary seed dormancy in *Arabidopsis thaliana.* Plants 2020; 9(6): 749.
[http://dx.doi.org/10.3390/plants9060749] [PMID: 32549219]

[59]    Martel C, Blair LK, Donohue K. PHYD prevents secondary dormancy establishment of seeds exposed to high temperature and is associated with lower PIL5 accumulation. J Exp Bot 2018; 69(12): 3157-69.
[http://dx.doi.org/10.1093/jxb/ery140] [PMID: 29648603]

[60]    Leymarie J, Benech-Arnold RL, Farrant JM, Corbineau F. Thermodormancy and ABA metabolism in barley grains. Plant Signal Behav 2009; 4(3): 205-7.
[http://dx.doi.org/10.4161/psb.4.3.7797] [PMID: 19721750]

[61]    Chen J, Huang X, Xiao X, *et al.* Seed dormancy release and germination requirements of *Cinnamomum migao*, an endangered and rare woody plant in southwest china. Front Plant Sci 2022; 13: 770940.
[http://dx.doi.org/10.3389/fpls.2022.770940] [PMID: 35154219]

[62]    Silverstone AL, Chang C, Krol E, Sun T. Developmental regulation of the gibberellin biosynthetic gene *GA1* in *Arabidopsis thaliana.* Plant J 1997; 12(1): 9-19.
[http://dx.doi.org/10.1046/j.1365-313X.1997.12010009.x] [PMID: 9263448]

[63]    Gupta R, Chakrabarty SK. Gibberellic acid in plant. Plant Signal Behav 2013; 8(9): e25504.
[http://dx.doi.org/10.4161/psb.25504] [PMID: 23857350]

[64]    Ni BR, Bradford KJ. Germination and dormancy of abscisic acid- and gibberellin-deficient mutant tomato (*Lycopersicon esculentum*) seeds (sensitivity of germination to abscisic acid, gibberellin, and water potential). Plant Physiol 1993; 101(2): 607-17.
[http://dx.doi.org/10.1104/pp.101.2.607] [PMID: 12231716]

[65]    Medeiros MJ, Oliveira MT, Willadino L, Santos MG. Overcoming seed dormancy using gibberellic acid and the performance of young *Syagrus coronata* plants under severe drought stress and recovery. Plant Physiol Biochem 2015; 97: 278-86.
[http://dx.doi.org/10.1016/j.plaphy.2015.10.008] [PMID: 26509497]

[66]    Chen SY, Kuo SR, Chien CT. Roles of gibberellins and abscisic acid in dormancy and germination of red bayberry (*Myrica rubra*) seeds. Tree Physiol 2008; 28(9): 1431-9.
[http://dx.doi.org/10.1093/treephys/28.9.1431] [PMID: 18595855]

[67]    Graeber K, Linkies A, Steinbrecher T, *et al. Delay of germination 1* mediates a conserved coat-dormancy mechanism for the temperature- and gibberellin-dependent control of seed germination. Proc Natl Acad Sci USA 2014; 111(34): E3571-80.
[http://dx.doi.org/10.1073/pnas.1403851111] [PMID: 25114251]

# CHAPTER 8

# Genetic Transformation: Current Opinion And Future Prospect

**Damla Ekin Özkaya[1,2,*]**

[1] *Muğla Sıtkı Koçman University, Faculty of Science, Molecular Biology and Genetics Department, Menteşe, Muğla, Türkiye*

[2] *Okan University, Vocational School of Health Services, Medical Laboratory Techniques Department, 34959, Tuzla, İstanbul, Türkiye*

**Abstract:** Nowadays, it is possible to transfer desired foreign genes into the genomes of various plant species in a stable manner through genetic engineering applications. Through genetic transformation, plants can gain resistance to diseases, environmental pressures, and various chemical compounds such as herbicides and pesticides. Numerous genetic transformation procedures, primarily involving the transport of exogenous genes and the regeneration of transformed plants, have been continuously discovered and improved for high efficiency and convenient manipulation. The delivery of biomolecules by nanomaterials has piqued the curiosity of researchers in recent years. To improve crops, plant molecular biology breakthroughs must be translated into an effective genotype-independent plant transformation system. Improving the nutritional quality of plants is another important advantage of genetic transformation. Genetic transformation studies on plants started in the 1980s, and effective genetic transformation methods have been developed until today. These methods are generally divided into direct transformation and indirect transformation. Each of the developed techniques has some advantages and disadvantages. When deciding which method to use in transformation, it is very important to consider these advantages and disadvantages and to choose the most appropriate method for the plant to be gene transferred for a successful transformation. In this context, this chapter is aimed at explaining genetic transformation methods in plants in detail, the developments in genetic transformation from the past to the present, and the expectations about the genetic transformation process in the future.

**Keywords:** Agrobacterium, Genetic engineering, Genetic transformation, Genetic transformation, Transgenic plant.

\* **Corresponding author Damla Ekin Özkaya:** Muğla Sıtkı Koçman University, Faculty of Science, Molecular Biology and Genetics Department, Menteşe, Muğla, Türkiye; Okan University, Vocational School of Health Services, Medical Laboratory Techniques Department, 34959, Tuzla, İstanbul, Türkiye, E-mail: damlaekinn95@gmail.com

Ergun Kaya (Ed.)

# INTRODUCTION

Humans benefit from plants in many fields, such as food industry, animal feed, medicine, soft drinks, paints, cosmetics, perfume industry, and dye industry. Today, serious problems occur in plant resources, which are very important for living things, due to reasons such as the increasing world population and abiotic and biotic stress factors in agricultural areas. Classical breeding studies have been carried out for centuries to address these problems such as the slow pace of breeding cycles, and genetic linkage drag. With classical breeding methods, it is possible to select the desired traits from existing plants and combine these traits to obtain new plant varieties [1, 2].

However, the yield increase obtained from classical breeding methods is not sufficient today due to the decrease in arable land, the need for large plant populations, and therefore the increase in cost and labor requirements. In addition to the desired traits in classical plant breeding methods, undesirable traits also manifest themselves in the plant. This situation has revealed the necessity of investigating new technologies in plant breeding studies [3 - 6].

Genetic transformation methods are used to improve a plant with a certain trait without changing other traits [7, 8]. In molecular biology, the term transformation refers to the removal of a foreign gene from the cell membrane and incorporation into the genome of the host cell [9, 10]. The foreign gene transferred here is called a 'transgene', and organisms that have successfully transferred the gene are called 'transgenic' [11, 12]. With genetic transformation based on recombinant DNA technology, it is possible to increase the nutritional value of plants, provide resistance to biotic and abiotic stress factors, or study plant metabolism. Many parts of the plant, such as the stem, leaf, embryo, callus suspension, protoplast culture, cotyledon, or reproductive cells, can be used in transformation studies. Since the 1980s, this technology has been used in tomato [13], rice [14 - 19], wheat [20, 21], tobacco [22, 23], grape [24], maize [25 - 27], cassava [28], canola [29], and petunia [30]. The transferred genes reduce losses caused by stress factors and increase crop yields [31, 32, 12].

Transformation methods developed for transferring foreign DNA into target plant cells are generally divided into direct transformation and indirect transformation methods [33, 34]. In indirect methods, a bacterial cell is needed to transfer the foreign DNA into the plant cell, while in direct transformation methods, genetic transformation is performed without the need for an intermediary cell (Fig. **1**) [34, 35].

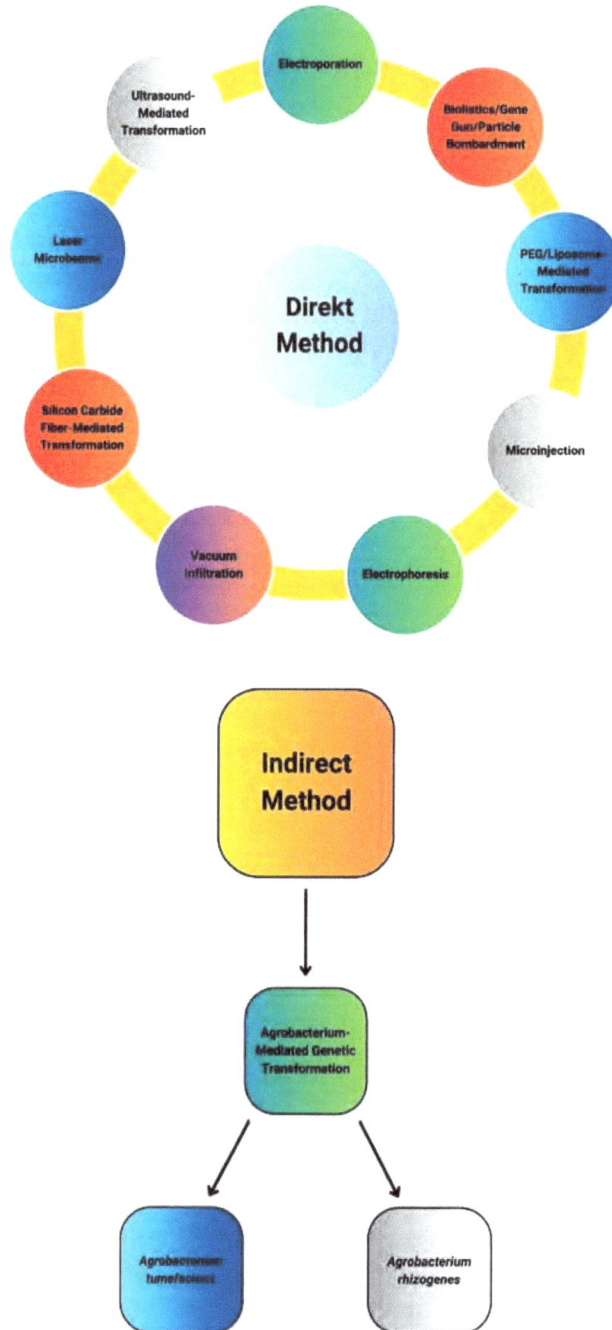

**Fig. (1).** Illustration of direct and indirect transformation methods.

**Indirect Methods**

*Agrobacterium-Mediated Transformation*

Indirect transformation is the transfer of the desired gene into the target plant cell by plasmid-containing bacteria such as *Agrobacterium tumefaciens* or *Agrobacterium rhizogenes* [33, 21, 36]. *A. tumefaciens* is a gram-negative bacterium carrying the Ti (tumor-inducing) plasmid that causes crown tumors in plants. Similarly, virulent strains of *A. rhizogenes* carry the Ri (root inducer) megaplasmid that promotes the development of hairy root disease (Fig. **2**). These megaplasmids are capable of generating pathogenicity and contain a region of T-DNA, the so-called transferred DNA. The T-DNA consists of the gene region responsible for opine production and the oncogenic region that causes tumor development. The oncogene is involved in tumor formation by producing enzymes that synthesize auxins and cytokinins. In addition, 25 bp repeat sequences located on the two sides of the T-DNA, called the left and right border (LB, RB), mediate the transfer of this region from the bacterium to the host cell [37, 38].

Oncogenic genes and opine catabolism genes are located inside the T-DNA region of the Ti plasmid, while virulence (*vir*) genes are located outside the T-DNA. A mutation in this region results in a loss of virulence, and thus these plasmids are used as a natural vector for plant transformation [39]. *A. tumefaciens*-mediated genetic transformation to plants consists of bacterial colonization, activation of the bacterial virulence system, formation of the T-DNA transfer complex, T-DNA transfer, and integration of T-DNA into the plant genome. The T-DNA transfer process is initiated by the recognition of various signals (*e.g.,* phenolic compounds) released by plant injury by the bacterium with the help of the products of the virulence gene (vir gene) and the chromosomal virulence gene (chv gene) encoded in the bacterial Ti plasmid (Fig. **3**). The vir region consists of seven major loci *virA, virB, virC, virD, virE, virF* and *virG*, which encode elements of the bacterial protein system. These are the loci required for the processing and transfer of T-DNA. The virA and virG regions activate other vir genes located in the Ti plasmid. *virB, virC, virD*, and *virE* enable the processing and transfer of T-DNA into the host plant cell and integrate the DNA into the plant genome [40]. *virC* and *virF* play a role in determining the plant species to be transformed by *Agrobacterium* [41].

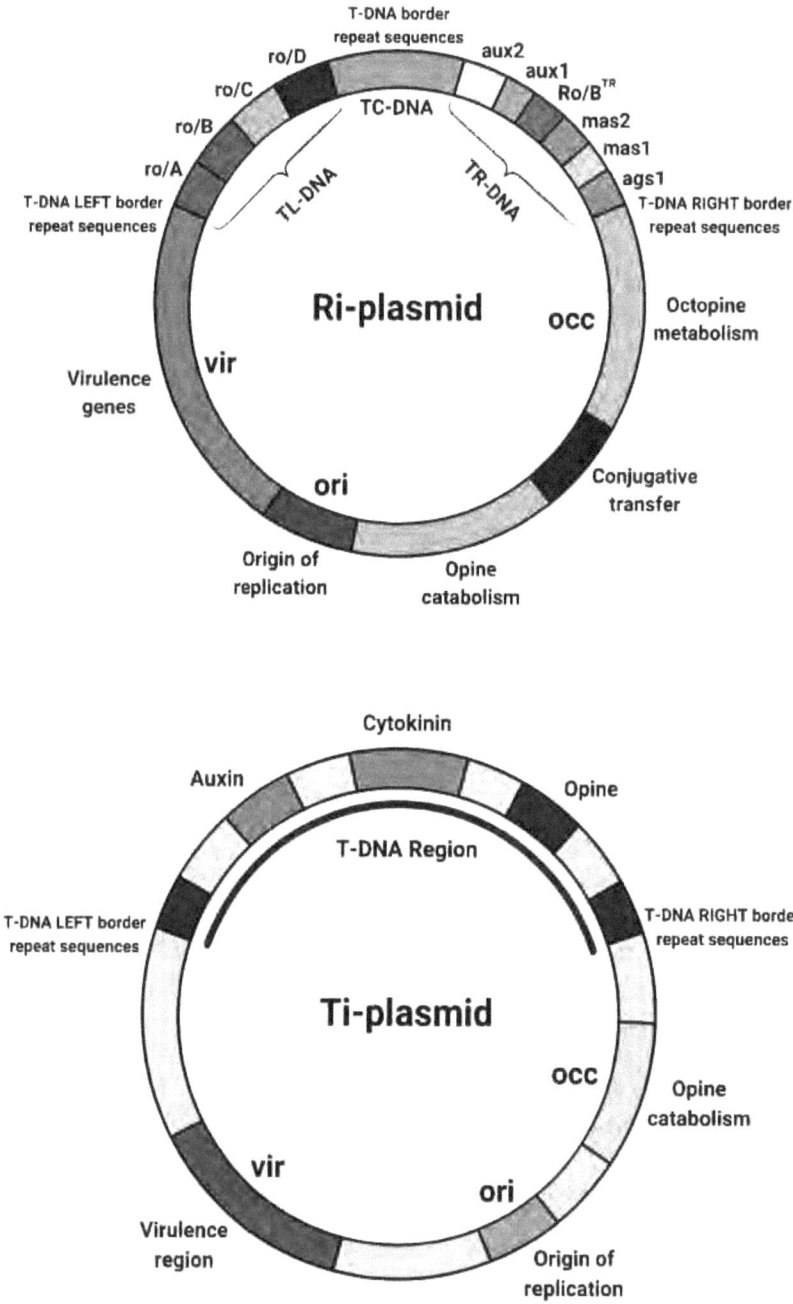

**Fig. (2).** Schematic of Ri-plasmid and Ti-plasmid Structures [47].

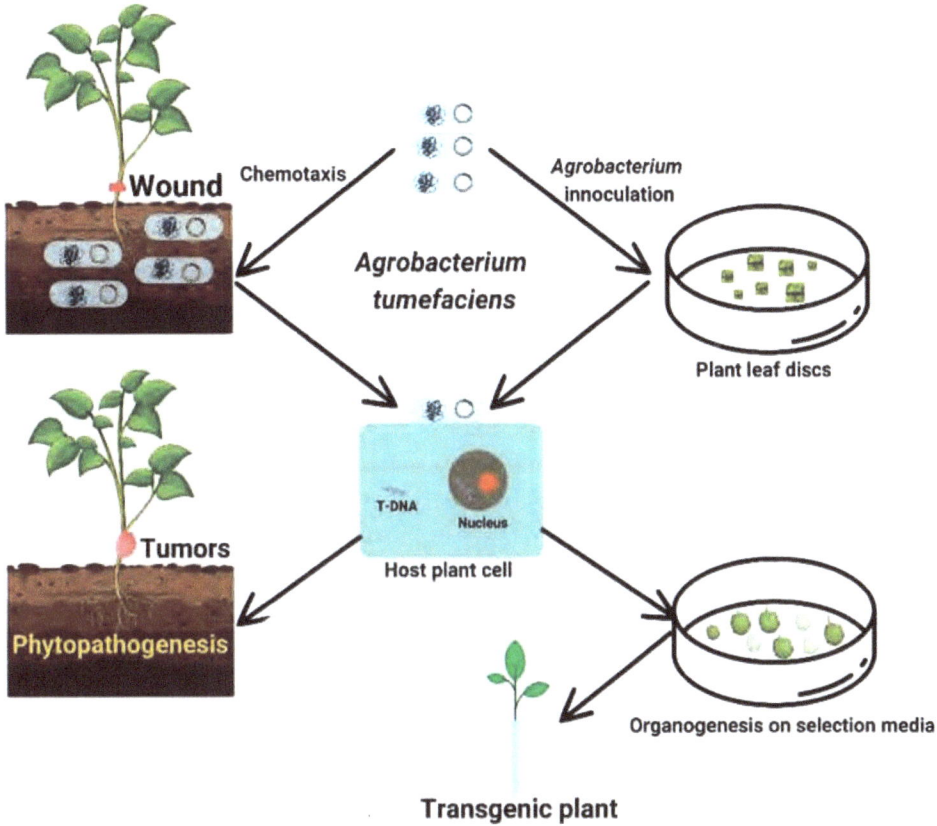

**Fig. (3).** *Agrobacterium*-mediated transformation. Natural infection of plants by Agrobacterium and crown gall disease formation and use of *Agrobacterium* species for plant transformation [48].

The basic steps in *A. tumefaciens*-mediated genetic transformation are as follows:

1. Some chemical signals secreted by wounded plants are detected by the *virA/virG* two-component system of *A. tumefaciens*, followed by transcription of vir gene promoters and thus the expression of the vir proteins.

2. T-DNA is notched at the left and right borders by *virD2/virD1*, resulting in a linear, single-stranded T-DNA molecule. *virD2* binds covalently to the 5' end of the single-stranded T-DNA (T-complex formation).

3. A. tumefaciens attaches to the plant and transfers the T-complex into the plant cell *via* the *virD4*/B T4SS transport system.

4. The T-complex is integrated into the nucleoplasm with the help of some proteins, and the T-DNA is integrated into the plant genome.

5. T-DNA genes are expressed in the plant cell, and the products of these genes induce tumor formation [42, 43].

There are many factors that need to be considered to perform transformation with *Agrobacterium* effectively. These factors include the *Agrobacterium* strain, the source of plant tissues, plant genotypes, the amount of salt and pH of the medium, the use of T-DNA promoting compounds, and the duration of tissue interactions with *Agrobacterium* [44]. In addition, various parts of the plant, such as immature or mature zygotic embryos, embryonic cultures, meristems, roots, stem tips, callus from matured seed, callus suspension cultures, cut leaf blades, primary leaf nodes, cotyledons (with or without nodal parts), and stem segments, can be used as target materials in this method [31].

Compared to other transformation methods, *Agrobacterium*-mediated transformation has some advantages. One of them is that it is a simple and highly efficient method. It is possible to produce a large number of stable transformants [45]. Large DNA fragments can be transferred into recipient cells, transferred genes can be passed on for generations, and transgenes with low copy numbers can be integrated into the genome [31]. In addition, the method allows the transfer of multiple genes linked to a T-DNA into a single plant tissue [46]. Furthermore, A. tumefaciens is preferred over A. rhizogenes in *Agrobacterium*-mediated transformation. This is because regeneration is difficult in plants infected by *A. rhizogenes* (fringe root transformed) [31, 6]. The application of *Agrobacterium* species for the transformation of monocot plants is limited. Therefore, the inability to use *Agrobacterium*-mediated transformation in every plant is a disadvantage of the method [40]. In addition, the expression level of transgenes is variable [41].

**Direct Method**

*Electroporation*

The electroporation method creates a pore on the cell surface by causing a polarity change in the cell membrane with electrical pulses, thus allowing DNA to enter the cell [49, 50, 6]. The device that generates the electrical pulse for genetic transformation is called an electroporator [51, 52, 32]. The electroporator device is an 80-800 μl structure with a 4 mm cavity containing two parallel plates of aluminum (Fig. **4**). The electrodes are in contact with the cells in suspension, and the liquid electrolyte containing the DNA is targeted for introduction into these cells. Each application consists of the application of an electrical pulse with a voltage ranging from 1.6 to 2.0 kV, lasting 10-6 to 10-2 s [6]. Under normal conditions (normal pressure and temperature conditions), when the potential difference of the applied voltage is greater than 0.5 V, an electrical imbalance occurs in the plasma membrane because the cell membrane voltage threshold is between 0.5-1 V and causes permeability in the cell membrane. Thus, the desired DNA can be introduced into these cells while maintaining cell viability [53]. The pores formed are temporary, and after a while they are closed and the cell membrane is restored [54, 55]. The membrane potential must be kept below the critical value for the pores to close. If the critical value is exceeded, the pores in the membrane become permanent, causing cell damage and cell death [56]. Many transgenic field crops such as rice, wheat, rye, maize, and carrots have been produced using electroporation [14, 19, 25, 49, 57 - 59]. However, due to the difficulty of the protocols for plant regeneration after transformation, they are usually applied to protoplasts [33, 60]. Despite these difficulties, it remains important because it can be used in monocotyl plants resistant to Agrobacterium transformation [49].

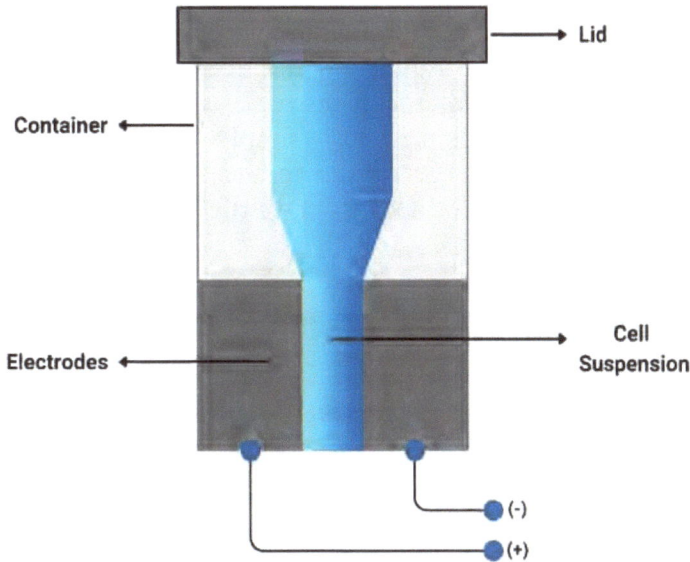

**Fig. (4).** Electroporator device [32].

Electroporation is frequently preferred in biotechnology, biology, and medicine due to its advantages, such as its applicability to many cell types, low cost, rapid technique, and high efficiency [61]. The disadvantages of this method are that it is difficult to apply to plant cells with thick cell walls and works with a limited number of receptor types [39]. In addition, strong electrical pulses can damage DNA and prevent the transformation from being carried out as desired [62].

## *Biolistics / Gene Gun / Particle Bombardment*

The method known as Biolistics, Gene Gun, or Particle Bombardment was developed in 1987 as an alternative way to transform resistant cereals [63]. In this technique, the DNA to be transferred is precipitated with spermidine and calcium chloride. The surface of the microcarriers (tungsten or gold particles about 0.6 - 1.0 mm in diameter) is then coated with this DNA and accelerated at high speed to reach the inside of the target cell. The microcarriers are spread evenly on a circular plastic film macrocarrier. The entire unit is then placed under the burst disk in the vacuum section of the biolistic device. A wire mesh (stop screen) to hold the macro carrier and allow the microcarriers to pass through it is positioned underneath the macro carrier, and the target tissue is positioned underneath the entire assembly (*i.e.,* underneath the launch unit). The device is then ignited, and the gas accelerator tube is filled with helium gas. The helium pressure pushes the macrocarriers containing the microcarriers forcefully towards the stop screen (Fig. **5**). Macrocarriers that cannot pass through the wire mesh are stopped there, while microcarriers pass through and hit the cell at high speed. Upon reaching the host cell through the cell wall, the microcarriers release the DNA they carry [31, 39]. The application of the biolistics method is simply shown in Fig. (**5**). In order to avoid contamination of the target tissue in the subsequent process of tissue culture, studies should be performed under sterile conditions [31].

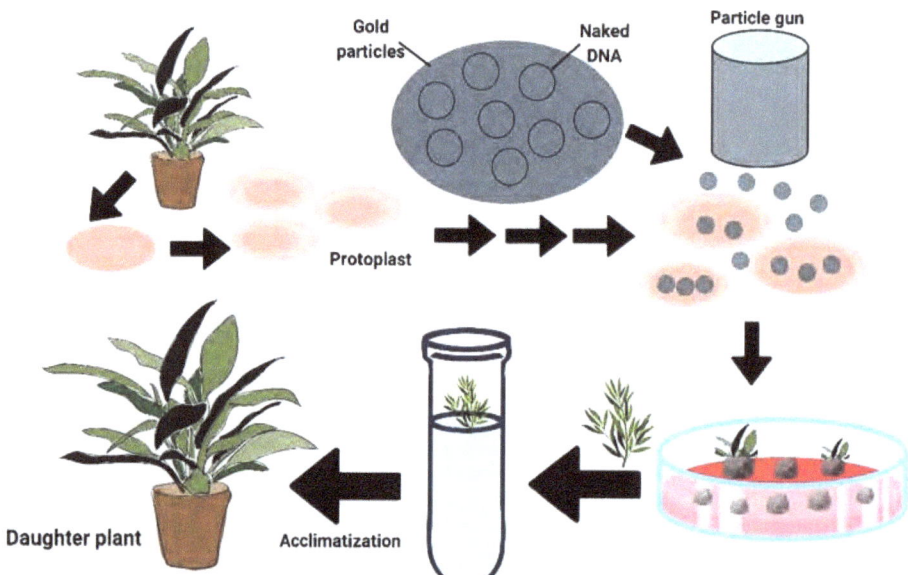

**Fig. (5).** Plant transformation by biolistics: (**1**) leaf protoplast isolation, (**2**) injection of DNA-coated gold particles into the protoplast with a gene gun, (**3**) transfer to solid medium, (**4**) acclimatization of the transgenic plant in a greenhouse [74, 10].

The most important advantages of the method are the absence of vectors, the absence of biological limits, the ability to use cells, meristems, protoplasts, calluses, or embryos as targets, and simultaneous multiple genetic transformation. In addition, it is a very popular method due to its ability to be used in many species, including bacteria, fungi, organelles, and animal cells [33], low cost, short transformation time, ease of transfer of multiple genes, or chimeric DNA (combining DNA from different organisms and non-homologous sources) [32, 6]. The efficiency of the biolistic technique is affected by factors such as the number of cells, the regeneration ability of the cells, the amount of DNA, the number of DNA-coated particles, and temperature. Furthermore, studies have shown that the probability of particles entering the cell increases in direct proportion to their kinetic energy [64]. Therefore, materials such as gold and tungsten are used to make particles [32, 64 - 70]. Gold particles are often preferred for their smaller size and efficiency, but tungsten particles are also preferred from a cost perspective [32, 70]. Although the integrity of the DNA to be transferred is a concern in this method, it has become easier to transfer large DNAs, such as yeast artificial chromosomes (YAC), into the plant genome [71, 72].

As with any technique, the biolistic method has some disadvantages. These include randomization of the intracellular target, vulnerability of DNA to damage, lower efficiency than Agrobacterium-mediated transformation, and higher cost [31]. Nevertheless, among the available transformation methods, biolistic and Agrobacterium-mediated transformation methods are currently the most widely used [73].

## *PEG/Liposome-Mediated Genetic Transformation*

Polyethylene glycol (PEG) is a chemical compound that enables direct DNA delivery to target protoplasts [39]. In other transformation methods, protoplasts are used in the PEG technique because the plant cell wall is a barrier to transferring foreign DNA into the cell [2]. The protoplasts used are usually isolated by enzymatic digestion of chloroplasts and chlorophyll-rich mesophyll cells of seedlings or leaves [75]. In this method, the DNA to be transferred is mixed with the protoplast, and PEG is added. DNA uptake of the protoplast occurs with the addition of PEG. Higher concentrations of PEG (15-25%) precipitate the DNA and stimulate DNA uptake by endocytosis without damaging the protoplasts [39, 76]. Thanks to PEG, there is a reversible increase in the permeability of the cell membrane, and the protoplast of the plant cell takes up the foreign DNA. Small gene sequences or large molecules, such as plasmids and micronuclei, can be easily taken into the cell through the cell membrane in the presence of PEG. The advantages, such as the ability to transfer molecules of different sizes, the ease of use of the method, and the fact that it does not require special equipment, have made it one of the most promising transformation methods. However, the method also has disadvantages, such as the low transformation frequency and the inability to form whole plants from protoplasts in many plant species. The fact that it is currently only used in plants whose protoplasts are efficiently regenerated has limited the application of this technique. When it is possible to regenerate protoplasts in other plant species to form whole plants, it may become a simple, reliable, effective, and inexpensive technique. Regeneration of the gene-transferred plant can be performed *in vitro* under optimized growth conditions [2, 77].

Related to the PEG method is the liposome-mediated transformation technique. Liposomes are microscopic, spherical lipid vesicles formed by the hydration of phospholipids and can contain a large number of plasmids. In this method, the desired DNA is loaded into the liposomes, and the liposomes are endocytosed by protoplasts. Since liposomes are positively charged, they tend to attract negatively charged DNA [31, 78]. This DNA is then released to integrate into the target plant's genome. This method has many advantages, such as the protection of DNA from degradation by nucleases, stability of nucleic acids encapsulated by liposomes, low cellular toxicity, and wide applicability. However, the fact that there are very few successful results in the literature regarding the application of this method to plant species, its laborious application, and its low efficiency limit the use of this method [31, 39].

## *Microinjection*

The method is called microinjection because DNA is introduced directly into the target plant cell using a 0.5-1.0 μm diameter glass microcapillary injection pipette containing mineral oil (Fig. **6**). In this technique, the target cells are immobilized under the microscope with a lubricated holding pipette and gentle suction, and the DNA is injected into the cell with the micropipette [33, 34, 79]. Low-melting agar is used to hold the cell during microinjection [2, 80]. The use of this method in plants is limited because the thick structure of the plant cell wall prevents the entry of microcapillary glass pipettes. Although this problem can be partially solved by the use of protoplasts, it is still not a complete solution [34]. The advantages of the technique are that it is efficient due to its precise delivery nature and allows the transfer of not only plasmids but also entire chromosomes into plant cells [6, 81]. However, microinjection is slow and laborious. It also requires an expensive micromanipulator device and experienced personnel [39].

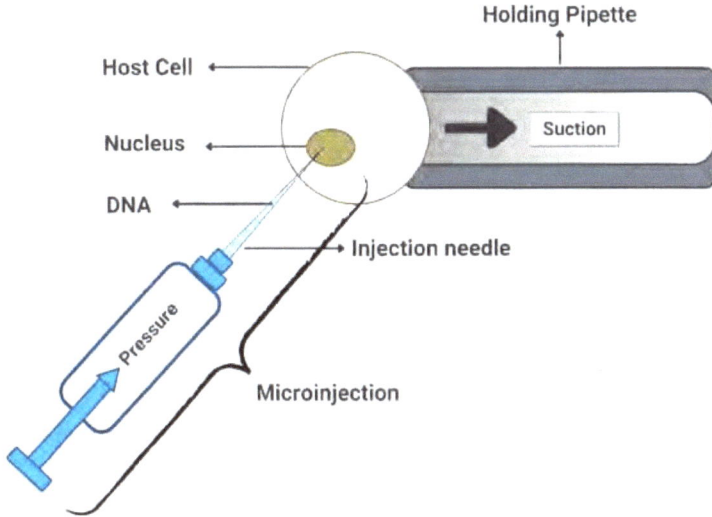

**Fig. (6).** Schematic representation of the microinjection method [82].

## Laser Microbeams

In the transformation method with laser microbeams (343 nm), micro holes are made in plant cells, and DNA is transferred inside (Fig. 7). Plant transformation methods usually use protoplasts, and it is difficult to create the whole plant from protoplasts. To avoid these difficulties, it is possible to use laser microbeams that make holes of about 0.5 μm in the cell wall, which are spontaneously restored [32]. In this way, the beams can be used to manipulate the nucleus and even organelles [83]. These microholes close in less than five seconds [32].

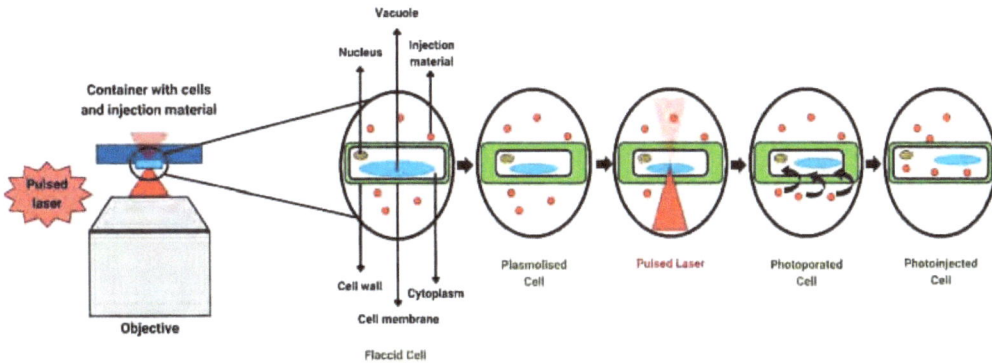

**Fig. (7).** Schematic representation of a standard laser microbeam experiment using an inverted microscopic setup and a pulsed laser source (plasmolysis of the plant cell prior to microbeam treatment promotes DNA uptake) [85].

The advantages of laser microbeam transformation are that a large number of cells can be transformed and that these DNA-transformed cells can be fully recovered. However, it is not a frequently used method due to the need for expensive equipment, the possibility of damage to the biological material by laser radiation, and the need for great care in handling [32, 84].

## Silicon Carbide-Mediated Transformation

Transformation *via* silicon carbide fibers was originally developed for DNA transfer into insect eggs but has also been successfully applied to plants [86]. In this method, silicon carbide fibers (10-80 mm in length and 0.6 mm in diameter) are used, and their sharp ends make holes in the cells that allow DNA transfer [33]. During the treatment, the target plant cells are first mixed with DNA in a suspension containing silicon carbide fibers. The DNA passes through the small holes formed as a result of collisions between the plant cells and the silicon carbide fibers and is transferred into the cell. The plant tissues to be transferred can be calluses, immature embryos, or cell piles [6, 87]. Factors affecting the efficiency of the method include the plant material, the thickness of the plant cell wall, the shape of the vessels used, and the size of the fibers. The transformation technique through silicon carbide fibers is a simple and inexpensive method, as it does not require any special technical equipment during its application [88]. However, the disadvantages of the method include low transformation efficiency and the fact that silicon carbide fiber fragments can cause serious harm to researchers if inhaled [89]. Considering all these advantages and disadvantages, Agrobacterium-mediated transformation is often preferred as an alternative method in cases where Agrobacterium-mediated transformation is not possible or a biolistic device is not available [33].

## Vacuum Infiltration

Vacuum infiltration is a method that promotes Agrobacterium-mediated plant transformation. Plants are placed upside down (with the floral parts in contact with the bacteria) in a container containing 5% sucrose suspension and Agrobacterium and exposed to a pressure of 0.05 bar for several minutes [33]. During the application of vacuum, a negative atmospheric pressure is created, which reduces the air spaces between the cells in the plant tissue. Thus, pathogenic bacteria enter these intercellular spaces (Fig. **8**). Adjusting the air spaces within the plant tissue to the desired degree depends on the duration of the vacuum and the pressure ratio. An increase in the pressure ratio allows the infiltration medium to settle into the plant tissue. It is very important to adjust the vacuum time correctly, as excessive hydration may occur if the time the plant tissue is exposed to the vacuum is prolonged [6, 32]. The promotion of

Agrobacterium-mediated transformation by vacuum infiltration was first realized in Arabidopsis plants in 1993 [90] and then applied to different plants. The main advantage of this technique is that it does not require cultures grown *in vitro* [33].

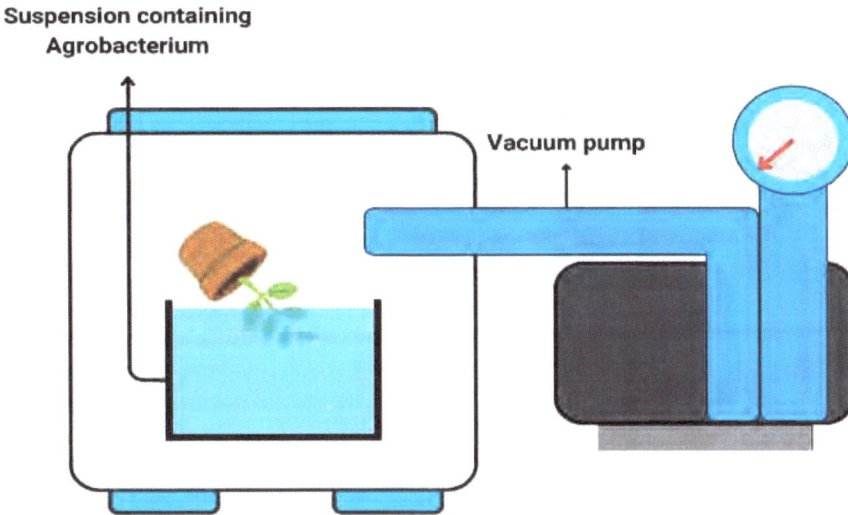

**Fig. (8).** Vacuum infiltration [32].

## *Ultrasonication-Mediated Transformation*

In this method, also called sonication, ultrasound (high frequency sound above 20 kHz) can temporarily alter the permeability of cellular membranes so that macromolecules such as DNA can be taken into the cell [91, 92]. Plant tissues to be used in this method are suspended in a few milliliters of sonication medium in a microcentrifuge tube. Plasmid DNA is then added to the medium, and the samples are prepared for sonication by rapidly mixing the medium. The ultrasonic waves used during transformation create cavitation and heating in cells and tissues. The heating is caused by both heat generation during cavitation and heating of the ultrasonic probe [93]. The cavitation state and the heating of the tissues up to several degrees above the biological temperature can lead to tissue perfusion [94]. In addition, there is a risk that the cavitation bubbles may damage the cell by completely breaking the cell membrane [6, 95]. Despite these disadvantages, ultrasonication-mediated transformation has been reported to increase transformation efficiency in somatic and zygotic embryos, embryonic suspension cells, leaf tissues, shoot tips, stems, roots, whole seedlings, and immature cotyledons. Furthermore, this method is often used in combination with sonication-assisted Agrobacterium-mediated transformation (SAAT) for the transformation of plant tissues [96, 97]. In this method, target plant tissues are again exposed to ultrasound in the presence of Agrobacterium, thereby increasing the transformation efficiency through the numerous micro-holes formed in the plant cells [39, 98 - 101].

## *Electrophoresis*

Gene transformation by electrophoresis is a simple and inexpensive method developed in the late 1980s. Embryos for genetic transformation are placed between the ends of two pipettes connected to electrodes. The narrow part of the pipette connected to the anode of the electrophoresis device contains agar, and the rest of the pipette contains electrophoresis buffer. The pipette connected to the cathode section contains agar mixed with DNA and an electrophoresis buffer. This pipette is in contact with the apical meristem of the embryo, and the second pipette is placed in the basal apical part of the embryo. The most commonly used parameters in the electrophoresis method are optimized as a voltage of 25 mV and a current of 0.5 mA for 15 minutes [6, 89]. Turning on the current causes a slow flow of DNA through the embryo, from the cathode to the anode (from the apical meristem to the basal part). The efficiency of the method depends on the electrophoresis time, the characteristics of the embryo, the applied electrical field, and the buffer concentration. Although it is a low-cost and easily applicable method, it is not a frequently preferred method due to the very low survival rate of embryos [33].

## ARTIFICIAL INTELLIGENCE IN GENETIC TRANSFORMATION

In genetic transformation studies, the use of artificial intelligence technologies to redesign some important proteins, especially those that play a role in metabolic processes, and to increase the efficiency of genetic transformation can be another perspective in today's technologies. By using an AI-based approach to protein design, researchers can achieve a higher degree of variation, reduce costs, and quickly create simulation scenarios before focusing on a specific subset of targets for experimental testing. On the other hand, traditional methods for designing new proteins can often involve extensive trial and error processes as well as increased costs.

Recombinant DNA technology is a field formed from a multidisciplinary perspective that forms the basis for genetic engineering and molecular biotechnology applications and covers fields such as molecular biology, biochemistry, immunology, genetics, chemical engineering and cell biology. The support of artificial intelligence-based technologies in all applications aimed at using prokaryotic and eukaryotic organisms in recombinant DNA applications, manipulating DNA and researching gene regulation or protein production by cloning will be able to accelerate the developments in these fields at an ultra-level. The contribution of artificial intelligence technologies can be great in adding an advanced technological dimension to methodical applications, especially cloning, and in more detail, DNA restriction and ligation of fragments, transformation of competent cells, detection of positive bacterial clones, PCR and site directed mutagenesis.

In this context, it brings plant physiologists face to face with a wider variety of models and algorithms in order to solve problems in the field of plant physiology on the basis of genetic transformation, to more deeply elucidate the metabolic pathways affecting growth and development, and to develop new protocols based on artificial intelligence by focusing on how to apply them. Thus, researchers will be able to focus on problems arising from bioinformatics, genomics and agriculture services with artificial intelligence-based applications. In the light of topics such as machine learning, data visualization, and the design of data-based learning algorithms, forward-looking applications can be carried out in fields such as synthetic biology and genetic algorithms.

## CONCLUSION

With the recent advances in genetic engineering, transformation methods have made it possible to transfer many traits to plants that could not be achieved by traditional breeding methods that have been practiced for centuries. Since the 1980s, many new and effective transformation methods have been developed for genetic transformation to different cell types and different plants. Due to the fact that not all transformation methods are suitable for all plant species, low efficiency, and problems during regeneration, researchers have searched for different methods. Therefore, despite the large number of methods available, Agrobacterium-mediated and biolistic-mediated transformation methods remain the two most preferred methods when comparing their advantages and disadvantages. Although the transformation of some plant species poses a challenge for researchers, new and improved methods are expected to be designed as the mechanisms of transformation and regeneration stages are elucidated. In addition, it holds great promise for meeting the ever-increasing food demand in the world by increasing the yield of plants with changes made only to the desired genes.

## REFERENCES

[1]    Robinson C. Genetic Modification Technology and Food: Consumer Health and Safety. Brussels, Belgium: ILSI Europe 2001.

[2]    Desai K, Solanki B, Mankad A, Pandya H. Genetic transformation of plants. IJSRR 2019; 8(2): 1792-806.

[3]    Canlı FA, Pektaş M, Temurtaş N. Advances in genetic transformation of apple. Tarım Bilimleri Araştırma Dergisi 2009; 2(1): 87-92.

[4]     Halford NG. Genetically Modified Crops. 2nd ed. London, UK: Imperial College Press 2012; pp. 1-4.

[5]     Safitri FA, Ubaidillah M, Kim KM. Efficiency of transformation mediated by *Agrobacterium tumefaciens* using vacuum infiltration in rice ( *Oryza sativa* L.). J Plant Biotechnol 2016; 43(1): 66-75.
[http://dx.doi.org/10.5010/JPB.2016.43.1.66]

[6]     Kalefetoğlu Macar T I, Macar O, Yalçın E, Çavuşoğlu K. Gene technology and plant genetic transformation methods. Afyon Kocatepe University Journal of Sciences and Engineering 2017; 17(2): 377-92.
[http://dx.doi.org/10.5578/fmbd.58669]

[7]     Zhu LH, Li XY, Ahlman A, Welander M. The rooting ability of the dwarfing pear rootstock BP10030 (*Pyrus communis*) was significantly increased by introduction of the rolB gene. Plant Sci 2003; 165(4): 829-35.
[http://dx.doi.org/10.1016/S0168-9452(03)00279-6]

[8]     Canli FA, Tian L. Regeneration of adventitious shoots from mature stored cotyledons of Japanese plum (*Prunus salicina* Lind1). Sci Hortic (Amsterdam) 2009; 120(1): 64-9. [Prunus salicina Lind I].
[http://dx.doi.org/10.1016/j.scienta.2008.09.017]

[9]     Wilson RHC, Coverley D. Transformation-induced changes in the DNA-nuclear matrix interface, revealed by high-throughput analysis of DNA halos. Sci Rep 2017; 7(1): 6475.
[http://dx.doi.org/10.1038/s41598-017-06459-7] [PMID: 28743923]

[10]    Atiq G, Nasrullah Khan N, Raheem MAR, Iqbal RK. Plant transformation in biotechnology. MEJAST 2019; 2(3): 103-23.

[11]    Babaoglu M, Davey MR, Power JB. Genetic engineering of grainlegumes: key transformation events. Agric Biotechnol 2000; 2: 1-12.

[12]    Keshavareddy G, Kumar ARV, Ramu VS. Methods of plant transformation-a review. Int J Curr Microbiol Appl Sci 2018; 7(7): 2656-68.
[http://dx.doi.org/10.20546/ijcmas.2018.707.312]

[13]    Horsch RB, Fry JE, Hoffmann NL, Eichholtz D, Rogers SG, Fraley RT. A simple and general method for transferring genes into plants. Science. 1985; 227(4691): 1229.
[http://dx.doi.org/10.1126/science.227.4691.1229]

[14]    Toriyama K, Arimoto Y, Uchimiya H, Hinata K. Transgenic rice plants after direct gene transfer into protoplasts. Nat Biotechnol 1988; 6(9): 1072-4.
[http://dx.doi.org/10.1038/nbt0988-1072]

[15]    Zhang W, Wu R. Efficient regeneration of transgenic plants from rice protoplasts and correctly regulated expression of the foreign gene in the plants. Theor Appl Genet 1988; 76(6): 835-40.
[http://dx.doi.org/10.1007/BF00273668] [PMID: 24232391]

[16]    Zhang HM, Yang H, Rech EL, *et al.* Transgenic rice plants produced by electroporation-mediated plasmid uptake into protoplasts. Plant Cell Rep 1988; 7(6): 379-84.
[http://dx.doi.org/10.1007/BF00269517] [PMID: 24240249]

[17]    Yang H, Zhang HM, Davey MR, Mulligan BJ, Cocking EC. Production of kanamycin resistant rice tissues following DNA uptake into protoplasts. Plant Cell Rep 1988; 7(6): 421-5.
[http://dx.doi.org/10.1007/BF00269528] [PMID: 24240260]

[18]    Shimamoto K, Terada R, Izawa T, Fujimoto H. Fertile transgenic rice plants regenerated from transformed protoplasts. Nature 1989; 338(6212): 274-6.
[http://dx.doi.org/10.1038/338274a0]

[19]    Tada Y, Sakamoto M, Fujimura T. Efficient gene introduction into rice by electroporation and analysis of transgenic plants: use of electroporation buffer lacking chloride ions. Theor Appl Genet 1990; 80(4): 475-80.
[http://dx.doi.org/10.1007/BF00226748] [PMID: 24221005]

[20]    Hess D, Dressler K, Nimmrichter R. Transformation experiments by pipetting Agrobacterium into the spikelets of wheat (*Triticum aestivum* L.). Plant Sci 1990; 72(2): 233-44.
[http://dx.doi.org/10.1016/0168-9452(90)90087-5]

[21]    Patnaik D, Khurana P. Wheat biotechnology: a minireview. Electron J Biotechnol 2001; 4(2): 38-66.

[22]    Bevan MW, Flavell RB, Chilton MD. A chimaeric antibiotic resistance gene as a selectable marker for plant cell transformation. Nature 1983; 304(5922): 184-7.
[http://dx.doi.org/10.1038/304184a0]

[23]    Herrera-Estrella L, Depicker A, Van Montagu M, Schell J. Expression of chimaeric genes transferred into plant cells using a Ti-plasmid-derived vector. Nature 1983; 303(5914): 209-13.
[http://dx.doi.org/10.1038/303209a0]

[24]    Perl A, Lotan O, Abu-Abied M, Holland D. Establishment of an Agrobacterium-mediated transformation system for grape (*Vitis vinifera* L.): The role of antioxidants during grape–Agrobacterium interactions. Nat Biotechnol 1996; 14(5): 624-8.
[http://dx.doi.org/10.1038/nbt0596-624] [PMID: 9630955]

[25]    Zimmermann U, Vienken J. Stable transformation of maize after gene transfer by electroporation. J Membr Biol 1982; 67: 165-82.
[http://dx.doi.org/10.1007/BF01868659] [PMID: 7050391]

[26]    Southgate EM, Davey MR, Power JB, Westcott RJ. A comparison of methods for direct gene transfer into maize (Zea mays L.). *In Vitro* Cell. Dev Biol Plant 1998; 34(3): 218-24.
[http://dx.doi.org/10.1007/BF02822711]

[27]    Armstrong CL. The first decade of maize transformation: a review and future perspective. Maydica 1999; 44(1): 101-9.

[28]    Zhang P, Jaynes JM, Potrykus I, Gruissem W, Puonti-Kaerlas J. Transfer and expression of an artificial storage protein (ASP1) gene in cassava (*Manihot esculenta* Crantz). Transgenic Res 2003; 12(2): 243-50.
[http://dx.doi.org/10.1023/A:1022918925882] [PMID: 12739891]

[29]    Moloney MM, Walker JM, Sharma KK. High efficiency transformation of *Brassica napus* using Agrobacterium vectors. Plant Cell Rep 1989; 8(4): 238-42.
[http://dx.doi.org/10.1007/BF00778542] [PMID: 24233146]

[30]    Fraley RT, Rogers SG, Horsch RB, *et al.* Expression of bacterial genes in plant cells. Proc Natl Acad Sci USA 1983; 80(15): 4803-7.
[http://dx.doi.org/10.1073/pnas.80.15.4803] [PMID: 6308651]

[31]    Barampuram S, Zhang ZJ. Recent Advances in Plant Transformation. In: Birchler J, Ed. Plant Chromosome Engineering: Methods and Protocols. Totowa, NJ: Humana Press 2011; pp. 1-35.
[http://dx.doi.org/10.1007/978-1-61737-957-4_1]

[32]    Rivera AL, Gómez-Lim M, Fernández F, Loske AM. Physical methods for genetic plant transformation. Phys Life Rev 2012; 9(3): 308-45.
[http://dx.doi.org/10.1016/j.plrev.2012.06.002] [PMID: 22704230]

[33]    Rakoczy-Trojanowska M. Alternative methods of plant transformation-a short review. Cell Mol Biol Lett 2002; 7(3): 849-58.
[PMID: 12378268]

[34]    Rao AQ, Bakhsh A, Kiani S, *et al.* RETRACTED: The myth of plant transformation. Biotechnol Adv 2009; 27(6): 753-63.
[http://dx.doi.org/10.1016/j.biotechadv.2009.04.028] [PMID: 19508888]

[35]    Tzfira T, Citovsky V. Agrobacterium-mediated genetic transformation of plants: biology and biotechnology. Curr Opin Biotechnol 2006; 17(2): 147-54.
[http://dx.doi.org/10.1016/j.copbio.2006.01.009] [PMID: 16459071]

[36]   Zupan J, Zambryski P, Citovsky V. The agrobacterium DNA transfer complex. Crit Rev Plant Sci 1997; 16(3): 279-95.
[http://dx.doi.org/10.1080/07352689709701951]

[37]   Oltmanns H, Frame B, Lee LY, *et al.* Generation of backbone-free, low transgene copy plants by launching T-DNA from the Agrobacterium chromosome. Plant Physiol 2010; 152(3): 1158-66.
[http://dx.doi.org/10.1104/pp.109.148585] [PMID: 20023148]

[38]   Mini P, Demurtas OC, Valentini S, *et al.* Agrobacterium-mediated and electroporation-mediated transformation of *Chlamydomonas reinhardtii*: a comparative study. BMC Biotechnol 2018; 18(1): 11.
[http://dx.doi.org/10.1186/s12896-018-0416-3] [PMID: 29454346]

[39]   Saeed T, Shahzad A. Basic Principles Behind Genetic Transformation in Plants. In: Shahzad A, Sharma S, Siddiqui S, Eds. Biotechnological Strategies for the Conservation of Medicinal and Ornamental Climbers. Cham: Springer 2016; pp. 327-50.
[http://dx.doi.org/10.1007/978-3-319-19288-8_13]

[40]   Hwang HH, Yu M, Lai EM. Agrobacterium-mediated plant transformation: biology and applications. Arabidopsis Book 2017; 15(15): e0186.
[http://dx.doi.org/10.1199/tab.0186] [PMID: 31068763]

[41]   Gelvin SB. Agrobacterium-mediated plant transformation: the biology behind the "gene-jockeying" tool. Microbiol Mol Biol Rev 2003; 67(1): 16-37.
[http://dx.doi.org/10.1128/MMBR.67.1.16-37.2003] [PMID: 12626681]

[42]   Guo M, Bian X, Wu X, Wu M. Agrobacterium-mediated genetic transformation: history and progress. In: Alvarez M, Ed. Genetic Transformation. InTech 2011.
[http://dx.doi.org/10.5772/22026]

[43]   Desai K, Modi N. Genetic Modification of Plants: An Emerging Technology. Int J Agric Syst 2000; 8(2): 64-76.
[http://dx.doi.org/10.20956/ijas.v8i2.2100]

[44]   Opabode JT. Agrobacterium-mediated transformation of plants: emerging factors that influence efficiency. Biotechnol Mol Biol Rev 2006; 1(1): 12-20.

[45]   Michielse CB, Hooykaas PJJ, van den Hondel CAMJJ, Ram AFJ. Agrobacterium-mediated transformation as a tool for functional genomics in fungi. Curr Genet 2005; 48(1): 1-17.
[http://dx.doi.org/10.1007/s00294-005-0578-0] [PMID: 15889258]

[46]   Naqvi S, Farré G, Sanahuja G, Capell T, Zhu C, Christou P. When more is better: multigene engineering in plants. Trends Plant Sci 2010; 15(1): 48-56.
[http://dx.doi.org/10.1016/j.tplants.2009.09.010] [PMID: 19853493]

[47]   Mukherjee E, Gantait S. Genetic transformation in sugar beet (*Beta vulgaris* L.): technologies and applications. Sugar Tech 2023; 25(2): 269-81.
[http://dx.doi.org/10.1007/s12355-022-01176-6]

[48]   Rahman SU, Khan MO, Ullah R, Ahmad F, Raza G. Agrobacterium-mediated transformation for the development of transgenic crops;p and future prospects. Mol Biotechnol 2023; 1-17.
[http://dx.doi.org/10.1007/s12033-023-00826-8] [PMID: 37573566]

[49]   Fromm ME, Taylor LP, Walbot V. Stable transformation of maize after gene transfer by electroporation. Nature 1986; 319(6056): 791-3.
[http://dx.doi.org/10.1038/319791a0] [PMID: 3005872]

[50]   Saulis G, Venslauskas MS, Naktinis J. Kinetics of pore resealing in cell membranes after electroporation. Bioelectrochem Bioenerg 1991; 26(1): 1-13.
[http://dx.doi.org/10.1016/0302-4598(91)87029-G]

[51]   Okada K, Nagata T, Takebe I. Introduction of functional RNA into plant protoplasts by electroporation. Plant Cell Physiol 1986; 27(4): 619-26.

[http://dx.doi.org/10.1093/oxfordjournals.pcp.a077141]

[52]    Speyer JF. A simple and effective electroporation apparatus. Biotechniques 1990; 8(1): 28-30.
        [PMID: 2182075]

[53]    Djuzenova CS, Zimmermann U, Frank H, Sukhorukov VL, Richter E, Fuhr G. Effect of medium
        conductivity and composition on the uptake of propidium iodide into electropermeabilized myeloma
        cells. Biochim Biophys Acta Biomembr 1996; 1284(2): 143-52.
        [http://dx.doi.org/10.1016/S0005-2736(96)00119-8] [PMID: 8914578]

[54]    Lee EW, Gehl J, Kee ST. Introduction to Electroporation. In: Lee EW, Gehl J, Kee ST, Eds. Clinical
        Aspects of Electroporation. New York: Springer 2011; pp. 3-7.
        [http://dx.doi.org/10.1007/978-1-4419-8363-3_1]

[55]    Hussain J, Manan S, Ahmad R, Ahmed T, Shah MM. Biotechnilogies used in genetic transformation
        of Triticumaestivum: a mini overview. FUUAST J Biol 2013; 3(2): 105.

[56]    Kandušer M, Miklavčič D. Electroporation in Biological Cell and Tissue: An Overview. In: Nikolai L,
        Eugene V, Eds. Electrotechnologies for Extraction from Food Plants and Biomaterials. New York:
        Springer 2009; pp. 1-37.
        [http://dx.doi.org/10.1007/978-0-387-79374-0_1]

[57]    Fromm M, Taylor LP, Walbot V. Expression of genes transferred into monocot and dicot plant cells
        by electroporation. Proc Natl Acad Sci USA 1985; 82(17): 5824-8.
        [http://dx.doi.org/10.1073/pnas.82.17.5824] [PMID: 3862099]

[58]    de la Peña A, Lörz H, Schell J. Transgenic rye plants obtained by injecting DNA into young floral
        tillers. Nature 1987; 325(6101): 274-6.
        [http://dx.doi.org/10.1038/325274a0]

[59]    Sorokin AP, Ke XY, Chen DF, Elliott MC. Production of fertile transgenic wheat plants *via* tissue
        electroporation. Plant Sci 2000; 156(2): 227-33.
        [http://dx.doi.org/10.1016/S0168-9452(00)00260-0] [PMID: 10936530]

[60]    Rao AQ, Bakhsh A, Kiani S, *et al*. Retracted: The myth of plant transformation. Biotechnol Adv 2019;
        37(5): 827.
        [http://dx.doi.org/10.1016/j.biotechadv.2019.06.003]

[61]    Kar S, Loganathan M, Dey K, *et al*. Single-cell electroporation: current trends, applications and future
        prospects. J Micromech Microeng 2018; 28(12): 123002.
        [http://dx.doi.org/10.1088/1361-6439/aae5ae]

[62]    Yan Y, Zhu X, Yu Y, Li C, Zhang Z, Wang F. Nanotechnology strategies for plant genetic
        engineering. Adv Mater 2022; 34(7): 2106945.
        [http://dx.doi.org/10.1002/adma.202106945] [PMID: 34699644]

[63]    Sanford JC, Klein TM, Wolf ED, Allen N. Delivery of substances into cells and tissues using a particle
        bombardment process. Particul Sci Technol 1987; 5(1): 27-37.
        [http://dx.doi.org/10.1080/02726358708904533]

[64]    Anderson BR, Boynton JE, Dawson J, *et al*. Sub-micron gold particles are superior to larger particles
        for efficient biolistic transformation of organelles and some cell types. Bulletin 2015; 1-3.

[65]    Oard JH, Paige DF, Simmonds JA, Gradziel TM. Transient gene expression in maize, rice, and wheat
        cells using an airgun apparatus. Plant Physiol 1990; 92(2): 334-9.
        [http://dx.doi.org/10.1104/pp.92.2.334] [PMID: 16667278]

[66]    Frame BR, Zhang H, Coccicolone SM, *et al*. Production of transgenic maize from bombarded type II
        callus: Effect of gold particle size and callus morphology on transformation efficiency. *In Vitro* Cell.
        Dev Biol Plant 2000; 36(1): 21-9.
        [http://dx.doi.org/10.1007/s11627-000-0007-5]

[67]    Altpeter F, Baisakh N, Beachy R, *et al*. Particle bombardment and the genetic enhancement of crops:

myths and realities. Mol Breed 2005; 15(3): 305-27.
[http://dx.doi.org/10.1007/s11032-004-8001-y]

[68] Taylor NJ, Fauquet CM. Microparticle bombardment as a tool in plant science and agricultural biotechnology. DNA Cell Biol 2002; 21(12): 963-77.
[http://dx.doi.org/10.1089/104454902762053891] [PMID: 12573053]

[69] Sanford JC. The development of the biolistic process. *In Vitro* Cell. Dev Biol Plant 2000; 36(5): 303-8.
[http://dx.doi.org/10.1007/s11627-000-0056-9]

[70] Southgate EM, Davey MR, Power JB, Marchant R. Factors affecting the genetic engineering of plants by microprojectile bombardment. Biotechnol Adv 1995; 13(4): 631-51.
[http://dx.doi.org/10.1016/0734-9750(95)02008-X] [PMID: 14536367]

[71] Pawlowski WP, Somers DA. Transgene inheritance in plants genetically engineered by microprojectile bombardment. Mol Biotechnol 1996; 6(1): 17-30.
[http://dx.doi.org/10.1007/BF02762320] [PMID: 8887358]

[72] Kohli A, Gahakwa D, Vain P, Laurie DA, Christou P. Transgene expression in rice engineered through particle bombardment: molecular factors controlling stable expression and transgene silencing. Planta 1999; 208(1): 88-97.
[http://dx.doi.org/10.1007/s004250050538]

[73] Dai S, Zheng P, Marmey P, *et al.* Comparative analysis of transgenic rice plants obtained by Agrobacterium mediated transformation and particle bombardment. Mol Breed 2001; 7(1): 25-33.
[http://dx.doi.org/10.1023/A:1009687511633]

[74] Narusaka Y, Narusaka M, Yamasaki S, Iwabuchi M. Methods to Transfer Foreign Genes to Plants. In: Ozden Ciftci Y, Ed. Transgenic Plants-Advances and Limitations. Rijeka: IntechOpen 2012.
[http://dx.doi.org/10.5772/32773]

[75] Wang H, Wang W, Zhan J, Huang W, Xu H. An efficient PEG-mediated transient gene expression system in grape protoplasts and its application in subcellular localization studies of flavonoids biosynthesis enzymes. Sci Hortic (Amsterdam) 2015; 191: 82-9.
[http://dx.doi.org/10.1016/j.scienta.2015.04.039]

[76] Davey MR, Anthony P, Power JB, Lowe KC. Plant protoplasts: status and biotechnological perspectives. Biotechnol Adv 2005; 23(2): 131-71.
[http://dx.doi.org/10.1016/j.biotechadv.2004.09.008] [PMID: 15694124]

[77] Binsfeld PC. Transgenic Sun ower: PEG-Mediated Gene Transfer. In: Jackson JF, Linskens HF, Eds. Genetic Transformation of Plants. Berlin, Heidelberg: Springer 2003; pp. 109-26.
[http://dx.doi.org/10.1007/978-3-662-07424-4_7]

[78] Gad AE, Rosenberg N, Altman A. Liposome mediated gene delivery into plant cells. Physiol Plant 1990; 79(1): 177-83.
[http://dx.doi.org/10.1111/j.1399-3054.1990.tb05883.x]

[79] Crossway A, Oakes JV, Irvine JM, Ward B, Knauf VC, Shewmaker CK. Integration of foreign DNA following microinjection of tobacco mesophyll protoplasts. Mol Gen Genet 1986; 202(2): 179-85.
[http://dx.doi.org/10.1007/BF00331634]

[80] Harwood WA, Davies DR. Protoplast Microinjection Using Agarose Microdrops. In: Pollard JW, Walker JM, Eds. Plant Cell and Tissue Culture Methods in Molecular Biology. Totowa, NJ: Humana Press 1990; pp. 323-33.
[http://dx.doi.org/10.1385/0-89603-161-6:323]

[81] Jones-Villeneuve E, Huang B, Prudhomme I, *et al.* Assessment of microinjection for introducing DNA into uninuclear microspores of rapeseed. Plant Cell Tissue Organ Cult 1995; 40(1): 97-100.
[http://dx.doi.org/10.1007/BF00041124]

[82] Sarı U, Tiryaki I. Genetic transformation techniques. Engineering and Architecture Sciences. 156.

[83]    Weber G, Monajembashi S, Wolfrum J, Greulich KO. Genetic changes induced in higher plant cells by a laser microbeam. Physiol Plant 1990; 79(1): 190-3.
[http://dx.doi.org/10.1111/j.1399-3054.1990.tb05885.x]

[84]    Lin PF, Ruddle FH. Photoengraving of coverslips and slides to facilitate monitoring of micromanipulated cells or chromosome spreads. Exp Cell Res 1981; 134(2): 485-8.
[http://dx.doi.org/10.1016/0014-4827(81)90452-3] [PMID: 7023960]

[85]    Heinemann D, Zabic M, Terakawa M, Boch J. Laser-based molecular delivery and its applications in plant science. Plant Methods 2022; 18(1): 82.
[http://dx.doi.org/10.1186/s13007-022-00908-9] [PMID: 35690858]

[86]    Kaeppler H, Gu W, Somers D, Rines H, Cockburn A. Silicon carbide fiber-mediated DNA delivery into plant cells. Plant Cell Rep 1990; 9(8): 415-8.
[http://dx.doi.org/10.1007/BF00232262] [PMID: 24227167]

[87]    Wang K, Drayton P, Frame B, Dunwell J, Thompson J. Whisker-mediated plant transformation: An alternative technology. *In Vitro* Cell. Dev Biol Plant 1995; 31(2): 101-4.
[http://dx.doi.org/10.1007/BF02632245]

[88]    Joung YH, Choi PS, Kwon SY, Harn CH. Plant Transformation Methods and Applications. In: Koh HJ, Kwon SY, Thomson M, Eds. Current Technologies in Plant Molecular Breeding. Dordrecht: Springer 2015; pp. 297-343.
[http://dx.doi.org/10.1007/978-94-017-9996-6_9]

[89]    Songstad DD, Somers DA, Griesbach RJ. Advances in alternative DNA delivery techniques. Plant Cell Tissue Organ Cult 1995; 40(1): 1-15.
[http://dx.doi.org/10.1007/BF00041112]

[90]    Bechtold N, Ellis J, Pelletier G. In-planta Agrobacterium-mediated gene-transfer by infiltration of adult *Arabidopsis thaliana* plants. N Bechtold - Sci. Paris. Life Sci 1993; 316(10): 1194-9.

[91]    Wyber JA, Andrews J, D'Emanuele A. The use of sonication for the efficient delivery of plasmid DNA into cells. Pharm Res 1997; 14(6): 750-6.
[http://dx.doi.org/10.1023/A:1012198321879] [PMID: 9210192]

[92]    Tachibana K, Uchida T, Ogawa K, Yamashita N, Tamura K. Induction of cell-membrane porosity by ultrasound. Lancet 1999; 353(9162): 1409.
[http://dx.doi.org/10.1016/S0140-6736(99)01244-1] [PMID: 10227224]

[93]    Miller DL, Pislaru SV, Greenleaf JF. Sonoporation: mechanical DNA delivery by ultrasonic cavitation. Somat Cell Mol Genet 2002; 27(1/6): 115-34.
[http://dx.doi.org/10.1023/A:1022983907223] [PMID: 12774945]

[94]    Zeqiri B. Exposure criteria for medical diagnostic ultrasound: II. Criteria based on all known mechanisms. Ultrasound Med Biol 2003; 29(12): 1809.
[http://dx.doi.org/10.1016/j.ultrasmedbio.2003.09.002]

[95]    Liu Y, Yang H, Sakanishi A. Ultrasound: Mechanical gene transfer into plant cells by sonoporation. Biotechnol Adv 2006; 24(1): 1-16.
[http://dx.doi.org/10.1016/j.biotechadv.2005.04.002] [PMID: 15935607]

[96]    Trick HN, Finer JJ. SAAT: sonication-assisted Agrobacterium-mediated transformation. Transgenic Res 1997; 6(5): 329-36.
[http://dx.doi.org/10.1023/A:1018470930944]

[97]    Weber S, Friedt W, Landes N, *et al.* Improved Agrobacterium -mediated transformation of sunflower ( *Helianthus annuus* L.): assessment of macerating enzymes and sonication. Plant Cell Rep 2003; 21(5): 475-82.
[http://dx.doi.org/10.1007/s00299-002-0548-7] [PMID: 12789451]

[98]    Liu Z, Park BJ, Kanno A, Kameya T. The novel use of a combination of sonication and vacuum

infiltration in Agrobacterium-mediated transformation of kidney bean (*Phaseolus vulgaris* L.) with lea gene. Mol Breed 2005; 16(3): 189-97.
[http://dx.doi.org/10.1007/s11032-005-6616-2]

[99]    Paliwal S, Mitragotri S. Ultrasound-induced cavitation: applications in drug and gene delivery. Expert Opin Drug Deliv 2006; 3(6): 713-26.
[http://dx.doi.org/10.1517/17425247.3.6.713] [PMID: 17076594]

[100]   de Oliveira MLP, Febres VJ, Costa MGC, Moore GA, Otoni WC. High-efficiency Agrobacterium-mediated transformation of citrus *via* sonication and vacuum infiltration. Plant Cell Rep 2009; 28(3): 387-95.
[http://dx.doi.org/10.1007/s00299-008-0646-2] [PMID: 19048258]

[101]   Subramanyam K, Subramanyam K, Sailaja KV, Srinivasulu M, Lakshmidevi K. Highly efficient Agrobacterium-mediated transformation of banana cv. Rasthali (AAB) *via* sonication and vacuum infiltration. Plant Cell Rep 2011; 30(3): 425-36.
[http://dx.doi.org/10.1007/s00299-010-0996-4] [PMID: 21212957]

# CHAPTER 9

# Crispr-Cas Technology: Targeted Genome Editing in Plant Physiology

**Mohammad Mehdi Habibi**[1,*]

[1] *University of Tsukuba, Faculty of Life and Environmental Sciences, Tennodai, Tsukuba, Ibaraki, Japan*

**Abstract:** The phenomenon of global climate change poses a significant threat to global food security, primarily due to the limited adaptability of major staple crops and plant species to the changing climatic conditions. This poses a significant challenge for farmers, agricultural experts, and policymakers worldwide as they seek to develop sustainable solutions to ensure adequate food supply in the face of climate change-induced threats. Significant improvement has been made to preserve crop yield, employing traditional breeding methods and cutting-edge molecular techniques to enhance the procedure. The utilization of CRISPR/Cas technology has recently gained traction as a viable alternative to transgenic methods in plant breeding. Our study in this chapter, for the first time, delves into the advantages of the CRISPR/Cas system in plant physiology, exploring key areas such as its impact on environmental factors, the underlying mechanisms of the CRISPR/Cas system, enhanced quality and yield, mitigation of biotic and abiotic stresses, ethical considerations, and regulatory issues, as well as the future prospects of this method.

**Keywords:** CRISPR/Cas system, Genome editing, Genetically modified organisms (GMO).

## INTRODUCTION

Traditional plant breeding methods no longer suffice to address the challenge of feeding a growing population in a sustainable and productive manner. These methods are limited by reduced plant diversity, high costs, and time-consuming breeding processes [1]. Therefore, modern biotechnology and genetic engineering must be combined with traditional plant breeding. Genetic engineering is a complex area of biotechnology that involves the selection, location, isolation, purification, multiplication, and transfer of genes to create targeted changes and transformations in the genomes of plants and animals [2, 3]. While modern biotechnology holds tremendous promise for medicine, agriculture, and industry,

---

[*] **Corresponding author Mohammad Mehdi Habibi:** University of Tsukuba, Faculty of Life and Environmental Sciences, Tennodai, Tsukuba, Ibaraki, Japan; E-mail: mehdihabibi9@gmail.com

**Ergun Kaya (Ed.)**

safety concerns regarding genetically modified organisms (GMOs) and their products must be taken into account [4 - 6].

The practice of targeted genome editing in plant physiology holds vast potential for improving plant breeding. One such application is the modification of genes involved in growth and development, including those that affect the physiology of plants, which can increase crop yield [7 - 12]. Additionally, genes linked to disease resistance can be edited to bolster plants' natural defenses against pathogens, thus reducing the necessity for chemical pesticides [13 - 15]. Moreover, targeted genome editing presents an opportunity to enhance the nutritional quality of plants. By modifying the genes responsible for synthesizing crucial nutrients like vitamins, crop plants can be engineered to produce higher levels of these nutrients, which can help address nutritional deficiencies in human diets [16 - 20]. Generally, targeted genome editing has the potential to transform agriculture by enabling the creation of crops with improved traits, greater resistance to environmental pressures, and increased nutritional value [21 - 24]. However, it is crucial to approach this technology with responsibility, considering ethical and environmental concerns and implementing strict regulations to prevent unintended impacts on ecosystems and biodiversity.

New methods for improving gene targeting techniques, known as gene editing technologies, have recently been developed. These include zinc-finger nucleases (ZFNs), transcription activator-like effector nucleases (TALENs), and CRISPR (Clustered Regularly Interspaced Short Palindromic Repeats) systems [25 - 27]. Among these, based on advanced research, the CRISPR/Cas system is the most efficient, cost-effective, and dependable for gene editing purposes.

The CRISPR/Cas system is an advanced method of transgenic plant production that has been developed to address concerns related to crop quality and quantity (Fig. 1). This genome editing technique is capable of creating mutations in multiple gene locations and producing large deletions, thus enhancing the function and activity of plant genes and creating new traits [28, 29]. The CRISPR/Cas system has been widely used in animals and plants for gene silencing, gene replacement, multiple gene editing, identification of gene function, and regulation of the transcription process [30 - 32]. Studies on the CRISPR/Cas system have demonstrated its potential to accelerate plant breeding by boosting the enhancement of high-performance crops.

The utilization of CRISPR/Cas in plant physiology has numerous benefits, including creating disease-resistant crops and enhancing crop adaptability and resilience to varying environmental conditions like drought, extreme temperatures, and salinity. Moreover, it can improve crop nutritional value, quality, and shelf life. Our study aims to provide a comprehensive view of how CRISPR/Cas9 is used in targeted genome editing of plants and its impact on plant physiology and transgenic plant development.

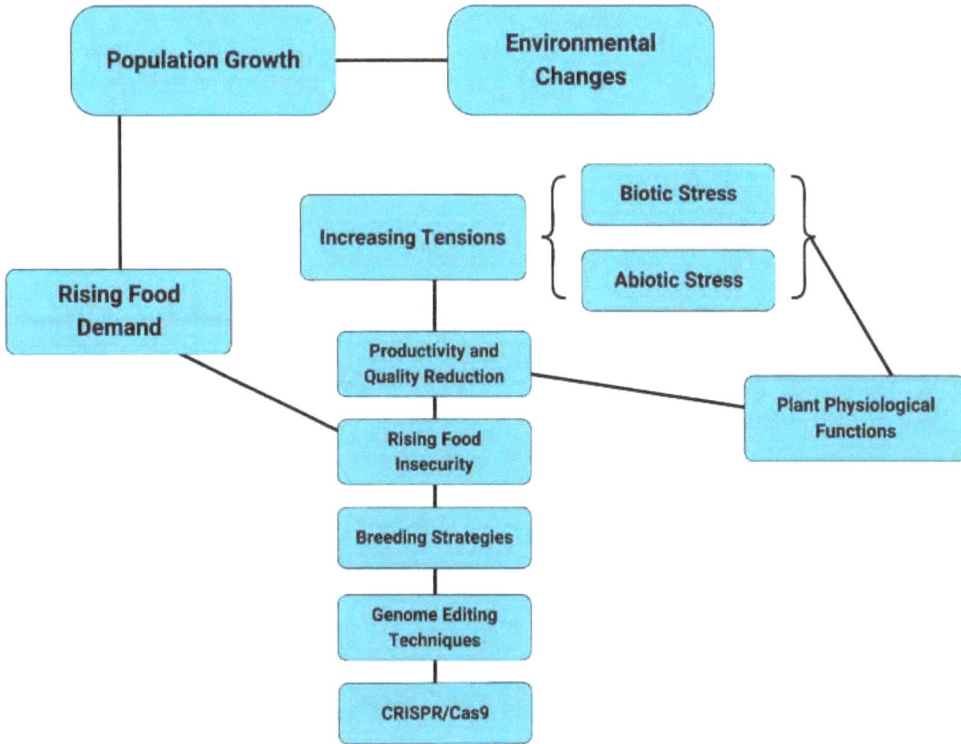

**Fig. (1).** Schematic overview of the CRISPR/CAS system's importance in targeted genome editing for plant physiology.

## Precision Editing

Enhanced and accurate editing tools, such as CRISPR/Cas, which utilize numerous host DNA-repair paths to edit the genome in host cells, are needed to produce a wide range of plants with modified cells. However, achieving the highest efficiency and eliminating off-target effects poses a challenge. CRISPR/Cas system allows scientists to precisely target and modify specific DNA sequences within a plant's genome. It utilizes guide RNA to direct the Cas protein to the desired location, where it induces changes such as insertions, deletions, or substitutions [33, 34].

The system known as CRISPR/Cas was initially discovered in 1987 in the genome of *Escherichia coli*, functioning as an acquired immune system that evolved in bacteria and archaea. CRISPR gene families (Cas) were subsequently identified in all prokaryotes in the year 2000 [35, 36]. The CRISPR system consists of two regions, including genes encoding Cas nuclease enzymes and the gene locus of CRISPR arrays containing repeat sequences and spacer sequences between them. CRISPR/Cas systems can be divided into three types based on the partaking Cas proteins. Type I and type III for binding and targeting use a complex of multiple Cas proteins, whereas type II needs just a single protein (Cas-9). The system CRISPR/Cas9 operates based on RNA and can be modified with greater ease and efficiency compared to approaches that rely on proteins. Furthermore, this system permits the targeting of several sites [37 - 39]. The process of genome editing in the CRISPR/Cas9 system can be broken down into three key phases: recognition, cleavage, and repair.

### Recognition

Recognition, or targeting, is a crucial component of the CRISPR/Cas9 mechanism, as it ensures accurate identification of the target sequence for precise cleavage. During this step, the sgRNA-Cas9 complex is prepared and guides Cas9 to attach to the target sequence of interest anywhere on the gene [40]. The CRISPR/Cas9 system relies on a short sequence called PAM (protospacer adjacent motif) to ensure accuracy, as it is responsible for recognizing the protospacer selection domain. Cas-9 then unzips the DNA double helix, and sgRNA-Cas9 joins to the DNA through a complementary base pairing mechanism, forming a DNA-RNA hybrid structure and initiating DNA capture. Typically, the cas9 protein identifies the PAM sequence at 5'-NGG-3' [41, 42].

### Cleavage

The process of genome editing *via* CRISPR/Cas9 typically involves utilizing the double-strand break (DSB) of DNA and its two associated methods of repair, namely HDR and NHEJ. Once the binding process is complete and sequence matching has been achieved, the DNA is cut, and two nuclease domains (HNH and RuvC) of the Cas-9 enzyme are activated. These domains work together with the PAM and target sequences to effectively cleave the DNA, and the Cas-9 protein serves as a sort of molecular scissor during this critical step [43, 44].

### Repair

There are two mechanisms to repair the double standard breaks caused by the Cas-9 protein: Non-homologous end joining (NHEJ) and homology-directed repair (HDR). Normally, cells do not use a specific pattern to recombine broken DNA ends, which makes NHEJ more common but also poses a risk of errors in the target sequence and resulting mutations. In contrast, HDR is a precise mechanism with a high level of fidelity that utilizes a homologous DNA pattern to repair DNA breaks. Scientists can control the repair machinery system through HDR by providing a large quantity of donor homologous DNA templates with the desired sequence of interest. By adding a homologous DNA template with the desired sequence at the double-standard break spot, HDR can achieve exact gene insertion or replacement [45, 46].

## Exploration of Gene Functions

The CRISPR/Cas system is a mechanism for researchers to study gene functions, as it enables them to create precise and targeted mutations. This approach is particularly useful in gaining insights into gene roles in plant growth, development, and responses to environmental stimuli. With its ability to manipulate genes with high specificity, the CRISPR/Cas9 system has established a new route for research in the field of genetics and genomics.

The CRISPR/Cas9 system provides tools for manipulating genes, enabling researchers to study gene function, introduce genetic modifications, and regulate gene expression in various organisms [47]. Advances in gene editing, such as knock-out, knock-in, and gene expression studies, have revolutionized molecular biology and hold great potential for genetic research and development [48 - 51].

### *Knock-out*

CRISPR/Cas9 can inactivate (silencing) or "knock-out" specific genes in an organism. By designing guide RNAs (gRNAs) that target the desired gene, the Cas protein can be guided to the location of the gene and induce a double-strand break (DSB) in the DNA. The cell's DNA repair machinery then repairs this DSB by inserting or deleting a couple of nucleotides (NHEJ). However, it often leads to errors, resulting in frameshift mutations or gene disruption [52, 53]. This methodology can be effectively employed to investigate the functionality and involvement of individual genes in cellular mechanisms. Specifically, it helps eliminate undesirable traits in plants ruled by a particular gene [54, 55]. An example of the knock-out function of the CRISPR/Cas9 system is the enhancement of resistance to biotic stresses, such as pests and diseases (Table **1**).

**Table 1. Application of CRISPR/Cas genome editing in a gene functional study on various plants.**

| Plant | Gen Function | Targeted Genes | Gene Reg. | Results | Refs. |
|---|---|---|---|---|---|
| **Cotton (*Gossypium hirsutum* L.)** | Knockout, | *GhCLA1, GhVP* | Deletions Insertion Substation | Establishment of genome editing for agronomic trait enhancement | [268] |

*(Table 1) cont.....*

| Plant | Gen Function | Targeted Genes | Gene Reg. | Results | Refs. |
|---|---|---|---|---|---|
| **Tetraploid Potato** | Knockout | *StPDS* | Deletions Insertion Replacement | Establishment of genome editing for agronomic trait enhancement | [268] |
| ***Petunia* cultivar 'Madness Midnight'** | Knockout | *F3HA, F3HB* | Deletions Insertion | Establishment of genome editing for agronomic trait enhancement (Flower color modification) | [269] |
| ***Camelina sativa*** | Knockout | *FAD2* | Deletions Insertion | Establishment of genome editing for agronomic trait enhancement (Increasing oleic acid content) | [270] |
| **Maize** | Knockout | *IPK* | Deletions Insertion | Gene editing efficiency in maize | [271] |
| **Melons (*Cucumis melo* L.)** | Knockout | *AtGRF5, AtPLT5* | Deletions Insertion replacement | Establishment of genome editing for agronomic trait enhancement | [272] |
| **Wheat Giza-168** | Knockout | *Sal1* | Deletions Insertion | Establishment of genome editing for agronomic trait improvement (Improve levels of drought stress tolerance) | [273] |
| **Rice (*O. sativa*)** | Knockout | *FAD2* | Deletions Insertion | Establishment of genome editing for agronomic trait enhancement (Increasing oleic acid content) | [274] |
| **Oil palm (*Elaeis guineensis* Jacq.)** | Knockout | *EgPAT, EgFAD2* | Deletions Insertion | Establishment of genome editing for agronomic trait enhancement (Increasing oleic acid content) | [275] |
| **Barley (*Hordeum vulgare* cv. "Golden Promise")** | Knockout | *ENGase* | Deletions | Establishment of genome editing for agronomic trait improvement (Study the modification of N-glycans in cereal grains) | [276] |
| ***Arabidopsis thaliana* Col-0** | Knockout Expression | GLV | Deletions Insertion | Gene editing efficiency in Arabidopsis | [277] |
| **Petunia (*Petunia hybrid*)** | Knockout | *PDS* | Deletions | Establishment of genome editing for agronomic trait improvement | [278] |
| **Soybean** | Knock-out | *GmCPR5* | Deletions Insertion | Establishment of genome editing for agronomic trait enhancement | [279] |

*(Table 1) cont.....*

| Plant | Gen Function | Targeted Genes | Gene Reg. | Results | Refs. |
|---|---|---|---|---|---|
| *Arabidopsis* | Knock-out | *EC1.2* | Deletions Insertion | Gene editing efficiency in Arabidopsis | [280] |
| melon (*Cucumis melo*) | knockout | *CmPDS* | Deletions Insertion Replacement | Establishment of genome editing for agronomic trait improvement | [281] |
| grapevine (*Vitis vinifera*) | Knock-out, Expression | *VvPDS* | Insertion | To promote the development of functional genomics and breeding improvement in grape | [282] |
| *Nicotiana tabacum* cv. Petit Havana | Knock-out | *NtTOM1* | Deletions Insertion | Improve plant resistance to TMV virus | [283] |
| *Petunia hybrida* cv. 'Mirage Rose' | Knock-out | *ACO* | Deletions Insertion | Study gene function in plant (role of gen in seed germination) | [284] |
| Rice | Knock-out | *PDS* | Deletions Insertion | Establishment of genome editing for agronomic trait improvement | [285] |
| Pea (*Pisum sativum* L.) | Knock-out | *PsPDS* | Deletions | successful generation of gene-edited pea plants using the CRISPR/Cas9 system | [286] |
| Chickpea (*Cicer arietinum* L.) | Knock-out | *CaPDS* | Deletions Insertion | Establishment of genome editing for agronomic trait enhancement | [287] |
| *Indica* rice | Knock-out | *DEP1* | Deletions | Increasing grain number per panicle, and increasing grain yields | [288] |
| *Brassica oleracea* and *B. rapa* | Knock-out | *FRI, PDS* | Deletions Insertion | Establishment of genome editing for agronomic trait improvement | [289] |
| Tomato (*Solanum lycopersicum*) | Knock-out | *SlPDS, SlPIF4* | Deletions Insertion | Establishment of genome editing for agronomic trait improvement | [290] |
| Onion (*Allium cepa* L.) | Knock-out | *AcPDS* | Deletions Insertion Replacement | Establishment of genome editing for agronomic trait improvement | [291] |
| *Petunia hybrida* cv. Mirage Rose | Knock-out | *PhACO* | Deletions Insertion | Improvement of floricultural quality (enhanced flower longevity) | [292] |
| Tomato | Knock-out | *NPR1* | Deletions Insertion Replacement | Evaluate drought tolerance | [293] |

(Table 1) cont.....

| Plant | Gen Function | Targeted Genes | Gene Reg. | Results | Refs. |
|---|---|---|---|---|---|
| Melon (*Cucumis melo* L.) | Knock-out | *CmeIF4E* | Deletions | Induce resistance to virus | [294] |
| Soybean (*Glycine max* L. Merr.) | Knock-out | *PDS* | Deletions Insertion | Establishment of genome editing for agronomic trait improvement | [295] |
| Arabidopsis | Knock-out | *ABP1* | Deletions | Gene editing efficiency in Arabidopsis | [296] |
| Maize (*Zea mays*) | Expression | *Argonaute18* | Deletions Insertion | Gene editing efficiency in maize, response to abiotic stress | [297] |
| Pepper (*Capsicum annuum*) | Knock-out | *CaMLO2* | Deletions | Induce resistance against disease pathogen (powdery mildew) | [298] |
| Wheat (*Triticum aestivum* L.) | Knock-out | *PDS* | Deletions Insertion Replacement | Establishment of genome editing to improve the quality of crop | [299] |
| Rice (*Oryza sativa* L. ssp. *japonica*) | Knock-in | AOX1 OsBEL | Deletions | Induce resistance to biotic and abiotic stress | [300] |
| Rice | Knock-in | *ADHE* | Deletions Insertion | Establishment of genome editing for agronomic trait enhancement | [49] |

## Knock-in

CRISPR/Cas9 allows for the precise insertion of desired genetic material into specific genomic locations, known as knock-in. By co-delivering a DNA donor template with the gRNA and Cas protein, targeted DNA sequences (HDR) can be inserted into the genome by repairing the induced DSB. This function can introduce specific mutations, insert reporter genes, or introduce genes into specific genomic locations and regulate gene expression (overexpression) [56 - 58]. One of the key benefits of this function is the ability to improve and breed the desired agronomic traits of crops. This function can significantly increase crop yields, quality, and resilience to environmental stressors (Table **1**).

## Gene expression

The CRISPR/Cas9 system can also be utilized to modulate gene expression levels. By designing gRNAs that target specific regulatory regions of a gene, such as the promoter or enhancer regions, the Cas protein can be directed to these regions and induce modifications that alter the expression of the targeted gene. These modifications can include epigenetic changes, such as DNA methylation or histone modifications, or interference with transcription factor binding [59 - 62]. This feature provides a high level of accuracy in managing gene expression, which helps examine gene regulation and can significantly enhance the effectiveness of plant breeding initiatives (Table **1**).

## Accelerate Breeding Programs

In recent times, environmental issues like global warming, pollution, and other stressors have adversely affected plant growth and crop yield. To study the impact of biotic and abiotic stresses such as drought, salinity, and diseases on plant physiology, researchers have been focusing on these areas [63 - 66]. Additionally, innovative breeding techniques in crop production can enhance resilience against these stresses [67 - 69].

Traditionally, farmers have relied on basic methods to cultivate plants and propagate desirable traits. However, modern breeding techniques like molecular markers and genetic modification have revolutionized crop improvement [70 - 72]. CRISPR/Cas9 is a faster way to create desired plant traits by editing the genome directly. This new technique enhances resistance to diseases, pests, or environmental stressors and has opened up new possibilities for crop improvement [73, 74].

### *Biotic Stresses*

Biotic stresses such as bacterial, viral, and fungal diseases and pests cause significant losses during the production cycle of crops, posing a critical challenge worldwide. Combatting this challenge requires the development of disease-resistant plants that can effectively mitigate the impact of these stresses on agricultural yield. The use of CRISPR/Cas9 technology has shown remarkable success in creating resistant plants through several studies that have demonstrated the technology's efficacy in producing crops that are more resilient and sustainable.

In recent years, global warming has caused insect pest outbreaks, which have significantly reduced agricultural productivity. This negative impact is due to the invasion of fields and the spread of diseases, leading to changes in the quantity and quality of crops [75]. Plants have developed a range of morphological, biochemical, and molecular defense mechanisms to protect themselves from pests [76]. The induction of resistance in plants can be an effective approach to managing pests, which can significantly reduce the need for insecticides. The development of insect-resistant crops requires a comprehensive approach due to the complex relationships between pests and their hosts, as well as the genetic similarity of genes involved in regulating resistance to various stress factors [77,

78]. Recent advances in genome manipulation techniques, such as CRISPR/Cas9 genome editing, have provided a revolutionary means to alter plant genomes and achieve desired traits significantly [79 - 81]. The Asian corn borer is a highly destructive insect pest in Asia, causing significant damage to corn farms. To address concerns over excessive pesticide use and environmental impact, researchers have successfully implemented the Sterile Insect Technique (SIT) through the use of CRISPR/Cas9 as an alternative strategy [82]. *Spodoptera litura* is a highly damaging pest that can devastate several important crops, including tea, tobacco, and cabbage [83, 84]. In a recent study, the gene-editing tool CRISPR/Cas9 was successfully utilized to knock-out a specific gene responsible for male fertility in *S. litura* [85]. This breakthrough could have significant implications for controlling the reproductive capabilities of this pest species.

Fungi are one of the most significant factors that cause diseases in plants, leading to various physiological disorders and infections such as spots on leaves, flowers, and fruits, blights, and the formation of galls on shoots, roots, and leaves [86 - 88]. Fungal pathogens can infect plant tissues, which can have adverse effects on plant homeostasis and physiology, and in some cases, this can even lead to systemic damage [89]. Powdery mildew is the most prevalent fungal disease that damages several important crops every year [90]. To address the issues associated with synthetic chemicals, many researchers are exploring biological methods of controlling disease pathogens. Using genetic modification through genome editing is an effective and sustainable approach to managing plant pathogens. This method provides breeders with an expanded set of genetic tools for their use and allows for precise genetic alterations of single or multiple gene targets in plants. The potential of CRISPR/Cas9 for improving resistance to fungal diseases in plant varieties is highlighted. Apart from its applications in plants, CRISPR/Cas9 has also been used to target genes that encode proteins involved in interactions between host and fungal pathogens [19, 91 - 95]. This has supported an understanding of the molecular mechanism underlying host-pathogen recognition and in generating screening systems for disease resistance. Using CRISPR/Cas9, [96] authors conducted a study on tomatoes and successfully created a variety that is immune to powdery mildew fungal pathogens by deleting targeted genes. A recent study on cucumber plants demonstrated the efficacy of CRISPR/Cas9 technology in enhancing their resistance against powdery mildew. The study involved the generation of three different types of mutants using the CRISPR/Cas9 system, and the results showed a significant improvement in the plant's resistance against fungal disease [97].

Diseases caused by bacterial factors are classified into different categories based on the signs and physiological damage they can cause in crops, such as bacterial leaf blight on wheat and citrus canker [98 - 100]. Managing bacterial infections in plants is generally difficult because they are hard to detect, have concealed appearances, and there are no effective chemical pesticides available [101, 102]. Plants have developed various distinctive protection mechanisms to prevent diseases induced by different microbial pathogens. These mechanisms include the production of antimicrobial compounds to prevent pathogen invasion, the strengthening of cell walls to deter pathogen infiltration, and programmed cell death to limit the transmission of infection. Chemical pesticides are the primary method used to manage crop diseases. However, they are harmful to both humans and the environment. A sustainable and environmentally friendly approach to agriculture is developing genetically modified crops that are resistant to diseases. Genome editing breeding programs have become an effective way to prevent significant losses in crop yield [103, 104]. Using CRISPR/Cas9 to modify the genome of plants has become increasingly popular in recent years, enhancing their resistance to bacterial infections. CRISPR/Cas9 systems, originally developed in bacteria, have the potential to create bacteriophage-mediated delivery of antibacterial agents to penetrate the cell wall. The mechanism for the CRISPR systems against bacterial diseases includes reducing virulence aspects of bacteria, targeting genes liable for antibiotic resistance, and targeting genes of bacterial suppressors [105 - 107]. Pseudomonas syringae is a highly detrimental bacterial pathogen that causes disease in various important plant species [108]. For instance, it is responsible for bacterial speck disease in tomato plants. Recent research has shown promising results in using CRISPR/Cas9 technology to develop resistance against this pathogen in tomato plants [109]. Bacterial leaf blight (BLB) is a common bacterial disease in rice crops that can significantly impact crop yield and quality. However, with the advent of CRISPR/Cas9 technology, it is now possible to create resistance to BLB by modifying the rice plant's genetic material [110]. Yang *et al.* [111] conducted a study on rice (*Oryza sativa* L.) and found that the CRISPR/cas9 system produced mutants that exhibited a significantly enhanced resistance against rice blast and bacterial leaf, which are the main reasons for the reduction of productivity and quality of rice.

Numerous virus diseases pose a significant threat to several key crops worldwide, such as necrotic spot virus on pepper leaves, mosaic virus in roses, and barley yellow dwarf in wheat [112 - 115]. Viral diseases are immobile and can spread between plants through various carriers or vectors, including thrips and whiteflies. Furthermore, they can also be transmitted through fungi, infected plant organs, pollen, and seeds [116 - 118]. Therefore, it is crucial to study their control in plants to prevent any potential damage to the crops. Plant viruses need to hijack the functions of various host factors to complete their life cycle [119]. This requires the interaction and interference of viral components with host components, leading to physiological changes in the plant that result in the development of symptoms. Recent studies have shown that plant-virus interactions can affect a wide range of processes, such as hormonal regulation, cell cycle control, and macromolecule transport within the plant [120, 121]. There are multiple methods by which plants can impede viral replication and movement. These methods include gene silencing, activation of immune receptor signaling, protein degradation, metabolic adjustments, and defense *via* hormones [122 - 124]. Plant viruses cause a serious hazard to agriculture, leading to significant yield losses. In response, breeding programs are underway to create virus-resistant plant varieties. This is being achieved through genome editing, which enhances the genetic defenses of plants, and the creation of transgenic varieties with high resistance to viral diseases [125 - 127]. These approaches hold promise for effectively combatting plant viruses and securing food production. Recent research in plant genetics and breeding has focused on developing modified crops that exhibit excellent resistance to viral diseases. CRISPR/Cas9 plays a crucial role in the breeding and gene alteration of plants to improve their performance against viral DNA or RNA [128 - 130]. The CRISPR/Cas9 system is a powerful tool that can be employed to target specific viruses and host plant genomes. This is achieved through the precise interruption of viral cleavage or changing of the host genome [131 - 133]. As a result of these targeted modifications, the replication ability of the virus is significantly reduced [134, 135]. For instance, Kumar *et al.* [136] employed the CRISPR/Cas9 technique to create genetic mutations in tomatoes to develop resistance against potyvirus. The expressed genes of BCTIV in sugar beet leaves were targeted using gRNA/Cas9 constructs in a study, and their transient expression was evaluated for efficiency by Yıldırım *et al.* [137]. The CRISPR/Cas9 system utilized in this study demonstrated complete viral resistance, preventing systemic contamination. Similarly, Yin *et al.* [138] employed Cas9 to target two crucial regions of the single-stranded DNA genome of the cotton leaf curl Multan virus (CLCuMuV) in the Nicotiana benthamiana plant. This resulted in the cleavage of the corresponding DNA sequence, which, in turn, led to the interruption of viral infection. Transgenic plants that expressed Cas9 and dual gRNAs targeting different regions of the CLCuMuV genome were able to provide complete resistance to virus infection.

## *Abiotic Stresses*

Abiotic stresses such as drought, salinity, toxic substances, and drastic temperatures are major causes of reduced agricultural production [139]. In light of global warming and environmental pollution, abiotic stresses pose a serious challenge to food security. To address this challenge, it is necessary to develop crops with better performance, increased resistance, and adaptability to various environmental conditions [140]. Genetic engineering techniques have been employed to create transgenic crops that can adapt to changing environmental conditions. Scientists have been successful in the last two decades in developing transgenic plants by employing various transformation techniques to modify the expression of specific genes associated with desirable traits [141 - 144]. Biotechnology and molecular science offer promising solutions for generating more vigorous plants. CRISPR/Cas9, in particular, can play a key role in manipulating the genes that regulate compatibility in the face of abiotic stresses [145 - 147]. Several reports have highlighted the potential of this technology to produce different varieties of crops with improved traits.

Drought stress is a major concern for plants, as it can lead to significant changes in their physical, biochemical, and genetic structure, ultimately leading to decreased productivity [148]. Plants utilize a complex system of mechanisms to combat the adverse impacts of drought. Various stress response mechanisms trigger a series of complex molecular, physiological, and metabolic alterations that promote stress tolerance. These mechanisms involve modifications in the root system, adjustments in ionic levels, modulation of nutritional elements, and regulation of gene expression [81, 149 - 151]. The interplay of these mechanisms enables plants to manage this process effectively. Several studies have been conducted to explore the functional genes involved in stress response, with a focus on understanding the underlying mechanism of plant responses to water deficiency [152 - 156]. Genome editing methods, particularly the CRISPR/Cas9 system, have shown great promise in creating plant varieties that are better equipped to handle drought stress [157 - 160]. Addressing the issue, it is essential to develop new plant varieties that are well-suited to this stress. The utilization of the CRISPR/Cas9 genome editing technique in Arabidopsis was observed to cause a significant increase in the number of leaves as well as their leaf areas. This ultimately led to an improvement in the plant's tolerance towards drought conditions [161]. According to a study on rice, the result of applying CRISPR/Cas9 showed an improved reaction to ABA and enhanced drought stress through stomatal regulation [162].

Salinity is a significant abiotic stress factor that can trigger a series of alterations in plants. The high concentration of salt can significantly affect all aspects of plant growth and crop yield, including osmotic processes, nutrition, propagation, ion behavior, and oxidative stress [163 - 165]. Salinity can occur naturally or as a result of human intervention through soil management, fertilizer application, and irrigation. Adaptation or tolerance of plants to salinity tension depends on various aspects such as physiological characteristics, metabolic pathways, and molecular factors [166 - 170]. To develop plant species that are capable of thriving in salt-affected regions, it is essential to have a thorough understanding of how plants respond to salinity stress at different levels. It requires an integrated approach that combines molecular tools with physiological and biochemical techniques. Recent studies have shown that there are various adaptive responses to salinity stress at the molecular, metabolic, and physiological levels [171 - 174]. Recent studies indicate that breeding by gene-editing strategy is a promising system to enhance the ability of plants to withstand salinity-induced stress [175 - 178]. The application of CRISPR/Cas9 technology is crucial for creating plant breeds with regulated genes, which can enhance their resistance to salinity [26, 175, 176]. In research on rice, Targeted CRISPR/Cas9-mediated editing was employed to create a superior variety of rice that displayed remarkable tolerance to salinity [179]. According to Wang *et al.* [180], the utilization of CRISPR/Cas9 technology demonstrated a significant improvement in the plant's salinity tolerance in soybean.

Heavy metals are non-essential elements that can induce phytotoxicity by interfering with various physiological processes in plants. Cadmium (Cd) and lead (Pb) are two such heavy metals that are known to cause severe damage to plants due to their ability to accumulate in different plant tissues and disrupt various metabolic pathways. On the other hand, micronutrients like zinc (Zn) and copper (Cu) are beneficial to plant growth in small amounts [181]. Accumulation of heavy metals in plant organs can hinder normal growth processes and development, resulting in reduced productivity and food security concerns [182]. In response to heavy metal stress, plants have evolved two primary strategies: avoidance and tolerance. Avoidance involves the absorption and carrying of heavy metals to different plant organs, while tolerance relies on developing resistance against stress [183]. These strategies are crucial for the survival and growth of plants in environments contaminated with heavy metals. Advancements made recently suggest that genetic modification can be an effective approach to improve the stress tolerance of plants towards heavy metals. The use of CRISPR/Cas9 technology is vital in developing plant lines with regulated genes that can control the accumulation of toxic elements and ensure healthier plants with higher yields [184]. The utilization of crisper/Cas9 based on a study in rice was found to be effective in improving resistance against Cd stress [185]. This strategy has demonstrated promising results and holds potential for further research and application in the field of agricultural biotechnology.

The impact of temperature changes on agricultural productivity is significant. To ensure global food security, policymakers, researchers, and farmers must be conscious of the effects of climate change, especially high temperatures. Optimal temperature plays a critical role in influencing the physiological and biochemical processes of plants, ultimately determining their growth and development potential [186, 187]. Therefore, it is imperative to maintain the ideal temperature conditions to promote the maximum growth potential of plants. However, extreme temperatures significantly affect the physiological processes of plants, including respiration, transpiration, photosynthesis, and productivity [188 - 190]. One solution to the crisis of extreme temperatures in plant production is to develop temperature-tolerant cultivars. Gene-editing techniques, particularly the CRISPR/Cas system, have made it possible to identify temperature tolerance factors that can protect crops from the increasingly unpredictable challenges of climate change [26, 191 - 193]. A recent study investigated the expression of the SlMAPK3 gene in tomato plants after being subjected to various high-temperature treatments. It was found that CRISPR/Cas9-mediated mutants exhibited significantly higher tolerance to heat stress than wild-type plants. In another study, Yu *et al.* [27] identified OsVPE2 as a gene in rice that responds to cold temperatures. The knock-out function of the CRISPR/Cas9-generated mutants resulted in a significant enhancement of seedlings' ability to tolerate chilling without any detrimental effect on crop yield [25]. These findings suggest that genetic modifications could be employed to enhance the temperature tolerance of crops, which is especially relevant in the face of climate change and its associated temperature fluctuations.

## Enhanced Crop Yield and Quality

As the world's population continues to grow, food security faces several challenges, including crop production and environmental changes. When it comes to crop production, quality and yield are critical factors for ensuring food security. This issue can be tackled from various angles, such as methodology and obstacles [194, 195]. Breeding strategies should, therefore, focus on removing obstacles and assessing risks to increase production rates with high quality at a lower cost for farmers or a better definition of sustainability in production and a response to consumer demand. By editing genes responsible for yield, nutrient content, or stress tolerance, scientists can develop crops that are more productive, nutritious, and resilient in adverse conditions, thus addressing global food security challenges.

Genome editing technologies, such as CRISPR/Cas9, provide ideal genetic improvement methods. Through these strategies, breeders can modify plants with desired features in a short amount of time and at a lower cost compared to other breeding methods [196, 197]. In recent studies, CRISPR/Cas9 has shown a brilliant capacity to improve and produce varieties with high quality, involving improved nutrients, textures, and secondary metabolites in plants [197, 199].

The productivity of agricultural products is determined by various characteristics, such as plant height, seed weight or number, spike length, chlorophyll content, and photosynthetic rate [200 - 204]. Enhancing crop yield is a major goal of breeding programs. CRISPR/Cas9 technology has shown hopeful results in improving physiological traits in crops and boosting agriculture productivity, as demonstrated by several successful studies [205 - 207]. As a case in point, over the past few years, there has been a considerable demand for high-quality cereals as a staple food in many countries. Consequently, several researchers have employed the CRISPR/Cas9 genome editing strategy to investigate specific genes for breeding. This method has shown promising results in enhancing the performance and quality of cereals through targeted gene modifications. A study was conducted on the application of the CRISPR/Cas9 technology on rice to augment the grain size. The study successfully produced GS3 mutants, which demonstrated an increase in grain size without any changes in other agronomic characteristics [208]. A research project focused on enhancing the yield of bread wheat by manipulating the AP2/ERF gene, which controls the structure of spike inflorescences in bread wheat (*Triticum aestivum* L.). This was done through the use of the CRISPR/Cas9 system to create mutations that resulted in an increase in the number of spikelets, a rise in the number of grains per spike, and, most importantly, a boost in yield when tested under farm conditions [209].

Scientists have identified and extensively studied multiple genes that control different traits in crops, including their size, color, shape, protein content, starch content, carotenoid levels, and oil content [210 - 218]. Recent studies have explored the potential of using CRISPR/Cas9 to enhance the quality of different species.

Fruit color is an essential horticultural trait that greatly affects consumer purchase intentions. Yang *et al.* [219] used CRISPR/Cas9 in a study to develop a breeding method for producing tomato lines with varying fruit colors from red-fruited materials. This approach resulted in a range of tomato lines with distinct fruit colors. Manipulating the pigmentation of ornamental plants is a critical aspect of the horticultural industry, and breeders continuously strive to develop novel cultivars displaying a wide of colors. Breeders conducted a study on Poinsettia (*Euphorbia pulcherrima*) to alter the color of the red bracts. They applied the CRISPR/Cas9 technique to knock-out a specific flavonoid. The successful application of this method resulted in the production of poinsettias with orange bracts color [220].

Developing crops with high amylose and resistant starch (RS) content is crucial for producing healthier foods and preventing diabetes [221]. Modern breeding technologies, especially genome editing by the CRISPR/Cas9 system, can be utilized to achieve this goal. In a recent study, researchers utilized CRISPR/Cas9 to generate high-amylose wheat in two cultivars of winter wheat and spring wheat by targeted mutagenesis of TaSBEIIa. This successful approach provides a promising route for developing crops with improved nutritional value [222]. Likewise, Wang *et al.* [223] employed the CRISPR/Cas9 gene editing technique to target the OsSBEIIb gene in japonica rice. This resulted in the development of mutant rice lines that exhibit high levels of amylose and resistant starch.

The use of secondary metabolite derived from plants is gaining momentum across several industries, including pharmaceuticals, cosmetics, and food. Breeding strategies for plants must now consider the presence of these volatile fragrance compounds to fully utilize their benefits [224 - 226]. This approach could lead to a significant increase in both the quantity and quality of essential compounds. This is particularly important for medicinal plants, as it can aid in genetic conservation efforts and ultimately prevent extinction [227, 228]. To achieve this, the CRISPR/Cas9 system provides an advantageous tool for improving this trait in plants. For instance, breeders have applied CRISPR/Cas9 to produce aromatic maize plants by simultaneously knocking out the two BADH2 genes [229]. In another study, researchers used CRISPR/Cas9-mediated editing of genes to create aromatic two-line hybrid rice, where fragrance is a key factor in attracting consumers [230]. Additionally, in a study focused on improving secondary metabolite production, knocking out one allele of VvbZIP36 in grapevine with CRISPR/Cas9 promoted anthocyanin accumulation. According to research carried out by Tu *et al.* [231], the use of the CRISPR/Cas9 system to knock-out FAD2 and PAT genes resulted in an increased accumulation of oleic acid content in oil palms. These findings highlight the potential of using the CRISPR/Cas9 system to improve the production of essential oils and secondary metabolites in plants, which could have significant implications across various industries.

The loss of agricultural products can occur due to different factors, such as susceptibility to diseases, pests, and environmental stress, as well as physiological processes [232, 233]. Pioneer breeding strategies, including implementing the CRISPR/Cas9 gene editing system, can help breeders edit the genome of crops, reducing pre- or post-harvest processes that could lead to crop failure. In a study aimed at reducing the activity of PPO enzymes responsible for the oxidization of polyphenols leading to the browning of harvested eggplant berry flesh, Maioli *et al.* [234] successfully employed CRISPR/Cas9 to knock-out the PPO genes. In another study, breeders utilized the CRISPR/Cas9 system to target specific genes in potato varieties in order to prevent tuber browning [235]. The results showed successful editing of the genes responsible for browning, underlining the potential of CRISPR/Cas9 technology in improving the quality and shelf-life of crops.

**Reduce Environmental Impact**

Recent advancements in plant breeding technologies and strategies for producing high-quality and high-yield crops have delivered promising results, offering a significant opportunity for achieving sustainable agriculture and global food security. However, certain critical aspects still require greater attention in breeding programs, particularly in response to concerns surrounding the environmental impact of agriculture.

The production of crops has been shown to accelerate climate change, resulting in global warming, greenhouse gas emissions, increased herbicide application, fuel consumption, and soil pollution [236 - 238]. Therefore, plant breeding strategies must employ methods to reduce or eliminate barriers to meeting the growing demand for crop production while minimizing environmental impact. The utilization of CRISPR/Cas9 for product editing aimed at mitigating environmental impacts has exhibited promising outcomes or is currently underway.

The utilization of feedstock crops and advanced biofuel conversion technologies has lately become a topic of great interest owing to its immense potential in mitigating the world's reliance on fossil fuels [239, 240]. By generating biofuels from sustainable biomass, this approach can effectively contribute to the reduction of dependency on fossil fuels and thus help in achieving the goal of a cleaner and greener energy future [240, 241]. In a recent research study conducted on switchgrass, the highly efficient CRISPR/Cas9 gene-editing system was employed to reduce the lignin content in the plant. This resulted in a significant increase in the sugar release, which in turn led to improved bioethanol production [161]. In addition, Lee *et al.* [242] demonstrated that by using CRISPR/Cas9-mediated mutagenesis, the lignin content in barley can be reduced. This results in a higher recovery rate of fermentable glucose and improved efficiency in producing lignocellulosic biofuel.

Meeting the nutritious needs of the growing population is a critical challenge for food security. Livestock production is a major contributor to food supply, and forage production with better performance and quality is a key concern for breeders [243]. However, livestock production has a significant environmental impact, particularly in the case of greenhouse gas emissions and global warming [244]. Methane gas production by livestock is a well-known issue, and nitrogen excretion onto pastureland contributes to the creation of potent greenhouse gas nitrous oxide ($N_2O$). Nitrate pollution of land also leads to a loss of biodiversity [245, 246]. To meet these challenges, increasing nutrient utilization efficiency while reducing excretion is essential. The CRISPR/Cas9 system can be effectively applied to developing grasses and other forages, leading to significant contributions to achieving these goals. For example, Wolabu *et al.* [247] employed the CRISPR/Cas9 system to edit alfalfa. This resulted in a reduction of lignin content and an increase in crude protein (CP) and mineral contents. These changes are beneficial for forage digestibility and intake potential. Similarly, May *et al.* [248] used CRISPR/Cas9 to enhance the mutation efficiency of genome editing in ryegrass.

Pesticides are commonly employed in agriculture to mitigate crop losses and enhance the yield and quality of food. The demand for pesticides has increased over time due to the need to enhance food production and control diseases. However, the excessive use of pesticides has led to environmental pollution and harmful impacts on human health [249]. One of the significant adverse impacts of pesticides is the decline in biodiversity as all organisms are exposed to them in cultivated areas [250, 251]. Additionally, pesticides have contaminated soil and water over the long term. Pesticide residues found in many food products and the air pose a critical risk to human health [252, 253]. Agricultural production is often impacted by pests, weeds, and diseases, which can lead to a significant reduction in crop yield. To prevent such losses, the use of pesticides is necessary [254]. Therefore, it is crucial for researchers to search for solutions to address the challenges posed by pests and diseases. Scientists are currently engaged in researching the genetic alteration of crops to make them resistant to pests and diseases. This novel approach to pest management has the potential to significantly decrease the use of pesticides and their destructive effects on the environment [255]. The latest CRISPR gene-editing technology has simplified the process of altering the genetic makeup of any organism, thus making this solution a practical reality. In recent research, Williams *et al.* [256] successfully generated mutation in pyrethroid resistance by inducing a CRISPR/Cas9 genome modification. In another study, the CRISPR/Cas9 system was utilized to generate mutants for knock-down insecticide-resistance in Drosophila melanogaster [257].

## Potential Ethical and Regulatory Considerations

The use of CRISPR-Cas technology in agriculture has great potential, but it raises many ethical and regulatory concerns related to genetically modified organisms (GMOs) and their effects on both the environment and human health [258, 259]. Regulations for genome-edited crops vary from country to country, and the production of GM crops has been a controversial topic due to a lack of understanding among the public, fear-driven agricultural policies, ineffective information sharing by scientists, and inaccurate portrayals by some anti-GMO activists. Additionally, several issues need to be addressed, such as legal property rights, ethical considerations, standardization, and market viability [260 - 262]. From a technical perspective, there is a risk of non-plant DNA fragments infiltrating the genome of the parent plant, which could lead to gene escape and potentially undesirable transfer of the target gene [263, 264]. Moreover, specific genes, such as those responsible for antibiotic and herbicide resistance, raise further concerns [265, 266].

Despite the advantageous prospect of emerging technologies, various ethical considerations and challenges must be addressed to ensure their responsible and safe development. Ensuring the responsible use of CRISPR/Cas9 in plant physiology is crucial to mitigate unintended consequences and potential risks to biodiversity [267]. Regulatory frameworks and public debate should guide the deployment of this technology to ensure its benefits are distributed equitably and the environment is protected.

## FUTURE PROSPECTS

CRISPR/Cas9 technology has tremendous prospects for the future. Gene editing can revolutionize various fields, such as medicine, agriculture, and basic scientific research. CRISPR/Cas9 is a highly profitable tool for treating genetic diseases in the medical field. By making precise modifications to DNA sequences, scientists can correct mutations that cause disorders. This technology also offers new opportunities for pharmaceutical research.

The potential of CRISPR/Cas9 in agriculture is promising for the future. Continuous advancements in CRISPR-Cas technology aim to improve its efficiency, expand its applications, and address challenges such as off-target effects and delivery methods for larger-scale applications in agriculture. This technology can help solve various global challenges, such as population growth, climate change, and the decline of arable land. It allows for modifications to be made to crops, increasing their resilience to diseases, pests, and unfavorable climate conditions. Additionally, CRISPR/Cas9 can address malnutrition and food scarcity issues by enhancing crops' nutritional content and taste.

CRISPR/Cas9 has immense potential in advancing fundamental scientific research. The precision and simplicity with which it can modify DNA sequences enables scientists to explore gene functions more accurately and efficiently. Researchers can delve deeper into complex biological processes by studying gene interactions and their functions, potentially unlocking solutions.

## CONCLUSION

CRISPR/Cas9 technology holds immense potential for revolutionizing plant physiology and has paved the way for significant advancements in agriculture and crop improvement. With its precise gene-editing capabilities, this powerful tool can help create plants with desirable traits and characteristics, ultimately leading to increased crop yields and improved food security. Reliable rules and ethical dialogues will be needed to maximize the advantages while reducing the risks associated with this transformative technology as scientific investigations continue.

# REFERENCES

[1]     Wolter F, Schindele P, Puchta H. Plant breeding at the speed of light: the power of CRISPR/Cas to generate directed genetic diversity at multiple sites. BMC Plant Biol 2019; 19(1): 176.
[http://dx.doi.org/10.1186/s12870-019-1775-1] [PMID: 31046670]

[2]     Das DN, Paul D, Mondal S. Role of Biotechnology on Animal Breeding and Genetic Improvement. In: Mondal S, Singh RL, Eds. Emerging Issues in Climate Smart Livestock Production: Biological Tools and Techniques. London, United Kingdom: Academic Press 2022; pp. 317-37.
[http://dx.doi.org/10.1016/B978-0-12-822265-2.00015-6]

[3]     Rasmussen SK. Molecular genetics, genomics, and biotechnology in crop plant breeding. Agronomy (Basel) 2020; 10(3): 439.
[http://dx.doi.org/10.3390/agronomy10030439]

[4]     Ahmad N, Mukhtar Z. Genetic manipulations in crops: Challenges and opportunities. Genomics 2017; 109(5-6): 494-505.
[http://dx.doi.org/10.1016/j.ygeno.2017.07.007] [PMID: 28778540]

[5]     Dormatey R, Sun C, Ali K, *et al.* *ptxD/* Phi as alternative selectable marker system for genetic transformation for bio-safety concerns: a review. PeerJ 2021; 9: e11809.
[http://dx.doi.org/10.7717/peerj.11809] [PMID: 34395075]

[6]     Khalequzzaman M, Chowdhury A, Sonnino A. Biosafety of Genetically Modified Organisms: Basic concepts, methods and issues. Rome: Food Agric. Organ. United Nations 2009; p. 301.

[7]     Chen K, Wang Y, Zhang R, Zhang H, Gao C. CRISPR/Cas Genome Editing and Precision Plant Breeding in Agriculture. Annu Rev Plant Biol 2019; 70(1): 667-97.
[http://dx.doi.org/10.1146/annurev-arplant-050718-100049] [PMID: 30835493]

[8]     Moon TT, Maliha IJ, Khan AAM, *et al.* CRISPR-Cas genome editing for insect pest stress management in crop plants. Stresses 2022; 2(4): 493-514.
[http://dx.doi.org/10.3390/stresses2040034]

[9]     Lin X, Tang B, Li Z, Shi L, Zhu H. Genome-wide identification and expression analyses of CYP450 genes in sweet potato (*Ipomoea batatas* L.). BMC Genomics 2024; 25(1): 58.
[http://dx.doi.org/10.1186/s12864-024-09965-x] [PMID: 38218763]

[10]    Wang H, Hsu YC, Wu YP, Yeh SY, Ku MSB. Production of high amylose and resistant starch rice through targeted mutagenesis of starch branching enzyme iib by Crispr/cas9. Research Square 2021.
[http://dx.doi.org/10.21203/rs.3.rs-585478/v1]

[11]    Yu J, Cui J, Huang H, *et al.* Identification of flowering genes in *Camellia perpetua* by comparative transcriptome analysis. Funct Integr Genomics 2024; 24(1): 2.
[http://dx.doi.org/10.1007/s10142-023-01267-x] [PMID: 38066213]

[12]    Zhang C, Yun P, Xia J, *et al.* CRISPR/Cas9-mediated editing of *Wx* and *BADH2* genes created glutinous and aromatic two-line hybrid rice. Mol Breed 2023; 43(4): 24.
[http://dx.doi.org/10.1007/s11032-023-01368-2] [PMID: 37313522]

[13]    Bigini V, Camerlengo F, Botticella E, Sestili F, Savatin DV. Biotechnological resources to increase disease-resistance by improving plant immunity: a sustainable approach to save cereal crop production. Plants 2021; 10(6): 1146.
[http://dx.doi.org/10.3390/plants10061146] [PMID: 34199861]

[14]    Jacott C, Murray J, Ridout C. Trade-offs in arbuscular mycorrhizal symbiosis: disease resistance, growth responses and perspectives for crop breeding. Agronomy (Basel) 2017; 7(4): 75.
[http://dx.doi.org/10.3390/agronomy7040075]

[15]    Yuan M, Xin XF. Bacterial infection and hypersensitive response assays in Arabidopsis-*Pseudomonas syringae* pathosystem. Bio Protoc 2021; 11(24): e4268.
[http://dx.doi.org/10.21769/BioProtoc.4268] [PMID: 35087927]

[16]   Ding Z, Fu L, Wang B, *et al.* Metabolic GWAS-based dissection of genetic basis underlying nutrient quality variation and domestication of cassava storage root. Genome Biol 2023; 24(1): 289.
[http://dx.doi.org/10.1186/s13059-023-03137-y] [PMID: 38098107]

[17]   Gong C, Guo Z, Hu Y, *et al.* A horizontally transferred plant fatty acid desaturase gene steers whitefly reproduction. Adv Sci 2023; 2306653: 1-14.
[http://dx.doi.org/10.1002/advs.202306653] [PMID: 38145364]

[18]   Hossain MA, Tahjib-Ul-Arif M, Jahin SA, *et al.* 2023.
[http://dx.doi.org/10.1002/9781119906506.ch26]

[19]   Kumar A. 2024.Biotechnological Innnovations.

[20]   Wang Y, Zhang X, Yan Y, *et al.* GmABCG5, an ATP-binding cassette G transporter gene, is involved in the iron deficiency response in soybean. Front Plant Sci 2024; 14: 1289801.
[http://dx.doi.org/10.3389/fpls.2023.1289801] [PMID: 38250443]

[21]   Chen SH, Martino AM, Luo Z, *et al.* A high-quality pseudo-phased genome for *Melaleuca quinquenervia* shows allelic diversity of NLR-type resistance genes. Gigascience 2022; 12.
[http://dx.doi.org/10.1093/gigascience/giad102] [PMID: 38096477]

[22]   Li ZY, Ma N, Zhang FJ, *et al.* Functions of phytochrome interacting factors (PIFs) in adapting plants to biotic and abiotic stresses. Int J Mol Sci 2024; 25(4): 2198.
[http://dx.doi.org/10.3390/ijms25042198] [PMID: 38396875]

[23]   Loubet I, Meyer L, Michel S, *et al.* A high diversity of non-target site resistance mechanisms to acetolactate-synthase (ALS) inhibiting herbicides has evolved within and among field populations of common ragweed (*Ambrosia artemisiifolia* L.). BMC Plant Biol 2023; 23(1): 510.
[http://dx.doi.org/10.1186/s12870-023-04524-0] [PMID: 37875807]

[24]   Magarini A, Passera A, Ghidoli M, Casati P, Pilu R. Genetics and environmental factors associated with resistance to *Fusarium graminearum*, the causal agent of gibberella ear rot in maize. Agronomy (Basel) 2023; 13(7): 1836.
[http://dx.doi.org/10.3390/agronomy13071836]

[25]   Deng H, Cao S, Zhang G, *et al.* OsVPE2, a member of vacuolar processing enzyme family, decreases chilling tolerance of rice. Rice (N Y) 2024; 17(1): 5.
[http://dx.doi.org/10.1186/s12284-023-00682-9] [PMID: 38194166]

[26]   Shaheen N, Ahmad S, Alghamdi SS, *et al.* CRISPR-Cas system, a possible "Savior" of rice threatened by climate change: an updated review. Rice (N Y) 2023; 16(1): 39.
[http://dx.doi.org/10.1186/s12284-023-00652-1] [PMID: 37688677]

[27]   Yu G, Ma J, Jiang P, *et al.* The mechanism of plant resistance to heavy metal. IOP Conf Ser Earth Environ Sci 2019; 310(5): 052004.
[http://dx.doi.org/10.1088/1755-1315/310/5/052004]

[28]   Costigan R, Stoakes E, Floto RA, Parkhill J, Grant AJ. Development and validation of a CRISPR interference system for gene regulation in *Campylobacter jejuni.* BMC Microbiol 2022; 22(1): 238.
[http://dx.doi.org/10.1186/s12866-022-02645-4] [PMID: 36199015]

[29]   Liu J, Wang FZ, Li C, Li Y, Li JF. Hidden prevalence of deletion-inversion bi-alleles in CRISPR-mediated deletions of tandemly arrayed genes in plants. Nat Commun 2023; 14(1): 6787.
[http://dx.doi.org/10.1038/s41467-023-42490-1] [PMID: 37880225]

[30]   Bi D, Zhu Y, Gao Y, *et al.* A newly developed PCR-based method revealed distinct *Fusobacterium nucleatum* subspecies infection patterns in colorectal cancer. Microb Biotechnol 2021; 14(5): 2176-86.
[http://dx.doi.org/10.1111/1751-7915.13900] [PMID: 34309194]

[31]   Hao Y, Zong W, Zeng D, *et al.* Shortened snRNA promoters for efficient CRISPR/Cas-based multiplex genome editing in monocot plants. Sci China Life Sci 2020; 63(6): 933-5.
[http://dx.doi.org/10.1007/s11427-019-1612-6] [PMID: 31942685]

[32]   Nakamura S, Morohoshi K, Inada E, *et al.* Recent advances in *in vivo* somatic cell gene modification in newborn pups. Int J Mol Sci 2023; 24(20): 15301.
[http://dx.doi.org/10.3390/ijms242015301] [PMID: 37894981]

[33]   Charpentier E. CRISPR -Cas9: how research on a bacterial RNA -guided mechanism opened new perspectives in biotechnology and biomedicine. EMBO Mol Med 2015; 7(4): 363-5.
[http://dx.doi.org/10.15252/emmm.201504847] [PMID: 25796552]

[34]   Zhang D, Zhang Z, Unver T, Zhang B. CRISPR/Cas: A powerful tool for gene function study and crop improvement. J Adv Res 2021; 29: 207-21.
[http://dx.doi.org/10.1016/j.jare.2020.10.003] [PMID: 33842017]

[35]   Ishino Y, Shinagawa H, Makino K, Amemura M, Nakata A. Nucleotide sequence of the iap gene, responsible for alkaline phosphatase isozyme conversion in *Escherichia coli*, and identification of the gene product. J Bacteriol 1987; 169(12): 5429-33.
[http://dx.doi.org/10.1128/jb.169.12.5429-5433.1987] [PMID: 3316184]

[36]   Mojica FJM, Díez-Villaseñor C, Soria E, Juez G. Biological significance of a family of regularly spaced repeats in the genomes of Archaea, Bacteria and mitochondria. Mol Microbiol 2000; 36(1): 244-6.
[http://dx.doi.org/10.1046/j.1365-2958.2000.01838.x] [PMID: 10760181]

[37]   Bolotin A, Quinquis B, Sorokin A, Ehrlich SD. Clustered regularly interspaced short palindrome repeats (CRISPRs) have spacers of extrachromosomal origin. Microbiology (Reading) 2005; 151(8): 2551-61.
[http://dx.doi.org/10.1099/mic.0.28048-0] [PMID: 16079334]

[38]   Groenen PMA, Bunschoten AE, Soolingen D, Errtbden JDA. Nature of DNA polymorphism in the direct repeat cluster of *Mycobacterium tuberculosis*; application for strain differentiation by a novel typing method. Mol Microbiol 1993; 10(5): 1057-65.
[http://dx.doi.org/10.1111/j.1365-2958.1993.tb00976.x] [PMID: 7934856]

[39]   Makarova KS, Wolf YI, Alkhnbashi OS, *et al.* An updated evolutionary classification of CRISPR–Cas systems. Nat Rev Microbiol 2015; 13(11): 722-36.
[http://dx.doi.org/10.1038/nrmicro3569] [PMID: 26411297]

[40]   Malina A, Katigbak A, Cencic R, *et al.* Adapting CRISPR/Cas9 for functional genomics screens. Methods Enzymol 2014; 546(C): 193-213.
[http://dx.doi.org/10.1016/B978-0-12-801185-0.00010-6] [PMID: 25398342]

[41]   Doench JG, Hartenian E, Graham DB, *et al.* Rational design of highly active sgRNAs for CRISPR-Cas9–mediated gene inactivation. Nat Biotechnol 2014; 32(12): 1262-7.
[http://dx.doi.org/10.1038/nbt.3026] [PMID: 25184501]

[42]   Xu H, Xiao T, Chen CH, *et al.* Sequence determinants of improved CRISPR sgRNA design. Genome Res 2015; 25(8): 1147-57.
[http://dx.doi.org/10.1101/gr.191452.115] [PMID: 26063738]

[43]   Hsu PD, Scott DA, Weinstein JA, *et al.* DNA targeting specificity of RNA-guided Cas9 nucleases. Nat. Biotechnol. 201; 331(9): 827–832.
[http://dx.doi.org/10.1038/nbt.2647]

[44]   Jinek M, Chylinski K, Fonfara I, Hauer M, Doudna JA, Charpentier E. A programmable dual-RNA-guided DNA endonuclease in adaptive bacterial immunity. Science 2012; 337(6096): 816-21.
[http://dx.doi.org/10.1126/science.1225829] [PMID: 22745249]

[45]   Aguirre AJ, Meyers RM, Weir BA, *et al.* Genomic copy number dictates a gene-independent cell response to CRISPR/Cas9 targeting. Cancer Discov 2016; 6(8): 914-29.
[http://dx.doi.org/10.1158/2159-8290.CD-16-0154] [PMID: 27260156]

[46]   Makarova KS, Aravind L, Grishin NV, Rogozin IB, Koonin EV. A DNA repair system specific for thermophilic Archaea and bacteria predicted by genomic context analysis. Nucleic Acids Res 2002;

30(2): 482-96.
[http://dx.doi.org/10.1093/nar/30.2.482] [PMID: 11788711]

[47]   Qi LS, Larson MH, Gilbert LA, *et al.* Repurposing CRISPR as an RNA-guided platform for sequence-specific control of gene expression. Cell 2013; 152(5): 1173-83.
[http://dx.doi.org/10.1016/j.cell.2013.02.022] [PMID: 23452860]

[48]   Heidersbach AJ, Dorighi KM, Gomez JA, Jacobi AM, Haley B. A versatile, high-efficiency platform for CRISPR-based gene activation. Nat Commun 2023; 14(1): 902.
[http://dx.doi.org/10.1038/s41467-023-36452-w] [PMID: 36804928]

[49]   Lu Y, Tian Y, Shen R, *et al.* Targeted, efficient sequence insertion and replacement in rice. Nat Biotechnol 2020; 38(12): 1402-7.
[http://dx.doi.org/10.1038/s41587-020-0581-5] [PMID: 32632302]

[50]   Ranawakage DC, Okada K, Sugio K, *et al.* Efficient CRISPR-Cas9-mediated knock-in of composite tags in zebrafish using long ssDNA as a donor. Front Cell Dev Biol 2021; 8: 598634.
[http://dx.doi.org/10.3389/fcell.2020.598634] [PMID: 33681181]

[51]   Wang SW, Gao C, Zheng YM, *et al.* Current applications and future perspective of CRISPR/Cas9 gene editing in cancer. Mol Cancer 2022; 21(1): 57.
[http://dx.doi.org/10.1186/s12943-022-01518-8] [PMID: 35189910]

[52]   Chen S, Xie W, Liu Z, *et al.* CRISPR start-loss: a novel and practical alternative for gene silencing through base-editing-induced start codon mutations. Mol Ther Nucleic Acids 2020; 21: 1062-73.
[http://dx.doi.org/10.1016/j.omtn.2020.07.037] [PMID: 32854061]

[53]   Houdebine LM. Design of Vectors for Optimizing Transgene Expression. 2014.
[http://dx.doi.org/10.1016/B978-0-12-410490-7.00017-7]

[54]   Ferreira SS, Reis RS. Using CRISPR/Cas to enhance gene expression for crop trait improvement by editing miRNA targets. J Exp Bot 2023; 74(7): 2208-12.
[http://dx.doi.org/10.1093/jxb/erad003] [PMID: 36626564]

[55]   Zaman QU, Li C, Cheng H, Hu Q. Genome editing opens a new era of genetic improvement in polyploid crops. Crop J 2019; 7(2): 141-50.
[http://dx.doi.org/10.1016/j.cj.2018.07.004]

[56]   He X, Tan C, Wang F, *et al.* Knock-in of large reporter genes in human cells *via* CRISPR/Cas9-induced homology-dependent and independent DNA repair. Nucleic Acids Res 2016; 44(9): e85-5.
[http://dx.doi.org/10.1093/nar/gkw064] [PMID: 26850641]

[57]   Lau CH, Tin C, Suh Y. CRISPR-based strategies for targeted transgene knock-in and gene correction. Fac Rev 2020; 9(20): 20.
[http://dx.doi.org/10.12703/r/9-20] [PMID: 33659952]

[58]   Morgante CV, Arraes FBM, Moreira-Pinto CE, De Melo BP. Grossi-de-sa MF. 2021.

[59]   Available from: https://www.alice.cnptia.embrapa.br/bitstream/doc/1132006/1/CRISPR-Ing-Cap-4.Modulation-of-gene.pdf

[60]   Catarino RR, Stark A. Assessing sufficiency and necessity of enhancer activities for gene expression and the mechanisms of transcription activation. Genes Dev 2018; 32(3-4): 202-23.
[http://dx.doi.org/10.1101/gad.310367.117] [PMID: 29491135]

[61]   Kang JG, Park JS, Ko JH, Kim YS. Regulation of gene expression by altered promoter methylation using a CRISPR/Cas9-mediated epigenetic editing system. Sci Rep 2019; 9(1): 11960.
[http://dx.doi.org/10.1038/s41598-019-48130-3] [PMID: 31427598]

[62]   Pulecio J, Verma N, Mejía-Ramírez E, Huangfu D, Raya A. CRISPR/Cas9-based engineering of the epigenome. Cell Stem Cell 2017; 21(4): 431-47.
[http://dx.doi.org/10.1016/j.stem.2017.09.006] [PMID: 28985525]

[63]   Uddin F, Rudin CM, Sen T. CRISPR gene therapy: applications, limitations, and implications for the

future. Front Oncol 2020; 10(August): 1387.
[http://dx.doi.org/10.3389/fonc.2020.01387] [PMID: 32850447]

[64]   Dash PK, Rai R, Rai V, Pasupalak S. Drought induced signaling in rice: delineating canonical and non-canonical pathways. Front Chem 2018; 6: 264.
[http://dx.doi.org/10.3389/fchem.2018.00264] [PMID: 30258837]

[65]   dos Santos TB, Ribas AF, de Souza SGH, Budzinski IGF, Domingues DS. Physiological responses to drought, salinity, and heat stress in plants: a review. Stresses 2022; 2(1): 113-35.
[http://dx.doi.org/10.3390/stresses2010009]

[66]   Gull A, Ahmad Lone A, Ul Islam Wani N. Biotic and abiotic stresses in plants. In: de Oliveira AB, Ed. Abiotic and Biotic Stress in Plants. London, United Kingdom: IntechOpen 2019; pp. 1-19.
[http://dx.doi.org/10.5772/intechopen.85832]

[67]   Pandey P, Irulappan V, Bagavathiannan MV, Senthil-Kumar M. Impact of combined abiotic and biotic stresses on plant growth and avenues for crop improvement by exploiting physio-morphological traits. Front Plant Sci 2017; 8: 537.
[http://dx.doi.org/10.3389/fpls.2017.00537] [PMID: 28458674]

[68]   Benitez-Alfonso Y, Soanes BK, Zimba S, *et al.* Enhancing climate change resilience in agricultural crops. Curr Biol 2023; 33(23): R1246-61.
[http://dx.doi.org/10.1016/j.cub.2023.10.028] [PMID: 38052178]

[69]   Davies WJ, Ribaut JM. Stress resilience in crop plants: strategic thinking to address local food production problems. Food Energy Secur 2017; 6(1): 12-8.
[http://dx.doi.org/10.1002/fes3.105]

[70]   Razzaq A, Kaur P, Akhter N, Wani SH, Saleem F. Next-generation breeding strategies for climate-ready crops. Front Plant Sci 2021; 12: 620420.
[http://dx.doi.org/10.3389/fpls.2021.620420] [PMID: 34367194]

[71]   Ahmar S, Gill RA, Jung KH, *et al.* Conventional and molecular techniques from simple breeding to speed breeding in crop plants: recent advances and future outlook. Int J Mol Sci 2020; 21(7): 2590.
[http://dx.doi.org/10.3390/ijms21072590] [PMID: 32276445]

[72]   Moose SP, Mumm RH. Molecular plant breeding as the foundation for 21st century crop improvement. Plant Physiol 2008; 147(3): 969-77.
[http://dx.doi.org/10.1104/pp.108.118232] [PMID: 18612074]

[73]   Naqvi RZ, Siddiqui HA, Mahmood MA, *et al.* Smart breeding approaches in post-genomics era for developing climate-resilient food crops. Front Plant Sci 2021; 3: 1-19.
[http://dx.doi.org/10.3389/fpls.2022.972164] [PMID: 36186056]

[74]   Jaganathan D, Ramasamy K, Sellamuthu G, Jayabalan S, Venkataraman G. CRISPR for crop improvement: an update review. Front Plant Sci 2018; 9: 985.
[http://dx.doi.org/10.3389/fpls.2018.00985] [PMID: 30065734]

[75]   Tang Q, Wang X, Jin X, Peng J, Zhang H, Wang Y. CRISPR/Cas technology revolutionizes crop breeding. Plants 2023; 12(17): 3119.
[http://dx.doi.org/10.3390/plants12173119] [PMID: 37687368]

[76]   Skendžić S, Zovko M, Živković IP, Lešić V, Lemić D. The impact of climate change on agricultural insect pests. 2021.
[http://dx.doi.org/10.3390/insects12050440]

[77]   Belete T. Defense mechanisms of plants to insect pests: from morphological to biochemical approach. Trends in Technical & Scientific Research 2018; 2(2): 30-8.
[http://dx.doi.org/10.19080/TTSR.2018.02.555584]

[78]   Enders L, Begcy K. Unconventional routes to developing insect-resistant crops. Mol Plant 2021; 14(9): 1439-53.
[http://dx.doi.org/10.1016/j.molp.2021.06.029] [PMID: 34217871]

[79] Kumari P, Jasrotia P, Kumar D, *et al.* Biotechnological approaches for host plant resistance to insect pests. Front Genet 2022; 13: 914029.
[http://dx.doi.org/10.3389/fgene.2022.914029] [PMID: 35719377]

[80] Kumar R, Pandey MK. Advanced Technologies. In: Ghosh P, Ed. Insect Pest Management. 2023.

[81] Zhang C, Zhang Z, Yang JB, Meng WQ, Zeng LL, Sun L. Genome-wide identification and relative expression analysis of DGATs gene family in sunflower. Acta Agron Sin China 2023; 49(1): 73-85.
[http://dx.doi.org/10.3724/SP.J.1006.2023.14217]

[82] Singh S, Rahangdale S, Pandita S, *et al.* CRISPR/Cas9 for insect pests management: a comprehensive review of advances and applications. Agriculture 2022; 12(11): 1896.
[http://dx.doi.org/10.3390/agriculture12111896]

[83] Natikar PK, Balikai RA. Present status on bio-ecology and management of tobacco caterpillar, *Spodoptera litura* (Fabricius) – An update. Int J Plant Prot 2017; 10(1): 193-202.
[http://dx.doi.org/10.15740/HAS/IJPP/10.1/193-202]

[84] Sahu B, Pachori R, Navya RN, Patidar S. Extent of damage by *Spodoptera litura* on cabbage. J Entomol Zool Stud 2020; 8: 1153-6.

[85] Bi H, Xu X, Li X, Wang Y, Zhou S, Huang Y. CRISPR/Cas9-mediated Serine protease 2 disruption induces male sterility in *Spodoptera litura.* Front Physiol 2022; 13: 931824.
[http://dx.doi.org/10.3389/fphys.2022.931824] [PMID: 35991171]

[86] Fang X, Li X, Zhang Q, *et al.* Physiological and endophytic fungi changes in grafting seedlings of qi-nan clones (*Aquilaria sinensis*). Forests 2024; 15(1): 106.
[http://dx.doi.org/10.3390/f15010106]

[87] Varma A, Prasad R, Tuteja N. Mycorrhiza - Eco-Physiology, Secondary Metabolites, Nanomaterials. 4th ed. Springer 2017; pp. 1-334.
[http://dx.doi.org/10.1007/978-3-319-57849-1]

[88] Vergara C, Araujo KEC, Alves LS, *et al.* Contribution of dark septate fungi to the nutrient uptake and growth of rice plants. Braz J Microbiol 2018; 49(1): 67-78.
[http://dx.doi.org/10.1016/j.bjm.2017.04.010] [PMID: 28888828]

[89] Zeilinger S, Gupta VK, Dahms TES, *et al.* Friends or foes? Emerging insights from fungal interactions with plants. FEMS Microbiol Rev 2016; 40(2): 182-207.
[http://dx.doi.org/10.1093/femsre/fuv045] [PMID: 26591004]

[90] Vielba-Fernández A, Polonio Á, Ruiz-Jiménez L, de Vicente A, Pérez-García A, Fernández-Ortuño D. Fungicide resistance in powdery mildew fungi. Microorganisms 2020; 8(9): 1431.
[http://dx.doi.org/10.3390/microorganisms8091431] [PMID: 32957583]

[91] Makarova SS, Khromov AV, Spechenkova NA, Taliansky ME, Kalinina NO. Application of the CRISPR/Cas system for generation of pathogen-resistant plants. Biochemistry (Mosc) 2018; 83(12-13): 1552-62.
[http://dx.doi.org/10.1134/S0006297918120131] [PMID: 30878030]

[92] Park J, Lee HH, Moon H, *et al.* A combined transcriptomic and physiological approach to understanding the adaptive mechanisms to cope with oxidative stress in *Fusarium graminearum.* Microbiol Spectr 2023; 11(5): e01485-23.
[http://dx.doi.org/10.1128/spectrum.01485-23] [PMID: 37671872]

[93] Tyagi S, Kumar R, Kumar V, Won SY, Shukla P. Engineering disease resistant plants through CRISPR-Cas9 technology. GM Crops Food 2021; 12(1): 125-44.
[http://dx.doi.org/10.1080/21645698.2020.1831729] [PMID: 33079628]

[94] Zhang J, Zhang X, Liu X, Pai Q, Wang Y, Wu X. Molecular network for regulation of seed size in plants. Int J Mol Sci 2023; 24(13): 10666.
[http://dx.doi.org/10.3390/ijms241310666] [PMID: 37445843]

[95]    Nekrasov V, Wang C, Win J, Lanz C, Weigel D, Kamoun S. Rapid generation of a transgene-free powdery mildew resistant tomato by genome deletion. Sci Rep 2017; 7(1): 482.
[http://dx.doi.org/10.1038/s41598-017-00578-x] [PMID: 28352080]

[96]    Tek MI, Calis O, Fidan H, Shah MD, Celik S, Wani SH. CRISPR/Cas9 based mlo-mediated resistance against *Podosphaera xanthii* in cucumber (*Cucumis sativus* L.). Front Plant Sci 2022; 13: 1081506.
[http://dx.doi.org/10.3389/fpls.2022.1081506] [PMID: 36600929]

[97]    Dilarri G, de Lencastre Novaes LC, Jakob F, Schwaneberg U, Ferreira H. Bifunctional peptides as alternatives to copper-based formulations to control citrus canker. Appl Microbiol Biotechnol 2024; 108(1): 196.
[http://dx.doi.org/10.1007/s00253-023-12908-3] [PMID: 38153551]

[98]    Kazempour MN, Kheyrgoo M, Pedramfar H, Rahimian H. Isolation and identification of bacterial glum blotch and leaf blight on wheat (*Triticum aestivum* L.) in Iran. Afr J Biotechnol 2010; 9(20): 2860-5.

[99]    Prasad D, Singh RN. Major diseases of field and horticultural crops in Northern Bihar region of India. Int J Plant Sci 2022; 17(2): 180-90.
[http://dx.doi.org/10.15740/HAS/IJPS/17.2/180-190]

[100]   Ali S, Hameed A, Muhae-Ud-Din G, *et al.* Citrus canker: a persistent threat to the worldwide citrus industry—an analysis. Agronomy (Basel) 2023; 13(4): 1112.
[http://dx.doi.org/10.3390/agronomy13041112]

[101]   Brankova L, Shopova E, Ivanov S, *et al.* Involvement of oxidative stress in localization of bacterial spot infection in pepper plants. J Biosci 2023; 49(1): 2.
[http://dx.doi.org/10.1007/s12038-023-00390-y] [PMID: 38173312]

[102]   Langner T, Kamoun S, Belhaj K. CRISPR crops: plant genome editing toward disease resistance. Annu Rev Phytopathol 2018; 56(1): 479-512.
[http://dx.doi.org/10.1146/annurev-phyto-080417-050158] [PMID: 29975607]

[103]   Singh R. Genome Editing for Plant Disease Resistance. In: Singh KP, Jahagirdar S, Sarma BK, Eds. Emerging Trends in Plant Pathology. Singapore: Springer 2020; pp. 577-90.
[http://dx.doi.org/10.1007/978-981-15-6275-4_25]

[104]   Binnie A, Fernandes E, Almeida-Lousada H, de Mello RA, Castelo-Branco P. CRISPR-based strategies in infectious disease diagnosis and therapy. Infection 2021; 49(3): 377-85.
[http://dx.doi.org/10.1007/s15010-020-01554-w] [PMID: 33393066]

[105]   Prabhukarthikeyan SR, Parameswaran C, Keerthana U, *et al.* Understanding the plant-microbe interactions in CRISPR/Cas9 era: indeed a sprinting start in marathon. Curr Genomics 2020; 21(6): 429-43.
[http://dx.doi.org/10.2174/1389202921999200716110853] [PMID: 33093805]

[106]   Zhu X, Kuang Y, Chen Y, *et al.* miR2118 negatively regulates bacterial blight resistance through targeting several disease resistance genes in rice. Plants 2023; 12(22): 3815.
[http://dx.doi.org/10.3390/plants12223815] [PMID: 38005712]

[107]   O'Malley MR, Anderson JC. Regulation of the *Pseudomonas syringae* type iii secretion system by host environment signals. Microorganisms 2021; 9(6): 1227.
[http://dx.doi.org/10.3390/microorganisms9061227] [PMID: 34198761]

[108]   Ortigosa A, Gimenez-Ibanez S, Leonhardt N, Solano R. Design of a bacterial speck resistant tomato by CRISPR /Cas9-mediated editing of *Sl JAZ 2*. Plant Biotechnol J 2019; 17(3): 665-73.
[http://dx.doi.org/10.1111/pbi.13006] [PMID: 30183125]

[109]   Zhou J, Peng Z, Long J, *et al.* Gene targeting by the TAL effector PthXo2 reveals cryptic resistance gene for bacterial blight of rice. Plant J 2015; 82(4): 632-43.
[http://dx.doi.org/10.1111/tpj.12838] [PMID: 25824104]

[110] Yang T, Ali M, Lin L, *et al.* Recoloring tomato fruit by CRISPR/Cas9-mediated multiplex gene editing. Hortic Res 2023; 10(1): uhac214.
[http://dx.doi.org/10.1093/hr/uhac214] [PMID: 36643741]

[111] Alquicer G, Ibrahim E, Maruthi MN, Kundu JK. Identifying putative resistance genes for barley yellow dwarf Virus-PAV in wheat and barley. Viruses 2023; 15(3): 716.
[http://dx.doi.org/10.3390/v15030716] [PMID: 36992425]

[112] Rakhshandehroo F, Zamani Zadeh HR, Modarresi A, Hajmansoor S. Occurrence of *Prunus necrotic ringspot virus* and *Arabis mosaic virus* on Rose in Iran. Plant Dis 2006; 90(7): 975-5.
[http://dx.doi.org/10.1094/PD-90-0975B] [PMID: 30781055]

[113] Ranabhat NB, Bruce MA, Davis MA, Fritz AK, Zhang G, Rupp JL. Wheat (Triticum aestivum, 'multiple cultivars') Barley yellow dwarf; Barley yellow dwarf virus. Plant Disease Management Reports.2020.
[http://dx.doi.org/10.13140/RG.2.2.11280.35844]

[114] Torres R, Larenas J, Fribourg C, Romero J. Pepper necrotic spot virus, a new tospovirus infecting solanaceous crops in Peru. Arch Virol 2012; 157(4): 609-15.
[http://dx.doi.org/10.1007/s00705-011-1217-3] [PMID: 22218966]

[115] Ewald PW. Evolution of virulence. Infect Dis Clin North Am 2004; 18(1): 1-15.
[http://dx.doi.org/10.1016/S0891-5520(03)00099-0] [PMID: 15081500]

[116] McGavern DB, Kang SS. Illuminating viral infections in the nervous system. Nat Rev Immunol 2011; 11(5): 318-29.
[http://dx.doi.org/10.1038/nri2971] [PMID: 21508982]

[117] Rubio L, Galipienso L, Ferriol I. Detection of plant viruses and disease management: relevance of genetic diversity and evolution. Front Plant Sci 2020; 11: 1092.
[http://dx.doi.org/10.3389/fpls.2020.01092] [PMID: 32765569]

[118] Patarroyo C, Laliberté JF, Zheng H. Hijack it, change it: how do plant viruses utilize the host secretory pathway for efficient viral replication and spread? Front Plant Sci 2013; 3: 308.
[http://dx.doi.org/10.3389/fpls.2012.00308] [PMID: 23335933]

[119] Hashimoto M, Neriya Y, Yamaji Y, Namba S. Recessive resistance to plant viruses: potential resistance genes beyond translation initiation factors. Front Microbiol 2016; 7(OCT): 1695.
[http://dx.doi.org/10.3389/fmicb.2016.01695] [PMID: 27833593]

[120] Jiang T, Zhou T. Unraveling the mechanisms of virus-induced symptom. Plants 2023; 12(15): 2830.
[http://dx.doi.org/10.3390/plants12152830] [PMID: 37570983]

[121] Kwon MJ, Kwon SJ, Kim MH, *et al.* Visual tracking of viral infection dynamics reveals the synergistic interactions between cucumber mosaic virus and broad bean wilt virus 2. Sci Rep 2023; 13(1): 7261.
[http://dx.doi.org/10.1038/s41598-023-34553-6] [PMID: 37142679]

[122] Mandadi KK, Scholthof KBG. Plant immune responses against viruses: how does a virus cause disease? Plant Cell 2013; 25(5): 1489-505.
[http://dx.doi.org/10.1105/tpc.113.111658] [PMID: 23709626]

[123] Borrelli VMG, Brambilla V, Rogowsky P, Marocco A, Lanubile A. The enhancement of plant disease resistance using crispr/cas9 technology. Front Plant Sci 2018; 9: 1245.
[http://dx.doi.org/10.3389/fpls.2018.01245] [PMID: 30197654]

[124] Dong OX, Ronald PC. Genetic engineering for disease resistance in plants: recent progress and future perspectives. Plant Physiol 2019; 180(1): 26-38.
[http://dx.doi.org/10.1104/pp.18.01224] [PMID: 30867331]

[125] Romay G, Bragard C. Antiviral defenses in plants through genome editing. Front Microbiol 2017; 8: 47.

[http://dx.doi.org/10.3389/fmicb.2017.00047] [PMID: 28167937]

[126]  Ahmad M. Plant breeding advancements with "CRISPR-Cas" genome editing technologies will assist future food security. Front Plant Sci 2023; 14: 1133036.
[http://dx.doi.org/10.3389/fpls.2023.1133036]

[127]  Hussain B, Lucas SJ, Budak H. CRISPR/Cas9 in plants: at play in the genome and at work for crop improvement. Brief Funct Genomics 2018; 17(5): 319-28.
[http://dx.doi.org/10.1093/bfgp/ely016] [PMID: 29912293]

[128]  Tavakoli K, Pour-Aboughadareh A, Kianersi F, Poczai P, Etminan A, Shooshtari L. Applications of CRISPR-Cas9 as an advanced genome editing system in life sciences. BioTech 2021; 10(3): 14.
[http://dx.doi.org/10.3390/biotech10030014] [PMID: 35822768]

[129]  Fichter KM, Setayesh T, Malik P. Strategies for precise gene edits in mammalian cells. Mol Ther Nucleic Acids 2023; 32: 536-52.
[http://dx.doi.org/10.1016/j.omtn.2023.04.012] [PMID: 37215153]

[130]  Gaucherand L, Gaglia MM. The role of viral RNA degrading factors in shutoff of host gene expression. Annu Rev Virol 2022; 9(1): 213-38.
[http://dx.doi.org/10.1146/annurev-virology-100120-012345] [PMID: 35671567]

[131]  Rampersad S, Tennant P. Replication and expression strategies of viruses. Viruses 2018; 55-82.
[http://dx.doi.org/10.1016/B978-0-12-811257-1.00003-6]

[132]  Kuroiwa K, Danilo B, Perrot L, *et al.* An iterative gene-editing strategy broadens *eIF4E1* genetic diversity in *Solanum lycopersicum* and generates resistance to multiple potyvirus isolates. Plant Biotechnol J 2023; 21(5): 918-30.
[http://dx.doi.org/10.1111/pbi.14003] [PMID: 36715107]

[133]  Zafirov D, Giovinazzo N, Bastet A, Gallois JL. When a knockout is an Achilles' heel: Resistance to one potyvirus species triggers hypersusceptibility to another one in *Arabidopsis thaliana*. Mol Plant Pathol 2021; 22(3): 334-47.
[http://dx.doi.org/10.1111/mpp.13031] [PMID: 33377260]

[134]  Kumar S, Abebie B, Kumari R, *et al.* Development of PVY resistance in tomato by knockout of host eukaryotic initiation factors by CRISPR-Cas9. Phytoparasitica 2022; 50(4): 743-56.
[http://dx.doi.org/10.1007/s12600-022-00991-7]

[135]  Yıldırım K, Kavas M, Küçük İS, Seçgin Z, Saraç ÇG. Development of highly efficient resistance to beet curly top Iran virus (Becurtovirus) in sugar beet (*B. vulgaris*) *via* CRISPR/Cas9 System. Int J Mol Sci 2023; 24(7): 6515.
[http://dx.doi.org/10.3390/ijms24076515] [PMID: 37047489]

[136]  Yin K, Han T, Xie K, Zhao J, Song J, Liu Y. Engineer complete resistance to Cotton Leaf Curl Multan virus by the CRISPR/Cas9 system in *Nicotiana benthamiana*. Phytopathology Research 2019; 1(1): 9.
[http://dx.doi.org/10.1186/s42483-019-0017-7]

[137]  Wang W, Vinocur B, Altman A. Plant responses to drought, salinity and extreme temperatures: towards genetic engineering for stress tolerance. Planta 2003; 218(1): 1-14.
[http://dx.doi.org/10.1007/s00425-003-1105-5] [PMID: 14513379]

[138]  Nykiel M, Gietler M, Fidler J, Prabucka B, Labudda M. Abiotic stress signaling and responses in plants. 2023.
[http://dx.doi.org/10.3390/books978-3-0365-9254-1]

[139]  Khadka K, Raizada MN, Navabi A. Recent progress in germplasm evaluation and gene mapping to enable breeding of drought-tolerant wheat. Front Plant Sci 2020; 11: 1149.
[http://dx.doi.org/10.3389/fpls.2020.01149] [PMID: 32849707]

[140]  Saini R, Osorio-Gonzalez CS, Brar SK, Kwong R. A critical insight into the development, regulation and future prospects of biofuels in Canada. Bioengineered 2021; 12(2): 9847-59.
[http://dx.doi.org/10.1080/21655979.2021.1996017] [PMID: 34852717]

[141] Lee JH, Won HJ, Hoang Nguyen Tran P, Lee S, Kim HY, Jung JH. Improving lignocellulosic biofuel production by CRISPR/Cas9-mediated lignin modification in barley. Glob Change Biol Bioenergy 2021; 13(4): 742-52.
[http://dx.doi.org/10.1111/gcbb.12808]

[142] Mujeeb-Kazi A, Gul A, Ahmad I, *et al.* Genetic Resources for Some Wheat Abiotic Stress Tolerances. In: Ashraf M, Ozturk M, Athar HR, Eds. Salinity and Water Stress. Dordrecht: Springer 2009; pp. 149-63.
[http://dx.doi.org/10.1007/978-1-4020-9065-3_16]

[143] Trono D, Pecchioni N. Candidate genes associated with abiotic stress response in plants as tools to engineer tolerance to drought, salinity and extreme temperatures in wheat: an overview. Plants 2022; 11(23): 3358.
[http://dx.doi.org/10.3390/plants11233358] [PMID: 36501397]

[144] Basavaraj PS, Rane J, Boraiah KM, Gangashetty P, Harisha CB. Genetic analysis of tolerance to transient waterlogging stress in pigeonpea (*Cajanus cajan* L. Millspaugh). Indian J Genet Plant Breed 2023; 83(3): 316-25.
[http://dx.doi.org/10.31742/ISGPB.83.3.3]

[145] Khan S, Singh R, Kaur H, *et al.* Plant growth regulator-mediated response under abiotic stress: A review. J Appl Biol Biotechnol 2024; 12(2): 13-21.
[http://dx.doi.org/10.7324/JABB.2024.157299]

[146] Peer LA. Abiotic stress tolerance in common beans; a review. Int J Biol Pharm Allied Sci 2023; 12(11): 5289-303.
[http://dx.doi.org/10.31032/IJBPAS/2023/12.11.7592]

[147] Oguz MC, Aycan M, Oguz E, Poyraz I, Yildiz M. Drought stress tolerance in plants: interplay of molecular, biochemical and physiological responses in important development stages. Physiologia 2022; 2(4): 180-97.
[http://dx.doi.org/10.3390/physiologia2040015]

[148] Farooq M, Wahid A, Kobayashi N, Fujita D, Basra SMA. Plant drought stress: effects, mechanisms and management. Agron Sustain Dev 2009; 29(1): 185-212.
[http://dx.doi.org/10.1051/agro:2008021]

[149] Özkara A, Akyil D, Konuk M. Pesticides, Environmental Ppollution, and Health. In: Larramendy ML, Soloneski S, Eds. Environmental Health Risk - Hazardous Factors to Living Species. IntechOpen 2016.
[http://dx.doi.org/10.5772/63094]

[150] Omprakash Gobu R, Bisen P, Baghel M, Chourasia K. Resistance/Tolerance mechanism under water deficit (Drought) condition in plants. Int J Curr Microbiol Appl Sci 2017; 6(4): 66-78.
[http://dx.doi.org/10.20546/ijcmas.2017.604.009]

[151] Yadav S, Sharma KD. Molecular and morphophysiological analysis of drought stress in plants. In: Rigobelo EC, Ed. Plant Growth IntechOpen. 2016.
[http://dx.doi.org/10.5772/65246]

[152] Balaji B, Dharani E, Shricharan S, *et al.* Genome editing for speed breeding of horticultural crops. Journal of Agrisearch 2021; 9(3): 196-200.
[http://dx.doi.org/10.21921/jas.v9i03.11001]

[153] Chaudhary J, Khatri P, Singla P, *et al.* Advances in omics approaches for abiotic stress tolerance in tomato. Biology (Basel) 2019; 8(4): 90.
[http://dx.doi.org/10.3390/biology8040090] [PMID: 31775241]

[154] Rajarajan K, Sakshi S, Taria S, *et al.* Whole plant response of *Pongamia pinnata* to drought stress tolerance revealed by morpho-physiological, biochemical and transcriptome analysis. Mol Biol Rep 2022; 49(10): 9453-63.

[http://dx.doi.org/10.1007/s11033-022-07808-0] [PMID: 36057878]

[155] Salava H, Thula S, Mohan V, Kumar R, Maghuly F. Application of genome editing in tomato breeding: mechanisms, advances, and prospects. Int J Mol Sci 2021; 22(2): 682.
[http://dx.doi.org/10.3390/ijms22020682] [PMID: 33445555]

[156] Zhang S, Zhu H. Development and prospect of gene edited fruits and vegetables. Food Qual. Saf 2023; pp. 1-15.
[http://dx.doi.org/10.1093/fqsafe/fyad045]

[157] Ali Z, Rai SK, Jan S, Raina K. An assessment on CRISPR Cas as a novel asset in mitigating drought stress. Genet Resour Crop Evol 2022; 69(6): 2011-27.
[http://dx.doi.org/10.1007/s10722-022-01364-z]

[158] Saeed S, Ullah S, Afzal R, Umar F, Ali A. CRISPR-Cas Technologies for. CRISPRized Horticulture Crops. 1st ed., Elsevier 2023.
[http://dx.doi.org/10.1016/B978-0-443-13229-2.00019-3]

[159] Sami A, Xue Z, Tazein S, *et al.* CRISPR–Cas9-based genetic engineering for crop improvement under drought stress. Bioengineered 2021; 12(1): 5814-29.
[http://dx.doi.org/10.1080/21655979.2021.1969831] [PMID: 34506262]

[160] Wang T, Zhang H, Zhu H. CRISPR technology is revolutionizing the improvement of tomato and other fruit crops. Hortic Res 2019; 6(1): 77.
[http://dx.doi.org/10.1038/s41438-019-0159-x] [PMID: 31240102]

[161] Park JJ, Yoo CG, Flanagan A, *et al.* Defined tetra-allelic gene disruption of the 4-coumarate:coenzyme A ligase 1 (Pv4CL1) gene by CRISPR/Cas9 in switchgrass results in lignin reduction and improved sugar release. Biotechnol Biofuels 2017; 10(1): 284.
[http://dx.doi.org/10.1186/s13068-017-0972-0] [PMID: 29213323]

[162] Ogata T, Ishizaki T, Fujita M, Fujita Y. CRISPR/Cas9-targeted mutagenesis of OsERA1 confers enhanced responses to abscisic acid and drought stress and increased primary root growth under nonstressed conditions in rice. PLoS One 2020; 15(12): e0243376.
[http://dx.doi.org/10.1371/journal.pone.0243376] [PMID: 33270810]

[163] Hasegawa PM, Bressan RA, Zhu JK, Bohnert HJ. Plant cellular and molecular responses to high salinity. Annu Rev Plant Physiol Plant Mol Biol 2000; 51(1): 463-99.
[http://dx.doi.org/10.1146/annurev.arplant.51.1.463] [PMID: 15012199]

[164] Trotti J, Trapani I, Gulino F, *et al.* Physiological responses to salt stress at the seedling stage in wild (*Oryza rufipogon* Griff.) and cultivated (*Oryza sativa* L.) rice. Plants 2024; 13(3): 369.
[http://dx.doi.org/10.3390/plants13030369] [PMID: 38337902]

[165] Rishi A, Sneha S. Antioxidative defense against reactive oxygen species in plants under salt stress. Int J Curr Res 2013; 5(7): 1622-7.

[166] An J, Song A, Guan Z, *et al.* The over-expression of *Chrysanthemum crassum* CcSOS1 improves the salinity tolerance of chrysanthemum. Mol Biol Rep 2014; 41(6): 4155-62.
[http://dx.doi.org/10.1007/s11033-014-3287-2] [PMID: 24566689]

[167] dos Santos LB, de Lima Silva JR, Moreira AMT, *et al.* Response to carvacrol monoterpene in the emergence of *Allium cepa* L. seeds exposed to salt stress. Environ Sci Pollut Res Int 2024; 0123456789.
[http://dx.doi.org/10.1007/s11356-024-32048-z]

[168] Çamlıca M, Yaldız G. Effect of salt stress on seed germination, shoot and root length in Basil (*Ocimum basilicum*). International Journal of Secondary Metabolite 2017; 4(3): 69-76.
[http://dx.doi.org/10.21448/ijsm.356250]

[169] Chaudhary MT, Majeed S, Rana IA, *et al.* Impact of salinity stress on cotton and opportunities for improvement through conventional and biotechnological approaches. BMC Plant Biol 2024; 24(1): 20.
[http://dx.doi.org/10.1186/s12870-023-04558-4] [PMID: 38166652]

[170] Parvaiz A, Satyawati S. Salt stress and phyto-biochemical responses of plants - a review. Plant Soil Environ 2008; 54(3): 89-99.
[http://dx.doi.org/10.17221/2774-PSE]

[171] Ahmed R, Islam MM, Sarker HMMU, *et al.* Morphological responses of three contrasting Soybean (*Glycine max* (L.) Merrill) genotypes under different levels of salinity stress in the coastal region of Bangladesh. Journal of Plant Stress Physiology 2023; 9: 18-26.
[http://dx.doi.org/10.25081/jpsp.2023.v9.8595]

[172] Barbafieri M. Morphophysiological characterisation of Guayule (*Parthenium argentatum* A . Gray) in response to increasing NaCl arid and semiarid areas. 20241–23.

[173] Sneha S, Rishi A, Chandra S. Effect of short term salt stress on chlorophyll content, protein and activities of catalase and ascorbate peroxidase enzymes in pearl millet. Am J Plant Physiol 2013; 9(1): 32-7.
[http://dx.doi.org/10.3923/ajpp.2014.32.37]

[174] Zhou Y, Yin X, Wan S, *et al.* The *Sesuvium portulacastrum* plasma membrane Na+/H+ antiporter SpSOS1 complemented the salt sensitivity of transgenic arabidopsis sos1 mutant plants. Plant Mol Biol Report 2018; 36(4): 553-63.
[http://dx.doi.org/10.1007/s11105-018-1099-6]

[175] Gajardo HA, Gómez-Espinoza O, Boscariol Ferreira P, Carrer H, Bravo LA. The Potential of CRISPR/Cas technology to enhance crop performance on adverse soil conditions. Plants 2023; 12(9): 1892.
[http://dx.doi.org/10.3390/plants12091892] [PMID: 37176948]

[176] Han X, Chen Z, Li P, *et al.* Development of novel rice germplasm for salt-tolerance at seedling stage using CRISPR-Cas9. Sustainability (Basel) 2022; 14(5): 2621.
[http://dx.doi.org/10.3390/su14052621]

[177] Khalid MB, Haris M, Arslan M, *et al.* Use of CRISPR- CAS to save the climate affected rice: A review. Asian Journal of Biotechnology and Genetic Engineering 2023; 6(2): 179-90.
https://journalajbge.com/index.php/AJBGE/article/view/113

[178] Zhang A, Liu Y, Wang F, *et al.* Enhanced rice salinity tolerance *via* CRISPR/Cas9-targeted mutagenesis of the *OsRR22* gene. Mol Breed 2019; 39(3): 47.
[http://dx.doi.org/10.1007/s11032-019-0954-y] [PMID: 32803201]

[179] Sheng X, Ai Z, Tan Y, *et al.* Novel salinity-tolerant third-generation hybrid rice developed *via* CRISPR/Cas9-mediated gene editing. Int J Mol Sci 2023; 24(9): 8025.
[http://dx.doi.org/10.3390/ijms24098025] [PMID: 37175730]

[180] Wang T, Xun H, Wang W, *et al.* Mutation of GmAITR Genes by CRISPR/Cas9 genome editing results in enhanced salinity stress tolerance in soybean. Front Plant Sci 2021; 12: 779598.
[http://dx.doi.org/10.3389/fpls.2021.779598] [PMID: 34899806]

[181] Alcántara E, Romera FJ, Cañete M, De la Guardia MD. Effects of heavy metals on both induction and function of root Fe(lll) reductase in Fe-deficient cucumber ( *Cucumis sativus* L.) plants. J Exp Bot 1994; 45(12): 1893-8.
[http://dx.doi.org/10.1093/jxb/45.12.1893]

[182] Yu W, Wang L, Zhao R, *et al.* Knockout of SlMAPK3 enhances tolerance to heat stress involving ROS homeostasis in tomato plants. BMC Plant Biol 2019; 19(1): 354.
[http://dx.doi.org/10.1186/s12870-019-1939-z] [PMID: 31412779]

[183] Grandgirard J, Poinsot D, Krespi L, Nénon JP, Cortesero AM. Costs of secondary parasitism in the facultative hyperparasitoid *Pachycrepoideus dubius* : does host size matter? Entomol Exp Appl 2002; 103(3): 239-48.
[http://dx.doi.org/10.1046/j.1570-7458.2002.00982.x]

[184] Jogam P, Sandhya D, Alok A, *et al.* Editing of TOM1 gene in tobacco using CRISPR/Cas9 confers

resistance to Tobacco mosaic virus. Mol Biol Rep 2023; 50(6): 5165-76.
[http://dx.doi.org/10.1007/s11033-023-08440-2] [PMID: 37119416]

[185] Hussain SB, Manzoor A, Shoaib M, Zubair M, Gilani MM. Genome-wide identification and characterization of cytokinin metabolic gene families in chickpea (*Cicer arietinum*). Asian Plant Research Journal 2023; 11(6): 98-118.
[http://dx.doi.org/10.9734/aprj/2023/v11i6235]

[186] Li Z, Rao MJ, Li J, *et al.* CRISPR/Cas9 mutant rice Ospmei12 involved in growth, cell wall development, and response to phytohormone and heavy metal stress. Int J Mol Sci 2022; 23(24): 16082.
[http://dx.doi.org/10.3390/ijms232416082] [PMID: 36555723]

[187] Nievola CC, Carvalho CP, Carvalho V, Rodrigues E. Rapid responses of plants to temperature changes. Temperature 2017; 4(4): 371-405.
[http://dx.doi.org/10.1080/23328940.2017.1377812] [PMID: 29435478]

[188] Vollenweider P, Günthardt-Goerg MS. Diagnosis of abiotic and biotic stress factors using the visible symptoms in foliage. Environ Pollut 2005; 137(3): 455-65.
[http://dx.doi.org/10.1016/j.envpol.2005.01.032] [PMID: 16005758]

[189] Gray SB, Brady SM. Plant developmental responses to climate change. Dev Biol 2016; 419(1): 64-77.
[http://dx.doi.org/10.1016/j.ydbio.2016.07.023] [PMID: 27521050]

[190] Hasanuzzaman M, Nahar K, Alam M, Roychowdhury R, Fujita M. Physiological, biochemical, and molecular mechanisms of heat stress tolerance in plants. Int J Mol Sci 2013; 14(5): 9643-84.
[http://dx.doi.org/10.3390/ijms14059643] [PMID: 23644891]

[191] Liu X, Zhou Y, Xiao J, Bao F. Effects of chilling on the structure, function and development of chloroplasts. Front Plant Sci 2018; 9: 1715.
[http://dx.doi.org/10.3389/fpls.2018.01715] [PMID: 30524465]

[192] Blomme J, Develtere W, Köse A, *et al.* The heat is on: a simple method to increase genome editing efficiency in plants. BMC Plant Biol 2022; 22(1): 142.
[http://dx.doi.org/10.1186/s12870-022-03519-7] [PMID: 35331142]

[193] Custers R, Dima O. Genome-edited crops and 21st century food system challenges in-depth analysis. Panel Future Sci Technol 2022; 10: 290440.
[http://dx.doi.org/10.2861/290440]

[194] Schindele P, Puchta H. Engineering CRISPR/ *Lb* Cas12a for highly efficient, temperature-tolerant plant gene editing. Plant Biotechnol J 2020; 18(5): 1118-20.
[http://dx.doi.org/10.1111/pbi.13275] [PMID: 31606929]

[195] Kapusi E, Corcuera-Gómez M, Melnik S, Stoger E. Heritable genomic fragment deletions and small indels in the putative ENGase gene induced by CRISPR/Cas9 in barley. Front Plant Sci 2017; 8: 540.
[http://dx.doi.org/10.3389/fpls.2017.00540] [PMID: 28487703]

[196] Burey PP, Panchal SK, Helwig A. Sustainable food systems. In: Juliano P, Buckow R, Nguyen MH, Knoerzer K, Sellahewa J, Eds. Food Engineering Innovations Across the Food Supply Chain. Australia: Academic Press 2021; pp. 15-46.
[http://dx.doi.org/10.1016/B978-0-12-821292-9.00015-7]

[197] Rozaki Z, Meirani Rejeki T, Rahayu L, Fairuz Ramli M. Farmers' food security in forest and peatland fires prone areas of south kalimantan, Indonesia. Jurnal Sylva Lestari 2023; 11(3): 527-42.
[http://dx.doi.org/10.23960/jsl.v11i3.770]

[198] Anand A, Subramanian M, Kar D. Breeding techniques to dispense higher genetic gains. Front Plant Sci 2023; 13(January): 1076094.
[http://dx.doi.org/10.3389/fpls.2022.1076094] [PMID: 36743551]

[199] Xu Y, Li P, Zou C, *et al.* Enhancing genetic gain in the era of molecular breeding. J Exp Bot 2017; 68(11): 2641-66.

[http://dx.doi.org/10.1093/jxb/erx135] [PMID: 28830098]

[200] Sharma P, Pandey A, Malviya R, Dey S, Karmakar S, Gayen D. Genome editing for improving nutritional quality, post-harvest shelf life and stress tolerance of fruits, vegetables, and ornamentals. 2023.
[http://dx.doi.org/10.3389/fgeed.2023.1094965]

[201] Wan L, Wang Z, Tang M, *et al.* Crispr-cas9 gene editing for fruit and vegetable crops: strategies and prospects. Horticulturae 2021; 7(7): 193.
[http://dx.doi.org/10.3390/horticulturae7070193]

[202] Dong C, Fu Y, Liu G, Liu H. Growth, photosynthetic characteristics, antioxidant capacity and biomass yield and quality of wheat (*Triticum aestivum* L.) exposed to LED light sources with different spectra combinations. J Agron Crop Sci 2014; 200(3): 219-30.
[http://dx.doi.org/10.1111/jac.12059]

[203] Ghaffar A, Hussain N, Ajaj R, *et al.* Photosynthetic activity and metabolic profiling of bread wheat cultivars contrasting in drought tolerance. Front Plant Sci 2023; 14: 1123080.
[http://dx.doi.org/10.3389/fpls.2023.1123080] [PMID: 36844078]

[204] Kubar MS, Zhang Q, Feng M, *et al.* Growth, yield and photosynthetic performance of winter wheat as affected by co-application of nitrogen fertilizer and organic manures. Life (Basel) 2022; 12(7): 1000.
[http://dx.doi.org/10.3390/life12071000] [PMID: 35888089]

[205] Mathan J, Bhattacharya J, Ranjan A. Enhancing crop yield by optimizing plant developmental features. Development 2016; 143(18): 3283-94.
[http://dx.doi.org/10.1242/dev.134072] [PMID: 27624833]

[206] Richards RA. Selectable traits to increase crop photosynthesis and yield of grain crops. J Exp Bot 2000; 51(Spec No) (Suppl. 1): 447-58.
[http://dx.doi.org/10.1093/jexbot/51.suppl_1.447] [PMID: 10938853]

[207] Abdul Rahim A, Uzair M, Rehman N, *et al.* CRISPR/Cas9 mediated TaRPK1 root architecture gene mutagenesis confers enhanced wheat yield. J King Saud Univ Sci 2024; 36(2): 103063.
[http://dx.doi.org/10.1016/j.jksus.2023.103063]

[208] Han Y, Yang J, Wu H, Liu F, Qin B, Li R. Improving rice leaf shape using CRISPR/Cas9-mediated genome editing of SRL1 and characterizing its regulatory network involved in leaf rolling through transcriptome analysis. Int J Mol Sci 2023; 24(13): 11087.
[http://dx.doi.org/10.3390/ijms241311087] [PMID: 37446265]

[209] Santoso TJ, Trijatmiko KR, Char SN, Yang B, Wang K. Targeted mutation of GA20ox-2 gene using CRISPR/Cas9 system generated semi-dwarf phenotype in rice. IOP Conf Ser Earth Environ Sci 2020; 482(1): 012027.
[http://dx.doi.org/10.1088/1755-1315/482/1/012027]

[210] Usman B, Zhao N, Nawaz G, *et al.* CRISPR/Cas9 guided mutagenesis of grain size 3 confers increased rice (*Oryza sativa* L.) grain length by regulating cysteine proteinase inhibitor and ubiquitin-related proteins. Int J Mol Sci 2021; 22(6): 3225.
[http://dx.doi.org/10.3390/ijms22063225] [PMID: 33810044]

[211] Wang Y, Du F, Wang J, *et al.* Improving bread wheat yield through modulating an unselected AP2/ERF gene. Nat Plants 2022; 8(8): 930-9.
[http://dx.doi.org/10.1038/s41477-022-01197-9] [PMID: 35851621]

[212] Alvarez JM, Bueno N, Cañas RA, Avila C, Cánovas FM, Ordás RJ. Analysis of the Wuschel-related homeobox gene family in *Pinus pinaster*: New insights into the gene family evolution. Plant Physiol Biochem 2018; 123(123): 304-18.
[http://dx.doi.org/10.1016/j.plaphy.2017.12.031] [PMID: 29278847]

[213] Ku HK, Ha SH. Improving nutritional and functional quality by genome editing of crops: status and perspectives. Front Plant Sci 2020; 11: 577313.

[http://dx.doi.org/10.3389/fpls.2020.577313] [PMID: 33193521]

[214] Naeem M, Demirel U, Yousaf MF, Caliskan S, Caliskan ME, Wehling P. Overview on domestication, breeding, genetic gain and improvement of tuber quality traits of potato using fast forwarding technique (GWAS): A review. Plant Breed 2021; 140(4): 519-42.
[http://dx.doi.org/10.1111/pbr.12927]

[215] Niñoles R, Ruiz-Pastor CM, Arjona-Mudarra P, *et al.* Transcription Factor DOF4.1 regulates seed longevity in Arabidopsis *via* seed permeability and modulation of seed storage protein accumulation. Front Plant Sci 2022; 13: 915184.
[http://dx.doi.org/10.3389/fpls.2022.915184] [PMID: 35845633]

[216] Simkin AJ. Carotenoids and apocarotenoids in planta: their role in plant development, contribution to the flavour and aroma of fruits and flowers, and their nutraceutical benefits. Plants 2021; 10(11): 2321.
[http://dx.doi.org/10.3390/plants10112321] [PMID: 34834683]

[217] Vachon G, Tichtinsky G, Parcy F. LEAFY, le régulateur clé du développement de la fleur. Biol Aujourdhui 2012; 206(1): 63-7.
[http://dx.doi.org/10.1051/jbio/2012006] [PMID: 22463997]

[218] Xiao Q, Huang T, Cao W, *et al.* Profiling of transcriptional regulators associated with starch biosynthesis in sorghum (*Sorghum bicolor* L.). Front Plant Sci 2022; 13: 999747.
[http://dx.doi.org/10.3389/fpls.2022.999747] [PMID: 36110358]

[219] Yang W, Hu J, Behera JR, *et al.* A tree peony trihelix transcription factor PrASIL1 represses seed oil accumulation. Front Plant Sci 2021; 12: 796181.
[http://dx.doi.org/10.3389/fpls.2021.796181] [PMID: 34956296]

[220] Zhang Y, Chen S, Yang L, Zhang Q. Application progress of CRISPR/Cas9 genome-editing technology in edible fungi. Front Microbiol 2023; 14: 1169884.
[http://dx.doi.org/10.3389/fmicb.2023.1169884] [PMID: 37303782]

[221] Yang J, Fang Y, Wu H, *et al.* Improvement of resistance to rice blast and bacterial leaf streak by CRISPR/Cas9-mediated mutagenesis of Pi21 and OsSULTR3;6 in rice (*Oryza sativa* L.). Front Plant Sci 2023; 14: 1209384.
[http://dx.doi.org/10.3389/fpls.2023.1209384] [PMID: 37528980]

[222] Nitarska D, Boehm R, Debener T, Lucaciu RC, Halbwirth H. First genome edited poinsettias: targeted mutagenesis of flavonoid 3′-hydroxylase using CRISPR/Cas9 results in a colour shift. Plant Cell Tissue Organ Cult 2021; 147(1): 49-60.
[http://dx.doi.org/10.1007/s11240-021-02103-5] [PMID: 34776565]

[223] Huang L, Liu Q. High-resistant starch crops for human health. Proc Natl Acad Sci USA 2023; 120(22): e2305990120.
[http://dx.doi.org/10.1073/pnas.2305990120] [PMID: 37216520]

[224] Garcia-Ruiz H. Susceptibility genes to plant viruses. Viruses 2018; 10(9): 484.
[http://dx.doi.org/10.3390/v10090484] [PMID: 30201857]

[225] Li J, Jiao G, Sun Y, *et al.* Modification of starch composition, structure and properties through editing of *TaSBEIIa* in both winter and spring wheat varieties by CRISPR/Cas9. Plant Biotechnol J 2021; 19(5): 937-51.
[http://dx.doi.org/10.1111/pbi.13519] [PMID: 33236499]

[226] Wang Y, Liu X, Zheng X, *et al.* Creation of aromatic maize by CRISPR/Cas. J Integr Plant Biol 2021; 63(9): 1664-70.
[http://dx.doi.org/10.1111/jipb.13105] [PMID: 33934500]

[227] Coppen JJW. Flavour and fragrances of plant origin (Non-wood forest product 1). Rome: Natural Resources Institute 1995; p. 111.

[228] Mendes A, Oliveira A, Lameiras J, Mendes-Moreira P, Botelho G. Organic medicinal and aromatic plants: consumption profile of a portuguese consumer sample. Foods 2023; 12(22): 4145.

[http://dx.doi.org/10.3390/foods12224145] [PMID: 38002202]

[229] Samarth RM, Samarth M, Matsumoto Y. Medicinally important aromatic plants with radioprotective activity. Future Sci OA 2017; 3(4): FSO247.
[http://dx.doi.org/10.4155/fsoa-2017-0061] [PMID: 29134131]

[230] Chen SL, Yu H, Luo HM, Wu Q, Li CF, Steinmetz A. Conservation and sustainable use of medicinal plants: problems, progress, and prospects. Chin Med 2016; 11(1): 37.
[http://dx.doi.org/10.1186/s13020-016-0108-7] [PMID: 27478496]

[231] Khan S, Al-Qurainy F, Nadeem M. Biotechnological approaches for conservation and improvement of rare and endangered plants of Saudi Arabia. Saudi J Biol Sci 2012; 19(1): 1-11.
[http://dx.doi.org/10.1016/j.sjbs.2011.11.001] [PMID: 23961155]

[232] Wang Y, Wang S, Sun J, *et al.* Nanobubbles promote nutrient utilization and plant growth in rice by upregulating nutrient uptake genes and stimulating growth hormone production. Sci Total Environ 2021; 800: 149627.
[http://dx.doi.org/10.1016/j.scitotenv.2021.149627] [PMID: 34426308]

[233] Zhang P, Jialaliding Z, Gu J, Merchant A, Zhang Q, Zhou X. Knockout of ovary serine protease leads to ovary deformation and female sterility in the asian corn borer, *Ostrinia furnacalis.* Int J Mol Sci 2023; 24(22): 16311.
[http://dx.doi.org/10.3390/ijms242216311] [PMID: 38003502]

[234] Tu M, Fang J, Zhao R, *et al.* CRISPR/Cas9-mediated mutagenesis of VvbZIP36 promotes anthocyanin accumulation in grapevine (*Vitis vinifera*). Hortic. Res. 2022; 9(November 2021).
[http://dx.doi.org/10.1093/hr/uhac022]

[235] Avdiu V, Hodolli G, Dragusha B, Bunjaku K. The impact of abiotic and biotic factors on the productivity of the apple cultivars (*Malus domestica*). Pol J Environ Stud 2023; 32(4): 3025-31.
[http://dx.doi.org/10.15244/pjoes/163503]

[236] Teshome DT, Zharare GE, Naidoo S. The Threat of the combined effect of biotic and abiotic stress factors in forestry under a changing climate. Front Plant Sci 2020; 11: 601009.
[http://dx.doi.org/10.3389/fpls.2020.601009] [PMID: 33329666]

[237] Maioli A, Gianoglio S, Moglia A, *et al.* Simultaneous CRISPR/Cas9 editing of three PPO genes reduces fruit flesh browning in *Solanum melongena* L. Front Plant Sci 2020; 11: 607161.
[http://dx.doi.org/10.3389/fpls.2020.607161] [PMID: 33343607]

[238] González MN, Massa GA, Andersson M, *et al.* Reduced enzymatic browning in potato tubers by specific editing of a polyphenol oxidase gene *via* ribonucleoprotein complexes delivery of the CRISPR/Cas9 system. Front Plant Sci 2020; 10: 1649.
[http://dx.doi.org/10.3389/fpls.2019.01649] [PMID: 31998338]

[239] Fouli Y, Hurlbert M, Kröbel R. Greenhouse gas emissions from canadian agriculture: estimates and measurements SPP Briefing Paper. 2021. Available from: ssrn.com/abstract=4042259

[240] Kingston-Smith AH, Marshall AH, Moorby JM. Breeding for genetic improvement of forage plants in relation to increasing animal production with reduced environmental footprint. Animal 2013; 7 (Suppl. 1): 79-88.
[http://dx.doi.org/10.1017/S1751731112000961] [PMID: 22717231]

[241] Lynch J, Cain M, Frame D, Pierrehumbert R. Agriculture's contribution to climate change and role in mitigation is distinct from predominantly fossil CO2-emitting sectors. Front. Sustain. Food Syst. 20214; 4: 518039.
[http://dx.doi.org/10.3389/fsufs.2020.518039]

[242] Ambaye TG, Vaccari M, Bonilla-Petriciolet A, Prasad S, van Hullebusch ED, Rtimi S. Emerging technologies for biofuel production: A critical review on recent progress, challenges and perspectives. J Environ Manage 2021; 290: 112627.
[http://dx.doi.org/10.1016/j.jenvman.2021.112627] [PMID: 33991767]

[243]  Neupane D. Biofuels from renewable sources, a potential option for biodiesel production. Bioengineering (Basel) 2022; 10(1): 29.
[http://dx.doi.org/10.3390/bioengineering10010029] [PMID: 36671601]

[244]  Godde CM, Mason-D'Croz D, Mayberry DE, Thornton PK, Herrero M. Impacts of climate change on the livestock food supply chain; a review of the evidence. Glob Food Secur 2021; 28: 100488.
[http://dx.doi.org/10.1016/j.gfs.2020.100488] [PMID: 33738188]

[245]  Espinosa-Marrón A, Adams K, Sinno L, *et al.* Environmental impact of animal-based food production and the feasibility of a shift toward sustainable plant-based diets in the United States. Frontiers in Sustainability 2022; 3: 841106.
[http://dx.doi.org/10.3389/frsus.2022.841106]

[246]  Grossi G, Goglio P, Vitali A, Williams AG. Livestock and climate change: impact of livestock on climate and mitigation strategies. Anim Front 2019; 9(1): 69-76.
[http://dx.doi.org/10.1093/af/vfy034] [PMID: 32071797]

[247]  Lynch J, Pierrehumbert R. Climate impacts of cultured meat and beef cattle. Front Sustain Food Syst 2019; 3: 5.
[http://dx.doi.org/10.3389/fsufs.2019.00005] [PMID: 31535087]

[248]  Wolabu TW, Mahmood K, Jerez IT, *et al.* Multiplex CRISPR/Cas9-mediated mutagenesis of alfalfa *Flowering locus* Ta1(MsFTa1) leads to delayed flowering time with improved forage biomass yield and quality. Plant Biotechnol J 2023; 21(7): 1383-92.
[http://dx.doi.org/10.1111/pbi.14042] [PMID: 36964962]

[249]  Miladinovic D, Antunes D, Yildirim K, *et al.* Targeted plant improvement through genome editing: from laboratory to field. Plant Cell Rep 2021; 40(6): 935-51.
[http://dx.doi.org/10.1007/s00299-016-2062-3]

[250]  May D, Sanchez S, Gilby J, Altpeter F. Multi-allelic gene editing in an apomictic, tetraploid turf and forage grass (*Paspalum notatum* Flüggé) using CRISPR/Cas9. Front Plant Sci 2023; 14: 1225775.
[http://dx.doi.org/10.3389/fpls.2023.1225775] [PMID: 37521929]

[251]  Brühl CA, Zaller JG. Biodiversity decline as a consequence of an inappropriate environmental risk assessment of pesticides. Front Environ Sci 2019; 7: 177.
[http://dx.doi.org/10.3389/fenvs.2019.00177]

[252]  Ito HC, Shiraishi H, Nakagawa M, Takamura N. Combined impact of pesticides and other environmental stressors on animal diversity in irrigation ponds. PLoS One 2020; 15(7): e0229052.
[http://dx.doi.org/10.1371/journal.pone.0229052] [PMID: 32614853]

[253]  Damalas CA, Eleftherohorinos IG. Pesticide exposure, safety issues, and risk assessment indicators. Int J Environ Res Public Health 2011; 8(5): 1402-19.
[http://dx.doi.org/10.3390/ijerph8051402] [PMID: 21655127]

[254]  Pathak VM, Verma VK, Rawat BS, *et al.* Current status of pesticide effects on environment, human health and it's eco-friendly management as bioremediation: a comprehensive review. Front. Microbiol, 2022;m13: 1–29.
[http://dx.doi.org/10.3389/fmicb.2022.962619]

[255]  He DC, He MH, Amalin DM, Liu W, Alvindia DG, Zhan J. Biological control of plant diseases: an evolutionary and eco-economic consideration. Pathogens 2021; 10(10): 1311.
[http://dx.doi.org/10.3390/pathogens10101311] [PMID: 34684260]

[256]  Deguine JP, Aubertot JN, Flor RJ, Lescourret F, Wyckhuys KAG, Ratnadass A. Integrated pest management: good intentions, hard realities. A review. Agron Sustain Dev 2021; 41(3): 38.
[http://dx.doi.org/10.1007/s13593-021-00689-w]

[257]  Williams J, Cowlishaw R, Sanou A, Ranson H, Grigoraki L. *In vivo* functional validation of the V402L voltage gated sodium channel mutation in the malaria vector *An. gambiae*. Pest Manag Sci 2022; 78(3): 1155-63.

[http://dx.doi.org/10.1002/ps.6731] [PMID: 34821465]

[258] Kaduskar B, Kushwah RBS, Auradkar A, *et al.* Reversing insecticide resistance with allelic-drive in *Drosophila melanogaster.* Nat Commun 2022; 13(1): 291.
[http://dx.doi.org/10.1038/s41467-021-27654-1] [PMID: 35022402]

[259] Godheja J. Impact of GMO'S on environment and human health. Science and Technology. 2013; 5(5): 26–29. Available from: http://recent-science.com/

[260] Godheja J. Impact of GMO'S on environment and human health. Science and Technology. 2013; 5(5): 26–29. Available from: https://sitn.hms.harvard.edu/flash/2015/gmos-and-pesticides/

[261] Nuffield Council on Bioethics. Genetically modified crops : the ethical and. Bioethics. 1999. Available from: http://www.nuffieldbioethics.org/fileLibrary/pdf/gmcrop.pdf

[262] Genetic Inventions, Intellectual Property Rights and Licensing Practices: Evidence and Policies. Paris: OECD Publishing 2003.

[263] Sharma P, Singh SP, Iqbal HMN, Parra-Saldivar R, Varjani S, Tong YW. Genetic modifications associated with sustainability aspects for sustainable developments. Bioengineered 2022; 13(4): 9509-21.
[http://dx.doi.org/10.1080/21655979.2022.2061146] [PMID: 35389819]

[264] Bawa AS, Anilakumar KR. Genetically modified foods: safety, risks and public concerns—a review. J Food Sci Technol 2013; 50(6): 1035-46.
[http://dx.doi.org/10.1007/s13197-012-0899-1] [PMID: 24426015]

[265] Ryffel GU. Transgene flow: Facts, speculations and possible countermeasures. GM Crops Food 2014; 5(4): 249-58.
[http://dx.doi.org/10.4161/21645698.2014.945883] [PMID: 25523171]

[266] Bennett PM, Livesey CT, Nathwani D, Reeves DS, Saunders JR, Wise R. An assessment of the risks associated with the use of antibiotic resistance genes in genetically modified plants: report of the Working Party of the British Society for Antimicrobial Chemotherapy. J Antimicrob Chemother 2004; 53(3): 418-31.
[http://dx.doi.org/10.1093/jac/dkh087] [PMID: 14749339]

[267] Rhouma M, Archambault M, Butaye P. Antimicrobial use and resistance in animals from a one health perspective. Vet Sci 2023; 10(5): 319.
[http://dx.doi.org/10.3390/vetsci10050319] [PMID: 37235402]

[268] Ghimire BK, Yu CY, Kim WR, *et al.* Assessment of benefits and risk of genetically modified plants and products: current controversies and perspective. Sustainability (Basel) 2023; 15(2): 1722.
[http://dx.doi.org/10.3390/su15021722]

[269] Chen X, Lu X, Shu N, *et al.* Targeted mutagenesis in cotton (*Gossypium hirsutum* L.) using the CRISPR/Cas9 system. Sci Rep 2017; 7(1): 44304.
[http://dx.doi.org/10.1038/srep44304] [PMID: 28287154]

[270] Yu J, Tu L, Subburaj S, Bae S, Lee GJ. Simultaneous targeting of duplicated genes in Petunia protoplasts for flower color modification *via* CRISPR-Cas9 ribonucleoproteins. Plant Cell Rep 2021; 40(6): 1037-45.
[http://dx.doi.org/10.1007/s00299-020-02593-1] [PMID: 32959126]

[271] Morineau C, Bellec Y, Tellier F, *et al.* Selective gene dosage by CRISPR -Cas9 genome editing in hexaploid *Camelina sativa.* Plant Biotechnol J 2017; 15(6): 729-39.
[http://dx.doi.org/10.1111/pbi.12671] [PMID: 27885771]

[272] Sant'Ana RRA, Caprestano CA, Nodari RO, Agapito-Tenfen SZ. PEG-delivered CRISPR-Cas9 ribonucleoproteins system for gene-editing screening of maize protoplasts. Genes (Basel) 2020; 11(9): 1029.
[http://dx.doi.org/10.3390/genes11091029] [PMID: 32887261]

[273] Wan L, Wang Z, Zhang X, *et al.* Optimised agrobacterium-mediated transformation and application of developmental regulators improve regeneration efficiency in melons. Genes (Basel) 2023; 14(7): 1432.
[http://dx.doi.org/10.3390/genes14071432] [PMID: 37510336]

[274] Abdallah NA, Elsharawy H, Abulela HA, Thilmony R, Abdelhadi AA, Elarabi NI. Multiplex CRISPR/Cas9-mediated genome editing to address drought tolerance in wheat. GM Crops Food 2022; 00(00): 1-17.
[http://dx.doi.org/10.1080/21645698.2022.2120313] [PMID: 36200515]

[275] Bahariah B, Masani MYA, Rasid OA, Parveez GKA. Multiplex CRISPR/Cas9-mediated genome editing of the FAD2 gene in rice: a model genome editing system for oil palm. J Genet Eng Biotechnol 2021; 19(1): 86.
[http://dx.doi.org/10.1186/s43141-021-00185-4] [PMID: 34115267]

[276] Bahariah B, Masani MYA, Fizree MPMAA, Rasid OA, Parveez GKA. Multiplex CRISPR/Cas9 gene-editing platform in oil palm targeting mutations in EgFAD2 and EgPAT genes. J Genet Eng Biotechnol 2023; 21(1): 3.
[http://dx.doi.org/10.1186/s43141-022-00459-5] [PMID: 36630019]

[277] Jakobson L, Oney Birol S, Timofejeva L. (2024). Implementing Genome Editing in Barley Breeding. In: Ricroch A, Eriksson D, Miladinović D, Sweet J, Van Laere K, Woźniak-Gientka E. (eds) A Roadmap for Plant Genome Editing. Springer, Cham.
[http://dx.doi.org/10.1007/978-3-031-46150-7_10]

[278] Peterson BA, Haak DC, Nishimura MT, *et al.* Genome-wide assessment of efficiency and specificity in crispr/cas9 mediated multiple site targeting in arabidopsis. PLoS One 2016; 11(9): e0162169.
[http://dx.doi.org/10.1371/journal.pone.0162169] [PMID: 27622539]

[279] Zhang B, Yang X, Yang C, Li M, Guo Y. Exploiting the CRISPR/Cas9 system for targeted genome mutagenesis in petunia. Sci Rep 2016; 6(1): 20315.
[http://dx.doi.org/10.1038/srep20315] [PMID: 26837606]

[280] Subburaj S, Agapito-Tenfen SZ. Establishment of targeted mutagenesis in soybean protoplasts using CRISPR/Cas9 RNP delivery *via* electro−transfection. Front Plant Sci 2023; 14: 1255819.
[http://dx.doi.org/10.3389/fpls.2023.1255819] [PMID: 37841627]

[281] Wang ZP, Xing HL, Dong L, *et al.* Egg cell-specific promoter-controlled CRISPR/Cas9 efficiently generates homozygous mutants for multiple target genes in Arabidopsis in a single generation. Genome Biol 2015; 16(1): 144.
[http://dx.doi.org/10.1186/s13059-015-0715-0] [PMID: 26193878]

[282] Hooghvorst I, López-Cristoffanini C, Nogués S. Efficient knockout of phytoene desaturase gene using CRISPR/Cas9 in melon. Sci Rep 2019; 9(1): 17077.
[http://dx.doi.org/10.1038/s41598-019-53710-4] [PMID: 31745156]

[283] Ren F, Ren C, Zhang Z, *et al.* Efficiency optimization of CRISPR/CAS9-mediated targeted mutagenesis in grape. Front Plant Sci 2019; 10: 612.
[http://dx.doi.org/10.3389/fpls.2019.00612] [PMID: 31156675]

[284] Naing AH, Xu J, Kim CK. Editing of 1-aminocyclopropane-1-carboxylate oxidase genes negatively affects petunia seed germination. Plant Cell Rep 2022; 41(1): 209-20.
[http://dx.doi.org/10.1007/s00299-021-02802-5] [PMID: 34665313]

[285] Illa-Berenguer E, LaFayette PR, Parrott WA. 2023.
[http://dx.doi.org/10.3389/fgeed.2023.1074641]

[286] Li G, Liu R, Xu R, *et al.* Development of an Agrobacterium-mediated CRISPR/Cas9 system in pea (*Pisum sativum* L.). Crop J 2023; 11(1): 132-9.
[http://dx.doi.org/10.1016/j.cj.2022.04.011]

[287] Gupta SK, Vishwakarma NK, Malakar P, Vanspati P, Sharma NK, Chattopadhyay D. Development of an Agrobacterium-delivered codon-optimized CRISPR/Cas9 system for chickpea genome editing.

Protoplasma 2023; 260(5): 1437-51.
[http://dx.doi.org/10.1007/s00709-023-01856-4] [PMID: 37131068]

[288]  Wang Y, Geng L, Yuan M, *et al.* Deletion of a target gene in Indica rice *via* CRISPR/Cas9. Plant Cell Rep 2017; 36(8): 1333-43.
[http://dx.doi.org/10.1007/s00299-017-2158-4] [PMID: 28584922]

[289]  Murovec J, Guček K, Bohanec B, Avbelj M, Jerala R. DNA-free genome editing of *Brassica oleracea* and *B. rapa* protoplasts using CRISPR-cas9 ribonucleoprotein complexes. Front Plant Sci 2018; 9: 1594.
[http://dx.doi.org/10.3389/fpls.2018.01594] [PMID: 30455712]

[290]  Pan C, Ye L, Qin L, *et al.* CRISPR/Cas9-mediated efficient and heritable targeted mutagenesis in tomato plants in the first and later generations. Sci Rep 2016; 6(1): 24765.
[http://dx.doi.org/10.1038/srep24765] [PMID: 27097775]

[291]  Mainkar P, Manape TK, Satheesh V, Anandhan S. CRISPR/Cas9-mediated editing of phytoene desaturase gene in onion (*Allium cepa* L.). Front Plant Sci 2023; 14: 1226911.
[http://dx.doi.org/10.3389/fpls.2023.1226911] [PMID: 37701798]

[292]  Xu J, Kang BC, Naing AH, *et al.* CRISPR /Cas9-mediated editing of 1-aminocyclopropane-1-carboxylate oxidase1 enhances *Petunia* flower longevity. Plant Biotechnol J 2020; 18(1): 287-97.
[http://dx.doi.org/10.1111/pbi.13197] [PMID: 31222853]

[293]  Li R, Liu C, Zhao R, *et al.* CRISPR/Cas9-Mediated SlNPR1 mutagenesis reduces tomato plant drought tolerance. BMC Plant Biol 2019; 19(1): 38.
[http://dx.doi.org/10.1186/s12870-018-1627-4] [PMID: 30669982]

[294]  Shirazi Parsa H, Sabet MS, Moieni A, *et al.* CRISPR/Cas9-mediated cytosine base editing using an improved transformation procedure in Melon (*Cucumis melo* L.). Int J Mol Sci 2023; 24(13): 11189.
[http://dx.doi.org/10.3390/ijms241311189] [PMID: 37446368]

[295]  Lu QSM, Tian L. An efficient and specific CRISPR-Cas9 genome editing system targeting soybean phytoene desaturase genes. BMC Biotechnol 2022; 22(1): 7.
[http://dx.doi.org/10.1186/s12896-022-00737-7] [PMID: 35168613]

[296]  Gao X, Chen J, Dai X, Zhang D, Zhao Y. An effective strategy for reliably isolating heritable and Cas9-free arabidopsis mutants generated by CRISPR/Cas9-mediated genome editing. Plant Physiol 2016; 171(3): 1794-800.
[http://dx.doi.org/10.1104/pp.16.00663] [PMID: 27208253]

[297]  Char SN, Neelakandan AK, Nahampun H, *et al.* An *Agrobacterium* -delivered CRISPR /Cas9 system for high-frequency targeted mutagenesis in maize. Plant Biotechnol J 2017; 15(2): 257-68.
[http://dx.doi.org/10.1111/pbi.12611] [PMID: 27510362]

[298]  Kim H, Choi J, Won KH. A stable DNA-free screening system for CRISPR/RNPs-mediated gene editing in hot and sweet cultivars of *Capsicum annuum*. BMC Plant Biol 2020; 20(1): 449.
[http://dx.doi.org/10.1186/s12870-020-02665-0] [PMID: 31898482]

[299]  Tanaka J, Minkenberg B, Poddar S, Staskawicz B, Cho MJ. Improvement of gene delivery and mutation efficiency in the CRISPR-Cas9 wheat (*Triticum aestivum* L.) genomics system *via* biolistics. Genes (Basel) 2022; 13(7): 1180.
[http://dx.doi.org/10.3390/genes13071180] [PMID: 35885963]

[300]  Xu RF, Li H, Qin RY, *et al.* Generation of inheritable and "transgene clean" targeted genome-modified rice in later generations using the CRISPR/Cas9 system. Sci Rep 2015; 5(1): 11491.
[http://dx.doi.org/10.1038/srep11491] [PMID: 26089199]

# CHAPTER 10

# Bioinformatics Approaches in Plant Physiology

**Mehmet Emin Uras**[1,*]

[1] *Haliç University, Faculty of Arts and Sciences, Department of Molecular Biology and Genetics, Eyupsultan, Istanbul, Türkiye*

**Abstract:** Bioinformatics has proven to be a powerful tool in enhancing productivity across various fields, including plant biology. Bioinformatics provides significant capabilities for the acquisition, processing, analysis, and interpretation of large amounts of genomic data. With the help of next-generation sequencing technologies, large amounts of genetic data can be generated rapidly. The integration of bioinformatics tools into plant physiology allows the analysis of large amounts of genomic information, providing a better understanding of functional aspects of developmental, metabolic, and reproductive processes. Moreover, it offers a scientific framework for pre-experimental planning, in-experimental management, and post-experimental data analysis. The key applications of bioinformatics comprise gene and pathway identification, molecular docking, sequence analysis, RNA and protein sequence analysis and prediction, gene expression analysis, protein-protein interaction analysis, and statistical techniques that can be executed from genome to phenome. In order to enhance plants, bioinformatics may play a crucial role in encouraging the public release of all sequencing data through repositories, rationally annotating genes, proteins, and phenotypes, and elucidating links between the many components of the plant data. The integration of bioinformatics into plant physiology has the potential to facilitate crop improvement, identification, and/or development of new plant-based functional chemicals and biofortified functional foods and plants that are more resistant to stress conditions. Therefore, this contributes to a more comprehensive understanding in all areas of biology. These new approaches include pan-genomics, artificial intelligence, machine and deep learning applications, CRISPR technology and genome editing, single-cell RNA sequencing, third-generation sequencing systems, RNA engineering and post-transcriptional editing, and metagenomic studies. This chapter reviews the applications of bioinformatics methods in plant physiology and biological databases and their potential contributions to plant physiology.

**Keywords:** Artificial intelligence, Data analysis, Expression analysis, Genome, Metagenomics, Pan-genome, Protein-protein interaction, Single-cell RNA sequencing.

---

[*] **Corresponding author Mehmet Emin Uras:** Haliç University, Faculty of Arts and Sciences, Department of Molecular Biology and Genetics, Eyupsultan, Istanbul, Türkiye; E-mail: meminuras@halic.edu.tr

**Ergun Kaya (Ed.)**

# INTRODUCTION

Computational science supports almost all areas of biology, and plant biology is no exception. Bioinformatics has become an important tool for acquiring, processing, analyzing, and interpreting vast amounts of genomic data. High-throughput technologies generate huge amounts of data, and the integration of bioinformatics tools and the vast data on plant physiology will provide a better understanding of the functional aspects of plants.

Bioinformatics provides a scientific platform for analyzing data before and after laboratory experiments [1]. The main applications in bioinformatics are gene and biological pathway identification, gene annotation, gene sequence analysis, RNA and protein sequence analysis and prediction, and expression analysis; statistics can be performed at the genomic level up to phenomics [2, 3]. Biological pathway identification [4], ncRNA detection [5], plant microbiome analysis [6], integrated phenomics [7], and proteogenomic approach to plant genomics [8] are good examples of recent studies in plant bioinformatics.

Plant physiology can be defined as the science of plant functions, including the dynamic processes of metabolic responses, regulatory processes, plant pathology, growth, and reproduction in living plants [9 - 11]. Important topics, like plant responses to biotic and abiotic stresses [12], production of primary and secondary metabolites [13], and biofortified plant breeding with genetic and epigenetic alteration [14], have gained importance in the area of plant physiology. Plant physiology mainly depends on biophysics, biochemistry, and molecular biology. Bioinformatics and plant physiology meet in molecular applications. The integration of bioinformatic tools into plant physiology offers opportunities for crop improvement, the discovery and development of new plant-derived functional chemicals, biofortified functional foods, and the discovery and development of industrial and energy compounds [15]. This chapter is a discussion of how bioinformatics can be a versatile tool for plant physiology (Fig. 1).

# PLANT BIOINFORMATICS RESOURCES

The comprehension and regulation of plant physiological processes can be achieved through the use of biochemistry, biophysics, and molecular biology applications. However, integrating bioinformatic tools with these applications enables studies to become faster, more efficient, and more practical. One of the most important bioinformatics sources is databases, where genetic information is stored, and there are some tools for different analyses. This section provides an overview of key databases for molecular biology and bioinformatics research on plant physiology.

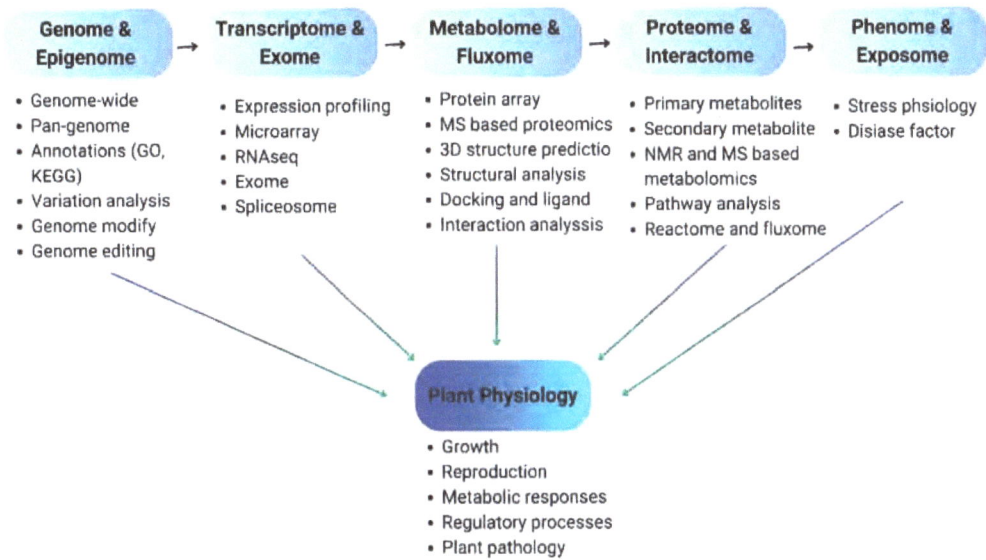

**Fig. (1).** Omics science and bioinformatics have the potential to advance research in plant physiology.

## Major Genetic Information Sources

The National Center for Biotechnology Information (NCBI) GenBank database is a publicly accessible repository of nucleotide and protein information and is managed by the National Library of Medicine of the USA. GenBank collaborates with the European Nucleotide Archive (ENA) and DNA DataBank of Japan (DDBJ) in the International Nucleotide Sequence Database Collaboration. The databases exchange sequencing data to provide a comprehensive and standardized collection of data [16]. The database deposits nucleotide and protein sequences that have been submitted by authors and institutions globally. The sequence size can vary from a single coding or non-coding DNA region to an entire genome (NCBI, 2023). Coding sequences can also be translated to protein sequences. GenBank comprises several sub-databases, including Nucleotide, Genome, Sequence Read Archive (SRA), Gene Expression Omnibus (GEO), Structural Variation (dbVar, exclusive to humans), Single Nucleotide Polymorphism database (dbSNP, exclusive to humans), Reference Sequence Database (RefSeq), Conserved Domain Database (CDD) and Protein Clusters, among others. Sequences in GenBank are submitted as either single/multiple sequences by authors or as bulk submissions from genome survey sequences (GSS), whole genome shotgun (WGS), expressed sequence tags (EST), transcriptome shotgun assembly (TSA), and other high-throughput sequencing studies (HTS) (Table **1**) [17].

**Table 1. Major genetic information databases regarding plant biology.**

| Database/Server | Information Content | Reference/Web |
|---|---|---|
| NCBI GenBank | The primary repositories of genomic information | https://www.ncbi.nlm.nih.gov/ |
| DNA Data Bank of Japan DDBJ | | https://www.ddbj.nig.ac.jp/ |
| European Nucleotide Archive - ENA | | https://www.ebi.ac.uk/ena/browser/home |
| Phytozome | A platform for plant genomes. | https://phytozome-next.jgi.doe.gov/ |
| Ensembl Plants | Includes automatically annotated genomic data of plants | https://plants.ensembl.org/ |
| PubChem (Protein Database) | Focused on chemical compounds | https://pubchem.ncbi.nlm.nih.gov/ |
| ChEBI | Focused on chemical compounds | https://www.ebi.ac.uk/chebi/ |
| UniProt (The Universal Protein Resource) | A comprehensive resource for proteins | https://www.uniprot.org/ |

GenBank is a crucial resource for researchers across all fields of biology, including plant physiology. The website's user-friendly interface and comprehensive search functions make it easy to navigate and retrieve relevant information. It serves as the primary repository of bioinformatics data, offering accessible information on sequences, sizes, chromosome positions, variations, mRNA information, organisms of origin, genes, proteins displaying noteworthy similarity, crucial gene and protein domains, and motifs for any physiological processes. It also provides some toolkits and online tools for analysis and annotation of the gene/protein of interest [16]. Additionally, GenBank and similar databases serve as a central repository for researchers to exchange their discoveries within the scientific community, facilitating accessibility for fellow academics.

Another significant major data source for plant genetics is the Phytozome database, which operates through the Data Portal of the U.S. Department of Energy (DOE) Joint Genome Institute (JGI) Data Portal. The database's main attraction is that Phytozome v13 contains 318 fully sequenced and extensively annotated plant genomes. All genes identified in the Phytozome database have been annotated using PFAM, KEGG, KOG, Panther, GO, the InterPro protein analysis tools, *etc*. Searching and visualization tools within the database enable quick and simple analysis of genes and proteins of interest [18]. The sequences deposited in the Phytozome database are generated exclusively by researchers from the JGI. Sequences from other sources are not accepted. The database offers comprehensive genetic data on plant species' metabolic and genetic characteristics that are of agronomic importance.

The European Molecular Biology Laboratory's European Bioinformatics Institute (EMBL-EBI) hosts an important plant genome database called EnsemblPlants, which is a part of the Ensembl Genomes Project [19]. There are currently more than 100 fully sequenced and annotated plant genomes present in the database. The database collaborates with other specialized plant genetic information databases, including Gramene, the Barley Genome Sequencing Consortium, and the PanOryza databases.

The other major key repository operated by the National Library of Medicine of the United States is PubChem. PubChem's focus is on chemicals like nucleotides, peptides, lipids, carbohydrates, plant primary and secondary metabolites, *etc.* It houses approximately 116 million compounds and 310,000 substances. The database provides essential data pertaining to the chemical and physical features, biological functions, molecular structures, hazardous profiles, descriptors, health aspects, patents, and safety concerns in the domain of plant physiology research [20]. The ChEBI database works as a complementary database to the PubChem database, and it focuses on the ontology of 'small' chemical compounds of biological importance. The ontology process of chemical entities is performed using two sub-ontologies: a chemical entity ontology that relies on chemical structure and a role ontology that is based on biological activity. ChEBI is hosted by the European Bioinformatics Institute (EBI) [21]. The Universal Protein Resource (UniProt) is a significant database consortium that offers high-quality data on protein annotation, sequence, and structure. The consortium hosts the UniProt Database (UniProtKB), the UniProt Archive (UniParc), and the UniProt Reference Clusters (UniRef) [22].

## Specialized Databases on Plant Biology

Specialized databases and consortia databases offer comprehensive information on particular subjects or species. The data is usually curated by field experts. Primary genetic information, annotations, expression profiles, and metabolic pathway models can be obtained from these databases on a species basis [23]. Some specialized databases are shown in Table **2**. Specialized databases are usually supported by consortia to consolidate standards within a specific field. The AgBioData consortium supports some databases specializing in plant genomics, particularly for industrial or agricultural plants. The consortium hosts and collaborates with specialized databases focused on various crops. The consortium supports genetic and breeding research, providing readily available genetic data to the public as a member of the scientific community FAIR (Findable, Accessible Interoperable, and Reusable) consortium [24, 25].

**Table 2. Some specialized databases and genetic information sources on plant genetics.**

| Database/Server | Information Content | Reference/Web |
|---|---|---|
| AgBioData | A consortium on Agriculturally important plant genetics | https://www.agbiodata.org/databases |
| FAIR | A consortium on genetics | https://fairsharing.org/ |
| The Arabidopsis Information Resource (TAIR) | Genomic information on the model plant *Arabidopsis thaliana* | https://www.arabidopsis.org |
| Plant Genome DataBase Japan (PGDBj) | A comprehensive Genetic Database for plants | http://pgdbj.jp |
| FLAGdb++ | Plant bioinformatics database for gene annotations | http://tools.ips2.u-psud.fr/FLAGdb |
| The Bio-Analytic Resource for Plant Biology (BAR) | Database of genetic information, mainly on plants | https://bar.utoronto.ca/ |
| miRBase | The archive for microRNA sequences and annotations | https://mirbase.org/ |
| Gramene | Database of genetic information, mainly on the Gramineae Family | https://www.gramene.org |
| PanOryza | Database of genetic information, mainly on the Genus Oryza | https://panoryza.org |
| GDR Rosa | Database of genetic information, mainly on the Rosaceae Family | https://www.rosaceae.org/ |
| Solanaceae Genomics Network (SGN) | Database of genetic information, mainly on the Solanaceae Family | https://solgenomics.net/ |
| MaizeGDB | Database of genetic information, mainly on Zea species | https://www.maizegdb.org/ |
| Brassicaceae Database (BRAD) | Database of genetic information, mainly on the Brassicaceae Family | http://brassicadb.cn/ |

The Arabidopsis Information Resource (TAIR) provides comprehensive information on the model plant *Arabidopsis thaliana* for genetic and molecular studies. Since the primary studies in plant physiology and genetics were mainly conducted on *A. thaliana*, it is regarded as a reference plant. The TAIR database offers significant annotations on genes and proteins that are valuable in plant physiology studies [26]. In addition to the TAIR database, Plant Genome DataBase Japan (PGDBj), FLAGdb++, Bio-Analytic Resource for Plant Biology (BAR), and miRbase databases offer significant information for plant improvement and breeding studies. PGDBj offers data on DNA markers, QTL listings, and orthologous genes, as well as cross-search against some other specialized databases [27]. FLAGdb++ database provides integrated structural and functional annotations of the gene and genomes on an integrative basis [28]. The Botany Array Resource (BAR) database offers comprehensive datasets on key plant species, including gene expression and protein analysis tools, mapping, and molecular marker tools, as well as visualization tools [29]. miRbase database provides micro-RNA sequences and annotations, which can be used in genome analysis, plant growth and reproduction, pathology, stress response, and metabolic pathway analysis [30].

Species-specific databases play a vital role in plant physiological studies, particularly in the areas of genetic and metabolic pathway analyses. These databases provide essential information and tools for researchers to conduct their investigations, allowing for a greater understanding of plant biology. Specialized databases are available for notable and significant family groups, especially for plants with commercial and agricultural importance. Gramene, PanOryza, GDR Rosa, Solanaceae Genomics Network (SGN), MaizeGDB, and Brassicaceae Database (BRAD) databases are the prominent species-specific databases.

Plant metabolomics plays an increasingly important role in managing and mitigating the effects of environmental stresses due to the changing environment. A holistic approach incorporating genomic, transcriptomic, proteomic, and metabolomic data in plant physiology provides a clearer understanding of physiological processes in plants (Table **3**) [31].

**Table 3. Major databases and sources for metabolites and metabolic pathways.**

| Database/Server | Information Content | Reference/Web |
|---|---|---|
| **GenomeNet** | A collection of databases, including KEGG and DBGET | https://www.genome.jp/ |
| **GO: The Gene Ontology Database** | Specialised in genomic ontology and annotations | https://geneontology.org/ |
| **PMN: Plant Metabolic Network** | Biochemical pathways of plant metabolism | https://www.plantcyc.org/ |
| **Plant Reactome** | Biochemical pathways of plant metabolism | https://plantreactome.gramene.org/ |
| **ModelSEED** | Plant metabolic models | https://modelseed.org/ |

There are some databases/servers specialized in metabolites, metabolic interactions, and pathways, as presented in Table **3**. These databases provide reliable information on genetic interactions and chemical pathways that are valuable for studies on plant physiology. The GenomeNet database houses the

DBGET, KEGG, varDB, and Community DBs databases, in addition to several bioinformatics tools. DBGET is a system for retrieving data from all sub and collaborated databases. KEGG is a specialized database on genetic connections of biochemicals and pathways [32, 33]. The Gene Ontology Database is a comprehensive resource focused on the functions of genes and their products, classified by molecular function, cellular component, and biological process [34]. The Plant Metabolic Network (PMN) provides vast information regarding genes, enzymes, reactions, metabolites, and pathways related to primary and secondary metabolism in plants on a species-specific basis [35]. The PLANT REACTOME database is another database that focuses on plant metabolic pathways. The database is open access, open sources, and peer-reviewed, and the data is curated manually by expert biologists [36]. Another database specializing in plant metabolic pathways is the ModelSEED database. The database has a semi-automated system for calculating genome-scale optimized metabolic models [37].

## GENOME-WIDE APPROACH IN PLANT PHYSIOLOGY

Plant physiological processes are the main basis for all plant ecological and agronomic traits. The production of primary and secondary metabolites depends on specialized metabolic pathways. Bioinformatic approaches have identified many metabolic gene clusters in major crops and continue to help in the discovery of novel enzymes and unknown pathways. In addition, there are some unresolved points about plant physiological processes related to plant defense against pathogens, adaptation to the environment, plant communication, and some other metabolic traits [38]. At this juncture, genome-wide (a genome-wide association study compares the genomes of numerous individuals in an effort to identify genetic markers linked to a specific phenotypic or illness risk) and pan-genomic methods (it encodes for every conceivable lifestyle that an organism could lead and defines the whole genomic repertoire of a particular evolutionary group) prove to be valuable tools for elucidating physiological processes at the molecular level. These methods may be utilized as pre- and post-analyses (Fig. **2**).

In the genome-wide approach, important plant metabolic pathway members are identified. Some critical members of plant metabolic pathways have been identified, including key regulators of the cell cycle, cyclin family proteins [41], important enzymes for structural maintenance, the lignification toolbox (monolignol synthesis) enzymes in *Arabidopsis* [42], pectin methylesterases in *Oryza sativa* [43], stress-protective agents, galactinol synthase proteins in tomato (*Solanum lycopersicum*) [44], and ascorbate and glutathione peroxidases [39]. QTL determination in C and N metabolic pathways in maize (*Zea mays* ssp. *mays*) [45] and elucidation of the genetic architecture of morphologic traits in Arabidopsis have also been identified [46]. In this approach, the investigation of certain features of the target gene is undertaken (Fig. **2**), namely the gene name and database ID, nucleotide sequence, physicochemical properties (such as position on the chromosome/s, exon/intron organization, and length), the domains and motifs contained within, expression profile, gene ontology (GO), cis-regulatory elements, transcripts, and polyadenylation signals. Some features of the protein have been investigated in genome-wide studies, including length, amino acid composition, molecular weight, pI, sub-cellular localization, N-glycosylation and phosphorylation positions, prediction and validation of 3D structure, and active sites.

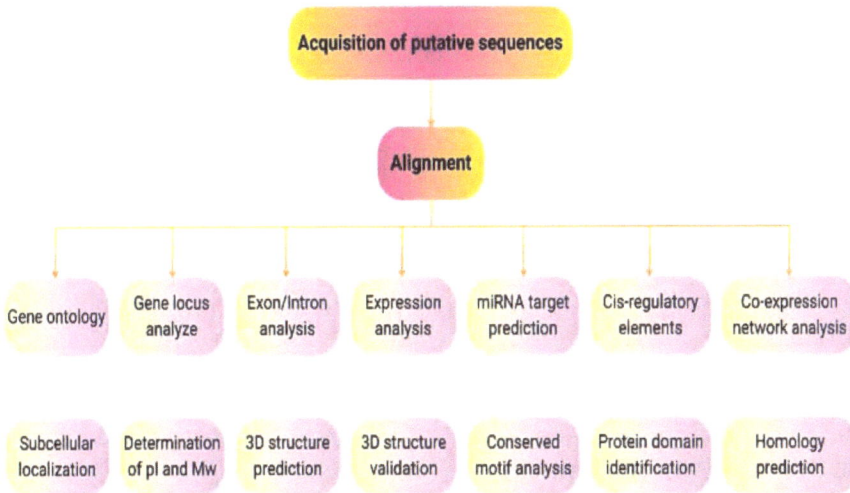

**Fig. (2).** Applications that can be used in a genome-wide approach. Adapted from [39] and [40].

## The Pan-Omics Approaches in Plant Physiology

The pan-genomic approach is a novel method that can be utilized to evaluate intraspecific diversity. Advances in analytical techniques and bioinformatics tools have facilitated the emergence of the pan-transcriptome and pan-epigenome fields, which may be developed by this method [47]. Integrating these approaches will lead to increased efficiency in studying topics such as intraspecific diversity, mutation, and alternative splicing [48]. The pangenome can be defined as the combined set of genomes from various lines, varieties, or population members within the same species [49]. The pan-genome approach was originally developed for prokaryotic genomes. Pan-genomes consist of two main parts: the core genome, which is present in all individuals, and the dispensable genome, which is not present in all individuals [50]. Using the pangenomes provides ease and reliability for detecting variations and understanding the interplay between genome and environment. In addition to single nucleotide polymorphism (SNP), presence-absence variation (PAV), small deletions, and small insertions, long fragment structural variation (SVs) can be effective in plant physiological variations. Structural variations may contribute to metabolic and phenotypic diversity, coping with stress conditions and mating systems. The pan-genomic approach helps to understand different physiological processes in ecotypes/varieties/lines of the same species [47].

Although some earlier studies on plant pan-genomes were conducted using partial genome data, the first whole genome assembly-based analysis of a plant pan-genome was published in 2014. In the study, wild soybean (*Glycine soja*) was used, and some variable genes were identified related to plant physiological characteristics, including the timing of flowering and maturity, biomass and organ size, disease resistance, and seed composition that are absent in cultivated soybean (*Glycine max* L.) [51, 52]. Subsequently, plant pan-genomes were constructed using *Arabidopsis thaliana* [53], *Brassica napus, B. napus, B. oleracea, Oryza sativa, O. rufipogon, Zea mays, Populus trichocarpa, Brachypodium distachyon, Medicago truncatula, Triticum aestivum, T. turgidum, Capsicum* spp., *Sesamum indicum, Helianthus annuus, Solanum lycopersicum, Juglans* spp., *Vitis vinifera, Musa acuminata, Ipomoea trifida, Cicer arietinum, Cucumis sativus, Malus domestica, Cocos nucifera, etc* [52, 54]. Pangenome investigations have primarily focused on plant species of agricultural importance due to whole genome sequencing studies being predominantly conducted on these species.

## Bioinformatics in Phenome Research

Second and third-generation sequencing technologies provide quick access to whole genomes or parts of genomes and without great investment and effort. Access to large amounts of data requires effective algorithms for analysis and data storage. In addition, with improved technology, some affordable equipment and set-ups are available that produce significant amounts of high-throughput phenomics data. By using bioinformatics tools to extract data from annotated genomes, it is generally easier to identify the physiological processes, enzymes, and other elements that determine phenotypic traits [55]. Some tools can be used to combine genomic and phenomics data. However, there are some challenges in combining the data sets. The extensive genomic and phenotypic data sets and specific tools for Arabidopsis are easy to combine; for crops and other plants, genomic data, phenotypic data, and combining tools are relatively scarce. For *Arabidopsis*, as a model organism, there are extensive databases such as easyGWAS (Genome-wide association studies) [56], AraGWAS [57], GWAPP [58], AraPheno, and AraRNASeq [59]. The alternative approach is to use machine learning methods with the limitation of focusing on only a few variables and excluding others. Over time, new databases will be created, and existing databases will be expanded to provide information for understanding the physiological process behind phenotypic traits from genome to phenome for plants other than Arabidopsis [55].

## SEQUENCE-BASED STUDIES IN PLANT PHYSIOLOGY

Next-generation and third-generation sequencing technologies have opened up an era of opportunity with more accurate, reliable, accessible, and affordable sequencing. Novel sequencing systems offer advantages for plant physiology and other related fields. These systems provide an objective and precise method for studying genetic material and analyzing biological processes. With their clear and concise data output, sequencing systems contribute to a logical and coherent flow of information, allowing for accurate scientific conclusions.

DNA and RNA sequencing are utilized in various studies, such as genomics, transcriptomics, and interactomes, to identify and annotate crucial genes and proteins of considerable physiological significance [60]. Recently, sequence-based genetic studies have become increasingly involved in the study of plant physiology. Sequence-based studies include RNA sequencing (RNA-Seq), which can be used to identify genes and regulators critical for development, stress response, and specific physiological processes, small RNA sequencing (sRNA-Seq), such as small interfering RNAs (siRNAs) and microRNAs (miRNAs), which can be used to characterize regulatory sRNAs in gene regulation, developmental stages and stress responses, epigenomics, which can be used to understand plant responses to environmental stressors such as salinity, drought, temperature, *etc* [61]. Additionally, to get a better understanding of the genetic and physiological traits, sequence-based analysis can be used for the identification of Quantitative Trait Loci (QTL) [62]. NGS-based methods such as restriction site-associated DNA sequencing (RAD-Seq), Genome-Wide Association Studies (GWAS) and Genotyping-by-Sequencing (GBS), and some other bioinformatics tools are employed for detecting QTL. A QTL is basically a specific genomic region/s that determines the phenotype associated with a crucial trait or disease [62]. Searching genomes for the QTL is a good example of using bioinformatics tools in plant physiology research.

RNA sequencing (RNA-seq) technology is a highly efficient method for detecting and analyzing both coding and non-coding RNAs in natural or experimental conditions. This technology has shown great potential for use in a wide range of research applications, including the quantification of gene expression levels, comparison of genomic information, and allele-specific expression with high accuracy and sensitivity [63]. Additionally, the technology is valuable for identifying splicing variations, annotating genes with low function and tissue specificity, and obtaining preliminary transcriptome-scale genomic data of newly discovered or orphan plant species [64, 65]. The overall process for transcriptome studies utilizing RNA-seq is illustrated in (Fig. **3**).

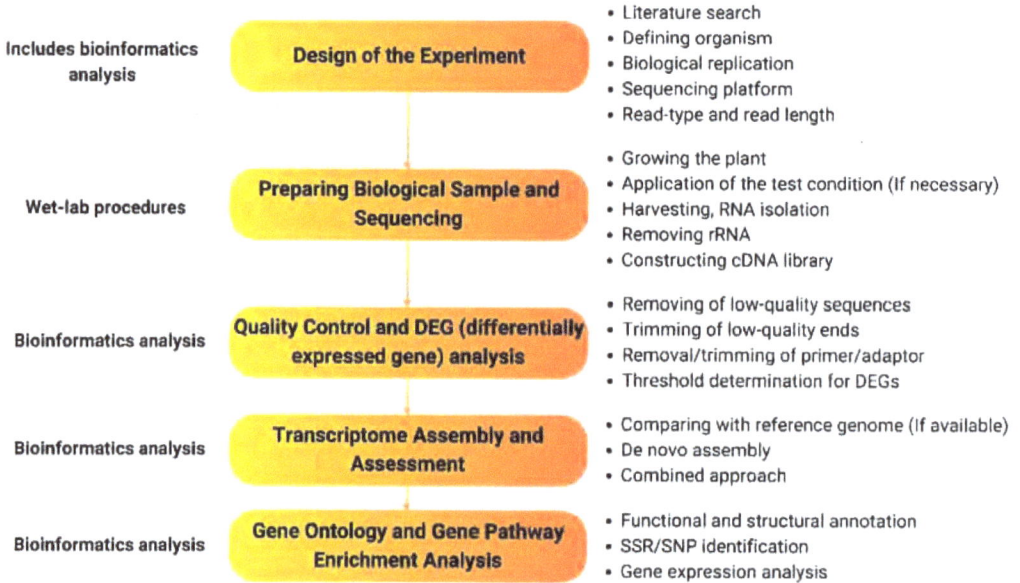

Fig. (**3**). General workflow for RNA-seq transcriptome analysis. Adapted from [66] and [67].

Bioinformatics tools can aid in designing experiments, validating findings, and solving problems during wet lab experimentation, as well as interpreting results afterward in studies on plant physiology. As a case study, Han *et al.* (2019) conducted a transcriptome study of cotton (*Gossypium hirsutum* L.) under cadmium (Cd) stress. By using the RNA-seq technology, the authors detected differentially expressed genes (DEGs), identified a novel gene that takes part in the Cd tolerance mechanism, and defined some important mechanisms in the interaction network of Cd stress in cotton. Bioinformatic tools were utilized in the study for data quality control, read alignment, transcript assembly, quantification, differential expression analysis, functional analysis, visualization, interpretation, data management, and integration steps [67]. While the study is mainly based on genetic and bioinformatic analysis, it has provided a comprehensive insight into cotton's physiological processes and metabolic responses under Cd stress.

Analyzing and interpreting high-throughput transcriptome data using expression networks is an effective approach to identifying physiologically significant gene interactions and previously unidentified genes. The identification process relies on three distinct applications: identification of regulators and their target molecules, prediction of metabolic pathways through structural gene discovery, and annotation transfer *via* comparative co-expression networks among plant species [68, 69]. To identify and analyze the neighboring metabolic genes, Toghe and Fernie (2020) analyzed various surveys that utilized bioinformatics tools for transcriptomics-based functional genomic research on plant metabolism. Neighboring metabolic genes are crucial to plant metabolic diversity and secondary metabolite production, and they are discovered as metabolic gene clusters and duplicated genes through neo-functionalization in plant genomes. The authors concluded that co-expression network analysis provides a powerful opportunity to explore such networks and interactions in plant genomes [70]. The bioinformatics workflow for the construction of co-expression networks is shown in Fig. (**4**).

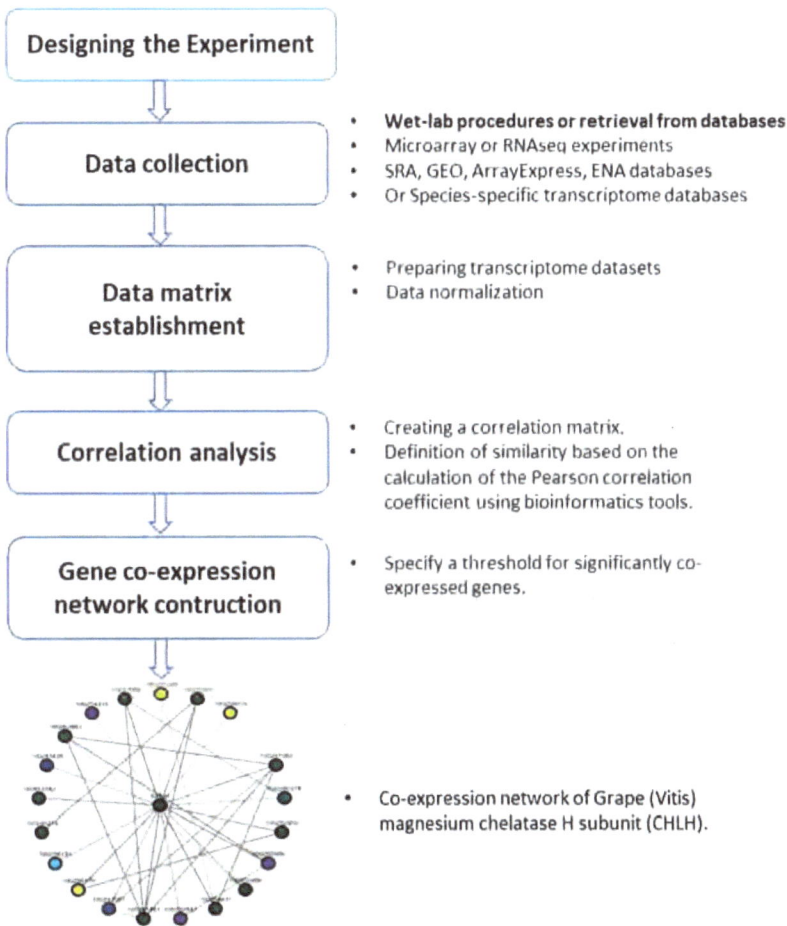

**Designing the Experiment**

**Data collection**
- **Wet-lab procedures or retrieval from databases**
- Microarray or RNAseq experiments
- SRA, GEO, ArrayExpress, ENA databases
- Or Species-specific transcriptome databases

**Data matrix establishment**
- Preparing transcriptome datasets
- Data normalization

**Correlation analysis**
- Creating a correlation matrix.
- Definition of similarity based on the calculation of the Pearson correlation coefficient using bioinformatics tools.

**Gene co-expression network contruction**
- Specify a threshold for significantly co-expressed genes.

- Co-expression network of Grape (Vitis) magnesium chelatase H subunit (CHLH).

**Fig. (4).** The workflow for the construction of co-expression networks. Adapted from [69]; a sample gene expression map was generated by using Attend II database [71].

Co-expression networks are constructed based on data obtained from wet lab methods such as microarrays and RNA-seq. Data can be retrieved from public transcriptome databases such as SRA, GEO, ArrayExpress, ENA, and species-specific databases. Easily accessible web-based and command-line tools are available for constructing co-expression networks that can be used to analyze plant psychological processes [69].

Understanding plant physiological changes and developmental stages through the processing of gene expression data using bioinformatics tools is a practical and efficient method. The utilization of bioinformatic tools has become indispensable for obtaining significant outcomes. In this context, a relatively new information source for plant physiology is single-cell sequencing, which involves analyzing the genetic information of an individual cell type at the multi-omics level. This technique provides a deeper understanding of cellular heterogeneity. Gene expression analysis and co-expression networks are based on the expression level of differentially expressed genes (DEGs) data. The data can be obtained from an organ, from a specific type of tissue, or from a single type of cell. Conventional sequencing studies involve isolated RNA from plant organs or bulk tissues without consideration of spatial and temporal resolution and tissue/cell-specific responses [72]. Single-cell sequencing offers detailed, high-resolution insight to uncover new genes by identifying unique expression patterns specific to individual cell types. The technique provides DEGs data considering the spatial distribution of the cells and their temporal state transition and developmental stages in plant cells [65, 72]. The process of single-cell transcriptomics, achieved through scRNA-seq, comprises steps akin to those taken in bulk RNA-seq analysis. Such steps include reading and aligning sequences, constructing an expression matrix, normalizing data, defining DEGs, and creating co-expression maps. However, there are further procedures and statistical analyses for reducing dimensionality and clustering cells that facilitate the elimination of low-quality and varying cell types from the dataset [73].

## ANALYSIS OF PROTEIN STRUCTURE AND INTERACTIONS RELATED TO PLANT PHYSIOLOGY

Plant metabolic pathways rely on a wide range of individual, complex, and molecular machinery proteins. A comprehensive understanding of the structural aspects and interactions between proteins is important in plant physiology, as well as in other areas of biology [74, 75]. Analysis of the 3D structure and physicochemical properties of proteins, protein-protein interactions in metabolic pathways, and protein responses to stress conditions can be evaluated using bioinformatics tools to contribute to plant physiological studies [76]. A further benefit of protein structure and interaction analysis is its ability to provide information on the interactions between the plant and its biotic environment [77].

The primary databases for protein analysis are the UniProt and RCSB Protein Data Bank (PDB). The UniProt database provides high-quality protein sequence and annotation information. The database includes BLAST search and alignment tools and new options for effective searching and analysis. Additionally, a powerful visualization tool was provided for analyzing the 3D structures of proteins. The database offers comprehensive information with the collaboration of multiple databases regarding name, taxonomy, subcellular localization, functional annotation, variants and isoforms, post-transcriptional modifications and/or processing, expression and interaction, 3D structure, and family and domain [22]. The PDB focuses on 3D protein structure and provides information on structure, drug design, and protein interactions [78].

Protein-protein interactions (PPI) are one of the main foundations of life's complexity [79]. For analysis of such interactions, there are some databases, including STRING, BioGRID, IntAct, DIP (Database of Interacting Proteins), KEGG, and Reactome. The databases provide information on interactions, biochemical reactions, and pathways. The databases have a significant limitation, as they contain insufficient plant PPI data. Typically, PPI data is readily available for model plants and economically significant plants, particularly *A. thaliana* [77]. There is the option of using software for manual analysis. The Cytoscape software is one the most convenient and flexible software for analyzing and visualizing large-scale PPI data [80].

## FUTURE PERSPECTIVES

The journey of molecular genetics began with Frederich Miescher's discovery of DNA in the cell nucleus. Over the past 150 years, the scientific community has succeeded in answering some fundamental questions, including elucidating the structure of nucleic acids and proteins, understanding how the genetic code works, amplifying and sequencing genomes and proteins, and investigating important molecular interactions at all omics levels. Now, we can edit genomes and create improved organisms. However, as our understanding expands, we encounter more significant obstacles and problems as we strive towards more substantial aspirations. In the study on plant physiology, bioinformatics tools are crucial in supporting cutting-edge research, and their ability to analyze complex data and identify patterns enables discoveries and advances in the understanding of the mechanisms that underpin the growth and development of plants.

Artificial intelligence (AI) has become a significant technology impacting various facets of our lives. Its applications include natural language processing, computer vision, text-to-speech, motion, and machine learning (ML). AI promises improvements in emerging and improved disciplines in biology, including structural biology [81], genetic engineering [82], drug design [83], and molecular interaction network analysis [65]. The most important improvement will be the application of quantum computing technology to AI and bioinformatics, which will take us into a new era [84].

The primary AI applications utilized in biological and molecular sciences are ML and DL technologies. ML and its subset, deep learning (DL), involve sophisticated statistical analysis to make predictions and decisions based on data [85]. The system seeks to predict outcomes based on the information available to it, and the more data we provide, the more accurate its predictions will be. While there is a more extensive dataset available for crucial agricultural models and plants, our knowledge of less significant plants is insufficient. To gain a better insight into the physiological aspects of less important and orphan crops, the dataset must be expanded. With the help of extensive data sets and the latest technologies, such as artificial intelligence, we can achieve more accurate results. A combined approach using single-cell sequencing and multiple omics, supported by AI, provides a comprehensive understanding of physiological processes in plants [65]. More detailed information on AI and ML can be found in Chapter 11.

As an important source of information and analytical tool, bioinformatics is involved in genome editing applications and RNAi technology. CRISPR-mediated genome editing and RNAi technologies are relatively new technologies where bioinformatics tools can be used effectively to contribute to plant physiology. The most important feature of CRISPR technology is its ability to intervene at precise locations in the genome. Bioinformatics tools are used to design the guide RNA (gRNA), identify potential target genes and elucidate functions related to plant physiology, predict off-target effects, ensure the efficiency and accuracy of CRISPR-mediated genome editing, and lead the development of improved crops [86]. RNAi technology is used to study gene function, identify target genes, and design functional small interfering RNA (siRNA) using bioinformatics tools. RNAi technology can be used to improve the response of plants to stress conditions in a non-transgenic approach [87, 88].

## CONCLUSION

With a changing environment and a growing population, there are emerging challenges in protecting vegetation and biodiversity and providing food for animals and humans. To overcome these challenges, plant genetics and physiological studies will be very crucial, and bioinformatics stands at the intersection of these two disciplines. Combining bioinformatics tools with studies related to genomics, proteomics, and metabolomics in plant physiology enhances outcomes effectively. The use of bioinformatics to understand the genetic basis of physiological processes provides significant insight into plant physiology. However, genetic data on plant genetics is relatively limited. While model and commercially significant plants have sufficient data in biological databases, a significant limitation lies in the lack of sufficient information for less important and orphan plants. To overcome these challenges, the generation of more information and interpretations by employing bioinformatics tools provides an excellent opportunity to cope with the agricultural and environmental problems that are likely to arise in the future.

# REFERENCES

[1]     Oliver GR, Hart SN, Klee EW. Bioinformatics for clinical next generation sequencing. Clin Chem 2015; 61(1): 124-35.
[http://dx.doi.org/10.1373/clinchem.2014.224360] [PMID: 25451870]

[2]     Ma X, Meng Y, Wang P, Tang Z, Wang H, Xie T. Bioinformatics-assisted, integrated omics studies on medicinal plants. Brief Bioinform 2020; 21(6): 1857-74.
[http://dx.doi.org/10.1093/bib/bbz132] [PMID: 32706024]

[3]     Pazhamala LT, Kudapa H, Weckwerth W, Millar AH, Varshney RK. Systems biology for crop improvement. Plant Genome 2021; 14(2): e20098.
[http://dx.doi.org/10.1002/tpg2.20098] [PMID: 33949787]

[4]     Zhang Y, Li D, Zhou R, *et al.* Transcriptome and metabolome analyses of two contrasting sesame genotypes reveal the crucial biological pathways involved in rapid adaptive response to salt stress. BMC Plant Biol 2019; 19(1): 66.
[http://dx.doi.org/10.1186/s12870-019-1665-6] [PMID: 30744558]

[5]     Chao H, Hu Y, Zhao L, *et al.* Biogenesis, functions, interactions, and resources of non-coding RNAs in plants. Int J Mol Sci 2022; 23(7): 3695.
[http://dx.doi.org/10.3390/ijms23073695] [PMID: 35409060]

[6]     Xu L, Pierroz G, Wipf HML, *et al.* Holo-omics for deciphering plant-microbiome interactions. Microbiome 2021; 9(1): 69.
[http://dx.doi.org/10.1186/s40168-021-01014-z] [PMID: 33762001]

[7]     Zhang Y, Zhang W, Cao Q, *et al.* WinRoots: a high-throughput cultivation and phenotyping system for plant phenomics studies under soil stress. Front Plant Sci 2022; 12: 794020.
[http://dx.doi.org/10.3389/fpls.2021.794020] [PMID: 35154184]

[8]     Chen MX, Zhu FY, Gao B, *et al.* Full-length transcript-based proteogenomics of rice improves its genome and proteome annotation. Plant Physiol 2020; 182(3): 1510-26.
[http://dx.doi.org/10.1104/pp.19.00430] [PMID: 31857423]

[9]     Berger S, Sinha AK, Roitsch T. Plant physiology meets phytopathology: plant primary metabolism and plant pathogen interactions. J Exp Bot 2007; 58(15-16): 4019-26.
[http://dx.doi.org/10.1093/jxb/erm298] [PMID: 18182420]

[10]    Taiz L, Zeiger E. Plant Physiology. 5$^{th}$ ed., California: The Benjamin Cummings Publishing Company 2015.

[11]    Ördög V, Moltar Z. Plant physiology. Debreceni Egyetem 2011; p. 1.

[12]    Aftab T, Roychoudhury A. Crosstalk among plant growth regulators and signaling molecules during biotic and abiotic stresses: molecular responses and signaling pathways. Plant Cell Rep 2021; 40(11): 2017-9.

[http://dx.doi.org/10.1007/s00299-021-02791-5] [PMID: 34561762]

[13]     Salam U, Ullah S, Tang ZH, *et al.* Plant metabolomics: an overview of the role of primary and secondary metabolites against different environmental stress factors. Life (Basel) 2023; 13(3): 706.
[http://dx.doi.org/10.3390/life13030706] [PMID: 36983860]

[14]     Li J, Scarano A, Gonzalez NM, *et al.* Biofortified tomatoes provide a new route to vitamin D sufficiency. Nat Plants 2022; 8(6): 611-6.
[http://dx.doi.org/10.1038/s41477-022-01154-6] [PMID: 35606499]

[15]     Oksman-Caldentey KM, Saito K. Integrating genomics and metabolomics for engineering plant metabolic pathways. Curr Opin Biotechnol 2005; 16(2): 174-9.
[http://dx.doi.org/10.1016/j.copbio.2005.02.007] [PMID: 15831383]

[16]     Benson DA, Cavanaugh M, Clark K, *et al.* GenBank. Nucleic Acids Res 2018; 46(D1): D41-7.
[http://dx.doi.org/10.1093/nar/gkx1094] [PMID: 29140468]

[17]     Sayers EW, Karsch-Mizrachi I. Using GenBank. In: Edwards D, Ed. Plant Bioinformatics: Methods and Protocols. New York: Humana Press 2016; pp. 1-22.
[http://dx.doi.org/10.1007/978-1-4939-3167-5_1]

[18]     Goodstein DM, Shu S, Howson R, *et al.* Phytozome: a comparative platform for green plant genomics. Nucleic Acids Res 2012; 40(D1): D1178-86.
[http://dx.doi.org/10.1093/nar/gkr944] [PMID: 22110026]

[19]     Bolser DM, Staines DM, Perry E, Kersey PJ. Ensembl plants: integrating tools for visualizing, mining, and analyzing plant genomic data. In: van Dijk ADJ, Ed. Plant Genomics Databases: Methods and Protocols. New York: Humana Press 2017; pp. 1-31.
[http://dx.doi.org/10.1007/978-1-4939-6658-5_1]

[20]     Kim S, Chen J, Cheng T, *et al.* PubChem 2023 update. Nucleic Acids Res 2023; 51(D1): D1373-80.
[http://dx.doi.org/10.1093/nar/gkac956] [PMID: 36305812]

[21]     Hastings J, Owen G, Dekker A, *et al.* ChEBI in 2016: Improved services and an expanding collection of metabolites. Nucleic Acids Res 2016; 44(D1): D1214-9.
[http://dx.doi.org/10.1093/nar/gkv1031] [PMID: 26467479]

[22]     Bateman A, Martin M-J, Orchard S, *et al.* UniProt: the üniversal protein knowledgebase in 2023. Nucleic Acids Res 2023; 51(D1): D523-31.
[http://dx.doi.org/10.1093/nar/gkac1052] [PMID: 36408920]

[23]     Xiong J. Essential bioinformatics. New York: Cambridge University Press 2006.
[http://dx.doi.org/10.1017/CBO9780511806087]

[24]     Harper L, Campbell J, Cannon EKS, *et al.* AgBioData consortium recommendations for sustainable genomics and genetics databases for agriculture. Database (Oxford) 2018; bay088.
[http://dx.doi.org/10.1093/database/bay088] [PMID: 30239679]

[25]     Reiser L, Harper L, Freeling M, Han B, Luan S. FAIR: a call to make published data more findable, accessible, interoperable, and reusable. Mol Plant 2018; 11(9): 1105-8.
[http://dx.doi.org/10.1016/j.molp.2018.07.005] [PMID: 30076986]

[26]     Berardini TZ, Reiser L, Li D, *et al.* The *Arabidopsis* information resource: Making and mining the "gold standard" annotated reference plant genome. Genesis 2015; 53(8): 474-85.
[http://dx.doi.org/10.1002/dvg.22877] [PMID: 26201819]

[27]     Asamizu E, Ichihara H, Nakaya A, *et al.* Plant genome dataBase Japan (PGDBj): a portal website for the integration of plant genome-related databases. Plant Cell Physiol. 2014; 1: 55(1): e8.
[http://dx.doi.org/10.1093/pcp/pct189]

[28]     Tamby JP, Brunaud V. FLAGdb++: A Bioinformatic Environment to Study and Compare Plant Genomes. In: van Dijk ADJ, Ed. Plant Genomics Databases: Methods and Protocols. New York: Humana Press 2017; pp. 79-101.

[http://dx.doi.org/10.1007/978-1-4939-6658-5_4]

[29]   Toufighi K, Brady SM, Austin R, Ly E, Provart NJ. The botany array resource: e-Northerns, expression angling, and promoter analyses. Plant J 2005; 43(1): 153-63.
[http://dx.doi.org/10.1111/j.1365-313X.2005.02437.x] [PMID: 15960624]

[30]   Kozomara A, Birgaoanu M, Griffiths-Jones S. miRBase: from microRNA sequences to function. Nucleic Acids Res 2019; 47(D1): D155-62.
[http://dx.doi.org/10.1093/nar/gky1141] [PMID: 30423142]

[31]   Piasecka A, Kachlicki P, Stobiecki M. Analytical methods for detection of plant metabolomes changes in response to biotic and abiotic stresses. Int J Mol Sci 2019; 20(2): 379.
[http://dx.doi.org/10.3390/ijms20020379] [PMID: 30658398]

[32]   Kanehisa M. Linking databases and organisms: GenomeNet resources in Japan. Trends Biochem Sci 1997; 22(11): 442-4.
[http://dx.doi.org/10.1016/S0968-0004(97)01130-4] [PMID: 9397687]

[33]   Fujibuchi W, Goto S, Migimatsu H, *et al.* DBGET/LinkDB: an integrated database retrieval system. InPac. Symp Biocomput 1998; 98: 683-94.

[34]   Aleksander SA, Balhoff J, Carbon S, *et al.* The gene ontology knowledgebase in 2023. Genetics 2023; 224(1): iyad031.
[http://dx.doi.org/10.1093/genetics/iyad031] [PMID: 36866529]

[35]   Hawkins C, Ginzburg D, Zhao K, *et al.* Plant Metabolic Network 15: A resource of genome-wide metabolism databases for 126 plants and algae. J Integr Plant Biol 2021; 63(11): 1888-905.
[http://dx.doi.org/10.1111/jipb.13163] [PMID: 34403192]

[36]   Naithani S, Gupta P, Preece J, *et al.* Plant Reactome: a knowledgebase and resource for comparative pathway analysis. Nucleic Acids Res 2019; 48(D1): gkz996.
[http://dx.doi.org/10.1093/nar/gkz996] [PMID: 31680153]

[37]   Henry CS, DeJongh M, Best AA, Frybarger PM, Linsay B, Stevens RL. High-throughput generation, optimization and analysis of genome-scale metabolic models. Nat Biotechnol 2010; 28(9): 977-82.
[http://dx.doi.org/10.1038/nbt.1672] [PMID: 20802497]

[38]   Schläpfer P, Zhang P, Wang C, *et al.* Genome-wide prediction of metabolic enzymes, pathways, and gene clusters in plants. Plant Physiol 2017; 173(4): 2041-59.
[http://dx.doi.org/10.1104/pp.16.01942] [PMID: 28228535]

[39]   Akbudak MA, Filiz E, Vatansever R, Kontbay K. Genome-wide identification and expression profiling of ascorbate peroxidase (APX) and glutathione peroxidase (GPX) genes under drought stress in Sorghum (*Sorghum bicolor* L.). J Plant Growth Regul 2018; 37(3): 925-36.
[http://dx.doi.org/10.1007/s00344-018-9788-9]

[40]   Neupane S, Schweitzer SE, Neupane A, *et al.* Identification and characterization of mitogen-activated protein kinase (MAPK) genes in sunflower (*Helianthus annuus* L.). Plants 2019; 8(2): 28.
[http://dx.doi.org/10.3390/plants8020028] [PMID: 30678298]

[41]   Wang G, Kong H, Sun Y, *et al.* Genome-wide analysis of the cyclin family in *Arabidopsis* and comparative phylogenetic analysis of plant cyclin-like proteins. Plant Physiol 2004; 135(2): 1084-99.
[http://dx.doi.org/10.1104/pp.104.040436] [PMID: 15208425]

[42]   Raes J, Rohde A, Christensen JH, Van de Peer Y, Boerjan W. Genome-wide characterization of the lignification toolbox in *Arabidopsis*. Plant Physiol 2003; 133(3): 1051-71.
[http://dx.doi.org/10.1104/pp.103.026484] [PMID: 14612585]

[43]   Jeong HY, Nguyen HP, Lee C. Genome-wide identification and expression analysis of rice pectin methylesterases: Implication of functional roles of pectin modification in rice physiology. J Plant Physiol 2015; 183: 23-9.
[http://dx.doi.org/10.1016/j.jplph.2015.05.001] [PMID: 26072144]

[44]    Filiz E, Ozyigit II, Vatansever R. Genome-wide identification of galactinol synthase (GolS) genes in *Solanum lycopersicum* and *Brachypodium distachyon*. Comput Biol Chem 2015; 58: 149-57.
[http://dx.doi.org/10.1016/j.compbiolchem.2015.07.006] [PMID: 26232767]

[45]    Zhang N, Gibon Y, Wallace JG, *et al*. Genome-wide association of carbon and nitrogen metabolism in the maize nested association mapping population. Plant Physiol 2015; 168(2): 575-83.
[http://dx.doi.org/10.1104/pp.15.00025] [PMID: 25918116]

[46]    Kooke R, Kruijer W, Bours R, *et al*. Genome-wide association mapping and genomic prediction elucidate the genetic architecture of morphological traits in Arabidopsis. Plant Physiol 2016; 170(4): 2187-203.
[http://dx.doi.org/10.1104/pp.15.00997] [PMID: 26869705]

[47]    Jayakodi M, Schreiber M, Stein N, Mascher M. Building pan-genome infrastructures for crop plants and their use in association genetics. DNA Res 2021; 28(1): dsaa030.
[http://dx.doi.org/10.1093/dnares/dsaa030] [PMID: 33484244]

[48]    Shen F, Hu C, Huang X, *et al*. Advances in alternative splicing identification: deep learning and pantranscriptome. Front Plant Sci 2023; 14: 1232466.
[http://dx.doi.org/10.3389/fpls.2023.1232466] [PMID: 37790793]

[49]    Gong Y, Li Y, Liu X, Ma Y, Jiang L. A review of the pangenome: how it affects our understanding of genomic variation, selection and breeding in domestic animals? J Anim Sci Biotechnol 2023; 14(1): 73.
[http://dx.doi.org/10.1186/s40104-023-00860-1] [PMID: 37143156]

[50]    Morgante M, Depaoli E, Radovic S. Transposable elements and the plant pan-genomes. Curr Opin Plant Biol 2007; 10(2): 149-55.
[http://dx.doi.org/10.1016/j.pbi.2007.02.001] [PMID: 17300983]

[51]    Li Y, Zhou G, Ma J, *et al*. De novo assembly of soybean wild relatives for pan-genome analysis of diversity and agronomic traits. Nat Biotechnol 2014; 32(10): 1045-52.
[http://dx.doi.org/10.1038/nbt.2979] [PMID: 25218520]

[52]    Bayer PE, Golicz AA, Scheben A, Batley J, Edwards D. Plant pan-genomes are the new reference. Nat Plants 2020; 6(8): 914-20.
[http://dx.doi.org/10.1038/s41477-020-0733-0] [PMID: 32690893]

[53]    Kang M, Wu H, Liu H, *et al*. The pan-genome and local adaptation of *Arabidopsis thaliana*. Nat Commun 2023; 14(1): 6259.
[http://dx.doi.org/10.1038/s41467-023-42029-4] [PMID: 37802986]

[54]    Guignon V, Toure A, Droc G, Dufayard J-F, Conte M, Rouard M. GreenPhylDB v5: a comparative pangenomic database for plant genomes. Nucleic Acids Res 2021; 49(D1): D1464-71.
[http://dx.doi.org/10.1093/nar/gkaa1068] [PMID: 33237299]

[55]    Bolger AM, Poorter H, Dumschott K, *et al*. Computational aspects underlying genome to phenome analysis in plants. Plant J 2019; 97(1): 182-98.
[http://dx.doi.org/10.1111/tpj.14179] [PMID: 30500991]

[56]    Grimm DG, Roqueiro D, Salomé PA, *et al*. easyGWAS: a cloud-based platform for comparing the results of genome-wide association studies. Plant Cell 2017; 29(1): 5-19.
[http://dx.doi.org/10.1105/tpc.16.00551] [PMID: 27986896]

[57]    Togninalli M, Seren Ü, Meng D, *et al*. The AraGWAS Catalog: a curated and standardized *Arabidopsis thaliana* GWAS catalog. Nucleic Acids Res 2018; 46(D1): D1150-6.
[http://dx.doi.org/10.1093/nar/gkx954] [PMID: 29059333]

[58]    Seren Ü, Vilhjálmsson BJ, Horton MW, *et al*. GWAPP: a web application for genome-wide association mapping in *Arabidopsis*. Plant Cell 2013; 24(12): 4793-805.
[http://dx.doi.org/10.1105/tpc.112.108068] [PMID: 23277364]

[59]    Togninalli M, Seren Ü, Freudenthal JA, *et al.* AraPheno and the AraGWAS Catalog 2020: a major database update including RNA-Seq and knockout mutation data for *Arabidopsis thaliana.* Nucleic Acids Res 2019; 48(D1): gkz925.
[http://dx.doi.org/10.1093/nar/gkz925] [PMID: 31642487]

[60]    Mohanta TK, Bashir T, Hashem A, Abd Allah EF. Systems biology approach in plant abiotic stresses. Plant Physiol Biochem 2017; 121: 58-73.
[http://dx.doi.org/10.1016/j.plaphy.2017.10.019] [PMID: 29096174]

[61]    Mekso MM, Feyissa T. RNA-seq as an effective tool for modern transcriptomics, a review-based study. Journal of Applied Research in Plant Sciences 2022; 3(2): 236-41.
[http://dx.doi.org/10.38211/joarps.2022.3.2.29]

[62]    Jamann TM, Balint-Kurti PJ, Holland JB. QTL mapping using high-throughput sequencing. In: Alonso AM, Stepanova AN, Eds. Plant Functional Genomics: Methods and Protocols. New York: Humana Press 2015; pp. 257-85.
[http://dx.doi.org/10.1007/978-1-4939-2444-8_13]

[63]    Garber M, Grabherr MG, Guttman M, Trapnell C. Computational methods for transcriptome annotation and quantification using RNA-seq. Nat Methods 2011; 8(6): 469-77.
[http://dx.doi.org/10.1038/nmeth.1613] [PMID: 21623353]

[64]    Mochida K, Shinozaki K. Advances in omics and bioinformatics tools for systems analyses of plant functions. Plant Cell Physiol 2011; 52(12): 2017-38.
[http://dx.doi.org/10.1093/pcp/pcr153] [PMID: 22156726]

[65]    Depuydt T, De Rybel B, Vandepoele K. Charting plant gene functions in the multi-omics and single-cell era. Trends Plant Sci 2023; 28(3): 283-96.
[http://dx.doi.org/10.1016/j.tplants.2022.09.008] [PMID: 36307271]

[66]    Garg R, Jain M. RNA-Seq for transcriptome analysis in non-model plants. In: Rose RJ, Ed. Legume Genomics: Methods and Protocols. Totowa: Humana Press 2013; pp. 43-58.
[http://dx.doi.org/10.1007/978-1-62703-613-9_4]

[67]    Han M, Lu X, Yu J, *et al.* Transcriptome analysis reveals cotton (*Gossypium hirsutum*) genes that are differentially expressed in cadmium stress toleranceInt. Int J Mol Sci 2019; 20(6): 1479.
[http://dx.doi.org/10.3390/ijms20061479] [PMID: 30909634]

[68]    Rao X, Dixon RA. Co-expression networks for plant biology: why and how. Acta Biochim Biophys Sin (Shanghai) 2019; 51(10): 981-8.
[http://dx.doi.org/10.1093/abbs/gmz080] [PMID: 31436787]

[69]    Zainal-Abidin RA, Harun S, Vengatharajuloo V, Tamizi AA, Samsulrizal NH. Gene co-expression network tools and databases for crop improvement. Plants 2022; 11(13): 1625.
[http://dx.doi.org/10.3390/plants11131625] [PMID: 35807577]

[70]    Tohge T, Fernie AR. Co-regulation of clustered and neo-functionalized genes in plant-specialized metabolism. Plants 2020; 9(5): 622.
[http://dx.doi.org/10.3390/plants9050622] [PMID: 32414181]

[71]    Obayashi T, Hibara H, Kagaya Y, Aoki Y, Kinoshita K. ATTED-II v11: a plant gene coexpression database using a sample balancing technique by subagging of principal components. Plant Cell Physiol 2022; 63(6): 869-81.
[http://dx.doi.org/10.1093/pcp/pcac041] [PMID: 35353884]

[72]    Seyfferth C, Renema J, Wendrich JR, *et al.* Advances and opportunities in single-cell transcriptomics for plant research. Annu Rev Plant Biol 2021; 72(1): 847-66.
[http://dx.doi.org/10.1146/annurev-arplant-081720-010120] [PMID: 33730513]

[73]    Shaw R, Tian X, Xu J. Single-cell transcriptome analysis in plants: advances and challenges. Mol Plant 2021; 14(1): 115-26.
[http://dx.doi.org/10.1016/j.molp.2020.10.012] [PMID: 33152518]

[74]    Schwacke R, Ponce-Soto GY, Krause K, *et al.* MapMan4: a refined protein classification and annotation framework applicable to multi-omics data analysis. Mol Plant 2019; 12(6): 879-92.
[http://dx.doi.org/10.1016/j.molp.2019.01.003] [PMID: 30639314]

[75]    Rasheed F, Markgren J, Hedenqvist M, Johansson E. Modeling to understand plant protein structure-function relationships - implications for seed storage proteins. Molecules 2020; 25(4): 873.
[http://dx.doi.org/10.3390/molecules25040873] [PMID: 32079172]

[76]    Hu J, Rampitsch C, Bykova NV. Advances in plant proteomics toward improvement of crop productivity and stress resistancex. Front Plant Sci 2015; 6: 209.
[http://dx.doi.org/10.3389/fpls.2015.00209] [PMID: 25926838]

[77]    Struk S, Jacobs A, Sánchez Martín-Fontecha E, Gevaert K, Cubas P, Goormachtig S. Exploring the protein–protein interaction landscape in plants. Plant Cell Environ 2019; 42(2): 387-409.
[http://dx.doi.org/10.1111/pce.13433] [PMID: 30156707]

[78]    Burley SK, Bhikadiya C, Bi C, *et al.* RCSB Protein Data Bank (RCSB.org): delivery of experimentally-determined PDB structures alongside one million computed structure models of proteins from artificial intelligence/machine learning. Nucleic Acids Res 2023; 51(D1): D488-508.
[http://dx.doi.org/10.1093/nar/gkac1077] [PMID: 36420884]

[79]    Szklarczyk D, Kirsch R, Koutrouli M, *et al.* The STRING database in 2023: protein–protein association networks and functional enrichment analyses for any sequenced genome of interest. Nucleic Acids Res 2023; 51(D1): D638-46.
[http://dx.doi.org/10.1093/nar/gkac1000] [PMID: 36370105]

[80]    Doncheva NT, Morris JH, Gorodkin J, Jensen LJ. Cytoscape StringApp: network analysis and visualization of proteomics data. J Proteome Res 2019; 18(2): 623-32.
[http://dx.doi.org/10.1021/acs.jproteome.8b00702] [PMID: 30450911]

[81]    Cheng F, Tuncbag N. Editorial overview: Artificial intelligence (AI) methodologies in structural biology. Curr Opin Struct Biol 2022; 74: 102387.
[http://dx.doi.org/10.1016/j.sbi.2022.102387] [PMID: 35589509]

[82]    Lee M. Deep learning in CRISPR-Cas systems: a review of recent studies. Front Bioeng Biotechnol 2023; 11: 1226182.
[http://dx.doi.org/10.3389/fbioe.2023.1226182] [PMID: 37469443]

[83]    Chen W, Liu X, Zhang S, Chen S. Artificial intelligence for drug discovery: Resources, methods, and applications. Mol Ther Nucleic Acids 2023; 31: 691-702.
[http://dx.doi.org/10.1016/j.omtn.2023.02.019] [PMID: 36923950]

[84]    Gill SS, Xu M, Ottaviani C, *et al.* AI for next generation computing: Emerging trends and future directions. Internet of Things 2022; 19: 100514.
[http://dx.doi.org/10.1016/j.iot.2022.100514]

[85]    Bhardwaj A, Kishore S, Pandey DK. Artificial intelligence in biological sciences. Life (Basel) 2022; 12(9): 1430.
[http://dx.doi.org/10.3390/life12091430] [PMID: 36143468]

[86]    Liu X, Wu S, Xu J, Sui C, Wei J. Application of CRISPR/Cas9 in plant biology. Acta Pharm Sin B 2017; 7(3): 292-302.
[http://dx.doi.org/10.1016/j.apsb.2017.01.002] [PMID: 28589077]

[87]    Touzdjian Pinheiro Kohlrausch Távora F, de Assis dos Santos Diniz F, de Moraes Rêgo-Machado C, *et al.* CRISPR/CAS- and Topical RNAi-based technologies for crop management and improvement: reviewing the risk assessment and challenges towards a more sustainable agriculture. Front Bioeng Biotechnol 2022; 10: 913728.
[http://dx.doi.org/10.3389/fbioe.2022.913728] [PMID: 35837551]

[88] Halder K, Chaudhuri A, Abdin MZ, Majee M, Datta A. RNA interference for improving disease resistance in plants and its relevance in this clustered regularly interspaced short palindromic repeats-dominated era in terms of dsRNA-based biopesticides. Front Plant Sci 2022; 13: 885128.
[http://dx.doi.org/10.3389/fpls.2022.885128] [PMID: 35645997]

**CHAPTER 11**

# Artificial Intelligence Technologies in Plant Physiology

**Mehmet Ali Balcı**[1,*] and **Ömer Akgüller**[1]

[1] *Muğla Sıtkı Koçman University, Faculty of Science, Department of Mathematics, Menteşe, Muğla, Türkiye*

**Abstract:** Analyzing phenotypic traits, diagnosing diseases, and anticipating yields are just a few of the many applications of plant organ segmentation in precision agriculture and plant phenotyping. Because plant structures are so varied and intricate, traditional methods have a hard time keeping up. By combining several data sources, such as images and point clouds, graph neural networks (GNNs) have completely altered crop organ segmentation. In this research, we present a new method for rethinking plant organ segmentation by using the powerful features of GNNs. The approach takes a look at point clouds of plant shoots and uses graph representations to capture deep structural intricacies and intricate spatial interactions. One important novelty is the use of betweenness centrality for weighting edges and vertex, which guarantees that the segmentation results are biologically significant. The model's ability to understand geometric and topological details is improved, leading to more accurate segmentation through dynamic computing and continuous updates of Forman-Ricci curvatures. This all-encompassing work opens new doors for plant phenotyping research by improving the accuracy of organ segmentation and facilitating the integration of complicated mathematical theories into biological analysis.

**Keywords:** Artificial intelligence, Graph neural networks, Gibberellic acid.

## INTRODUCTION

An essential part of precision agriculture and plant phenotyping, crop organ segmentation has recently been made much easier with the advent of Graph Neural Networks (GNNs). Using a variety of data sources, such as pictures and point clouds, crop organ segmentation identifies and categorizes distinct plant parts, including stems, leaves, and fruits [1 - 3]. Many applications rely on this method, such as phenotypic trait analysis, illness detection, and yield prediction. The complicated and diverse character of plant structures is often too much for traditional approaches, even though they are effective to some extent. GNNs offer

---

[*] **Corresponding author Mehmet Ali Balcı:** Muğla Sıtkı Koçman University, Faculty of Science, Department of Mathematics, Menteşe, Muğla, Türkiye; E-mail: mehmetalibalci@mu.edu.tr

**Ergun Kaya (Ed.)**

a potential answer to these problems because of their capacity to grasp the complex spatial correlations in data.

A group of characteristics developed through the ever-changing interplay between genes and the environment is the focus of plant phenotyping research [4, 5]. The focus of traditional phenotypic studies has been on plots and individual plant levels measured by hand. However, newer multidisciplinary research on genomics and phenomics has highlighted the need for more precise and high-throughput phenotypic acquisition, particularly for organ-level traits related to ideal plant architecture, such as leaf and stem traits [6 - 8]. New insights into high-throughput phenotyping have been provided by developments in image sensing and processing technologies [9 - 12], which in turn enhance the accuracy of phenotyping at the organ level. As a result, the development of methods for efficiently and accurately extracting organ-level traits is an important yet emerging field.

The octree algorithm, Difference of Normals, and 3D skeleton were traditionally used in traditional approaches to plant organ segmentation from 3D data, like LiDAR point clouds [13 - 16]. Although these methods are able to segment a variety of crop species with varying leaf shapes and canopy structures, they are still not generalizable beyond a small set of plants with simple structures that can be fine-tuned by time-consuming and labor-intensive parameter adjustment. The cutting edge of plant phenotyping research right now is the development of a universal 3D segmentation method that can be applied to a wide range of varieties throughout different phases of growth.

The recent advancements in 3D deep learning techniques, along with breakthroughs in neural network topologies and large amounts of data, have the potential to greatly enhance the accuracy and generalizability of organ segmentation [17, 18]. In an effort to gain a better grasp of 3D data, several studies have concentrated on Multi-view [19 - 22]. According to Jin *et al.* [23], in order to accomplish direct segmentation on the point cloud, it is necessary to first partition the point cloud into a large number of voxels. Then, a 3D convolution can be employed. However, this approach demands a lot of computing power. Among the first end-to-end deep learning networks to work directly on points, PointNet [24] and PointNet++ [25] can perform object classification and semantic segmentation on points at the same time. The SPGN [26] uses the similarity of each pair of points in the feature space, for instance, semantic segmentation, and many other researchers have used similar frameworks to optimize and improve the feature extraction modules, leading to improved network performance. To better enhance the connection between local features of the point cloud, researchers have resorted to Recurrent Neural Networks (RNN) [27 - 29],

Conditional Random Fields (CRFs) [30 - 34], and Graph Neural Networks [35 - 40].

In this comprehensive study, we embark on an innovative journey to redefine the landscape of plant organ segmentation by harnessing the advanced capabilities of Graph Neural Networks (GNNs). Our approach marks a significant departure from the conventional paradigms and methodologies that have dominated the fields of deep learning and computer vision as applied to botany. At the heart of our method lies the strategic extraction of graph representations from point clouds generated from plant shoots, executed through four meticulously designed methodologies. Each of these methodologies is tailored to meticulously capture and interpret the intricate spatial relationships and structural complexities inherent in plant organs. This paradigm shift, from viewing plant organs as isolated entities to treating them as interconnected systems, enables our GNN framework to delve into the depths of plant structure and functionality. The result is a groundbreaking pathway to achieving more precise and comprehensive plant phenotyping, thereby revolutionizing our understanding and analysis of plant biology.

Our novel strategy emphasizes the critical role of betweenness centrality in the weighting of edges and vertices within our graph-based model. This sophisticated approach does more than just enhance the segmentation process; it redefines the way we understand plant architecture by integrating the principles of graph theory into biological exploration. By leveraging betweenness centrality to assign weights, we ensure that our model doesn't merely recognize the physical structure of plant organs but also appreciates their biological importance. This methodological innovation allows us to achieve segmentation outcomes that are not only accurate from a computational perspective but also deeply meaningful within a biological framework, providing insights into the functional connectivity and significance of different parts of plant organs.

Expanding the horizon further, our research introduces the dynamic computation and continual updating of Forman Ricci curvatures for both edges and vertices as part of the convolutional steps within the GNN framework. This adaptive mechanism is designed to refine and enhance the model's grasp on the geometric and topological nuances of plant organs throughout the segmentation process. By updating the Forman Ricci curvatures with each convolution step, our model maintains an acute sensitivity to the minute variations in plant organ geometry, thus facilitating a segmentation accuracy and depth of insight that was previously beyond reach. This cutting-edge application of Forman Ricci curvature not only elevates the precision of plant organ segmentation to new heights but also paves the way for the integration of complex mathematical theories into the analysis of intricate biological structures. Through this pioneering approach, we are setting new benchmarks for what can be achieved in the field of plant phenotyping and opening up novel avenues for the exploration of advanced mathematical concepts in understanding the complexities of biological systems.

## METHODOLOGY

### Graphs from Point Cloud Data

Plant phenotypic analysis using graph structures produced from point clouds provides a multidimensional method for comprehending intricate biological systems. This approach allows for a quantitative analysis of morphology, branching patterns, and geometric aspects inside plants *via* a very detailed description of their spatial connections. Furthermore, it allows for the early diagnosis of illnesses and nutritional inadequacies by the recognition of distinguishable variations in structural patterns [41 - 45]. To facilitate more precise genetic engineering and breeding, 3D geometric data may be converted into graph structures to facilitate the mapping of genotype to phenotype. New possibilities for predictive analytics and automated trait assessment arise when graph-based representations are combined with machine learning algorithms. Graph analysis is at the forefront of breakthrough solutions in plant biology, agronomy, and resource optimization because of its scalability, capacity to unearth unique insights, and ability to allow multidisciplinary study. A more thorough and systematic investigation of plant structures and their interactions with the environment is made possible *via* its use in growth modeling, simulation, and resource allocation.

Point clouds are 3D data sets that generally depict the objects outside surfaces in a coordinate system. In mathematics, a point cloud $P$ is represented as a finite set $\{p_1,...,p_n\}$, where $p_i$ is a vector in $R^3$ with coordinates $\{x_i, y_i, z_i\}$. These points, which may be collected using techniques like 3D scanning or LiDAR technology, collectively describe the surface geometry of an item in three dimensions. Creating vertices and edges that connect these points is the first step in transforming this point cloud data into a graph structure. Point cloud $P$ may be represented as a graph $G = (V, R)$, where $V$ are each of the points and $E$ are edges representing spatial connections between them. In many cases, the edges and connections in a network are established according to a distance threshold. The spatial connections of the original point cloud are maintained in this graph-based representation, making it possible to use graph algorithms in further study of the structure.

Adjacency matrices are square matrices used to describe finite graphs in the mathematical subject of graph theory. The adjacency matrix $A = [a_{ij}]$ for an $n$-vertex graph $G$ is an $n \times n$ matrix with the entry $a_{ij}$ being 1 if there is an edge between vertices $i$ and $j$ and 0 otherwise. Both directed and undirected graphs may make use of the adjacency matrix, which represents the connectedness of the graph by means of its diagonal components, which indicate self-loops. However, the number of edges that are incident on a vertex is defined by its "vertex degree." In a directed graph, the degree is often broken down into the in-degree, which represents the number of edges entering the graph, and the out-degree, which represents the number of edges leaving the graph. For the sake of this research, we ignore directed graphs. The vertex degree and adjacency matrix are two fundamental ideas that help us understand the structure and features of the graph. In this study, we take a look at four different computational methods for extracting the underlying network structure from point cloud data.

A $\beta$-Skeleton is defined by connecting pairs of points $p$ and $q$ if there is no other point r within the sphere whose diameter is multiplied by $\beta$ and lies between $p$ and $q$. The constant variable Beta controls the sparsity of the graph, allowing for flexibility in determining edge connections. Depending on the method, computational complexity can range from low to very high. Gabriel Graphs, a special case of $\beta$-Skeletons with $\beta = 1$, connect points $p$ and $q$ if no other points lie within the circle whose diameter is the line segment connecting points $p$ and $q$. It is well known that Grabriel Graphs produce relatively sparse graphs. They can be constructed using Delaunay Triangulation in $O(n^2)$ time [46]. Delaunay triangulations are graphs that connect points in such a way that no point in the point cloud lies within the circumcircle of any triangle formed by connected points. In order to prevent "skinny" triangles from forming, Delaunay Triangulation generates a triangulation that optimizes the minimum of all the triangles' angles. Then, on average, its construction can be done in $O(\log n)$, but this may decline to $O(n^2)$ in the worst case [47]. Points $p$ and $q$ in a Relative Neighbor Graph (RNG) are connected if and only if there is no other point r in the circle whose diameter equals the line segment connecting $p$ and $q$. This is a particular case of $\beta$-Skeletons with $\beta = 2$. The crucial proximity edges are captured by RNGs, which also generate sparse graphs. They have a temporal complexity of $O(n^2)$, same as Gabriel Graphs [48].

The use of graph techniques, such as $\beta$-Skeletons, Gabriel Graphs, Delaunay Graphs, and the Relative Neighbor Graph, to represent and analyze point cloud data is of utmost importance in the field of plant phenotype. $\beta$-Skeletons, controlled by a continuous parameter $\beta$, allow for the capture of a wide range of spatial interactions inside complex plant structures, making them suitable for rigorous morphological analyses. In order to analyze crucial proximity interactions in plant architecture, the special instances of $\beta$-Skeletons known as Gabriel Graphs and Relative Neighbor Graphs provide sparser representations. In contrast, Delaunay graphs have an emphasis on surface triangulation and provide a strong foundation for surface reconstruction; this is particularly useful when trying to comprehend leaf and stem geometries. Fig. (**1**) is an illustration of obtaining graph structures. The distributions of vertex degrees and adjacency matrices of graphs are provided together with the 4095-point cloud sample acquired from a maize scan.

## Graph Neural Networks

Graph neural networks (GNNs) are a type of advanced deep learning algorithms designed to handle structured input in the form of graphs. Graph neural networks (GNNs) are specifically built to handle complicated data structures represented as graphs, which differ from standard grid-like structures like photos and sequential text. Unlike traditional neural networks, GNNs are adept at navigating and interpreting the intricate, often non-Euclidean topology of graph data. The versatility and significance of Graph Neural Networks (GNNs) in addressing real-world problems is evident in a variety of applications. For instance, GNNs are crucial in social network analysis, where individuals are represented as vertices and their relationships as edges. Similarly, in biochemical molecule interaction modeling, atoms can be seen as vertices and bonds as edges. This highlights the importance of GNNs in tackling a wide range of tasks [49 - 52].

GNNs distinguish themselves by their capacity to expand the neural network framework to fit the intrinsically irregular nature of graph data. Conventional deep learning models frequently face difficulties in handling irregularity, as they usually depend on a pre-established and consistent data structure for their operations. On the other hand, Graph neural networks (GNNs) excel in this setting by utilizing the distinct characteristics of graphs. They achieve this by employing a technique called message passing or neighborhood aggregation, wherein every vertex in the network gathers information from its adjacent vertices, therefore capturing both the local and global structural information inherent in the graph.

The basic objective of the GNN methodology is to acquire a high-dimensional representation, also known as an embedding, for every vertex in the graph. These embeddings are designed to capture a vast amount of information about each vertex, encompassing its own properties (such as features or labels linked to the vertex) as well as the qualities and impact of its nearby vertices. Moreover, these embeddings are specifically crafted to accurately represent the overall structure of the graph, encompassing an understanding of how vertices are coupled and how information propagates within the network. GNNs can effectively model intricate interactions and interdependencies between vertices due to their holistic perspective.

Gabriel Graph

Point Cloud of a Maize having 4095 points

Relative Neighbor Graph

β = 1.5 skeleton

Delaunay Triangulation

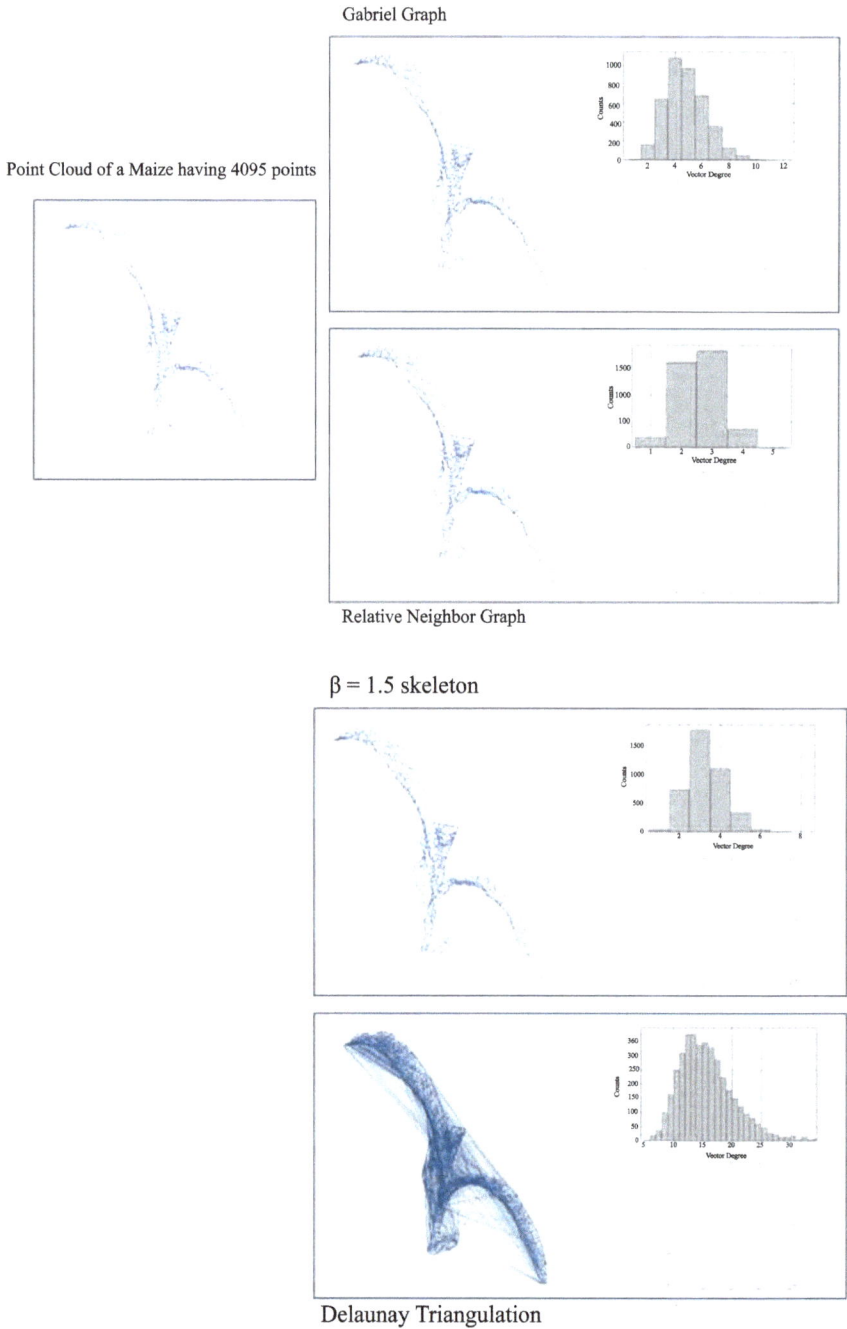

**Fig. (1).** β-Skeleton, Gabriel Graph, Delaunay Graph, and Relative Neighbor Graph emerging from a point cloud of a maize scan.

In order to accomplish this, Graph Neural Networks (GNNs) repeatedly enhance the representation of each vertex by merging its current attributes with aggregated data from its neighboring vertices, employing a range of advanced aggregation and update algorithms. The iterative method, also known as feature propagation, guarantees that the ultimate vertex embeddings contain abundant contextual information derived from both direct and indirect impacts from the neighborhood. The depth of these networks, usually indicated by the number of iterations or layers in the GNN, has a crucial impact on the amount of neighborhood information that is included in the embedding of each vertex. Deeper graph neural networks (GNNs) have the ability to capture more intricate and distant relationships between entities. However, this comes with the drawback of higher computational complexity and the risk of over-smoothing, which occurs when vertex representations become indistinguishable due to excessive aggregation of neighboring information.

Furthermore, graph neural networks (GNNs) are not uniform; a wide range of structures and approaches have been developed to enhance their efficiency for various graph types and workloads. Convolutional graph neural networks (GCNs) are influenced by the effectiveness of convolutional neural networks in image processing. They modify the convolutional procedures to operate on graphs by either defining them in the spectral domain or employing spatial approximations. Graph attention networks (GATs) incorporate an attention mechanism to dynamically assign weights to surrounding vertices, hence improving the model's capacity to concentrate on the most pertinent information throughout the aggregation phase.

The fundamental mechanism of GNNs is the message-passing framework, in which vertices repeatedly share information with their neighboring vertices. Mathematically, this process may be stated for every vertex v at every layer l of the GNN in the following manner:

Message Aggregation: Aggregate messages from the neighbors $N(v)$ for each vertex $v$. The aggregation function, denoted as $AGGREGATE^{(l)}$, has multiple options, including sum, mean, and max, which are commonly used. The combined message for vertex v at layer l is determined as follows:

$$m_v^{(l)} = AGGREGATE^{(l)} \left( h_u^{(l-1)} : u \in \mathcal{N}(v) \right), \tag{1}$$

where $h_u^{(l-1)}$ represents the neighbor u at the preceding layer $(l\text{-}1)$.

Update Function: The vertex subsequently modifies its representation by taking into account its current state and the combined message. The update function,

$UPDATE^{(l)}$, frequently incorporates neural network elements such as fully connected layers or more intricate designs like GRUs or LSTMs. The revised expression for the vertex v at layer l is as follows:

$$h_v^{(l)} = UPDATE^{(l)}\left(h_v^{(l-1)}, m_v^{(l)}\right). \tag{2}$$

Here, $h_v^{(l-1)}$ denotes the representation of vertex (*l*-1).

A notable example of GNNs is the Graph Convolutional Network (GCN), which simplifies the message-passing framework by using a normalized sum of neighbor features followed by a linear transformation and a non-linear activation function. The update rule for a vertex v in a GCN is:

$$h_v^{(l)} = \sigma\left(W^{(l)} \cdot MEAN\left\{h_u^{(l-1)}: u \in \mathcal{N}(v) \cup \{v\}\right\}\right), \tag{3}$$

where $\sigma$ is a non-linear activation function (*e.g.*, ReLU), $W^{(l)}$ is a learnable weight matrix for layer *l*, and *MEAN* denotes the mean aggregator. This formulation allows the GCN to effectively blend information from a vertex's neighbors and itself, leading to a powerful representation that captures both local structure and features.

Graph neural networks (GNNs) encounter several obstacles, such as effectively managing extensive graphs, acquiring knowledge from graphs with diverse architectures (*i.e.*, varying vertex and edge types), and capturing distant relationships inside graphs. In order to tackle these difficulties, several solutions have been suggested, including graph attention networks (GATs), which use attention processes to assess the significance of messages from neighboring vertices, and GraphSAGE, which selects a predetermined number of neighbors to handle huge networks efficiently [53, 54].

Applying vertex weighting techniques that rely on Forman Ricci curvature to tackle issues in GNNs introduces an innovative method that emphasizes utilizing inherent geometric characteristics of the graph.

An essential obstacle associated with GNNs is the effective management of extensive graphs. Conventional approaches frequently employ sampling strategies, like those utilized in GraphSAGE, or attention mechanisms, such as GATs, to decrease computational complexity. Nevertheless, these techniques may occasionally fail to include essential structural details, particularly in extensive and intricate networks.

By employing the Forman Ricci curvature to assign weights to vertices, one can establish a measure that inherently captures the topological and geometric characteristics of the network. The Forman Ricci curvature offers valuable insights into the geometric properties of the graph, enabling the identification of areas with high and low density, patterns of connectedness, and potential bottlenecks in the flow of information. The GNN may assign weights to vertices depending on their curvature, allowing it to prioritize vertices that are crucial for comprehending the overall structure of the graph. This could potentially decrease the requirement for intricate sampling or attention processes.

Graphs that have different designs, including different types of vertices and edges, present a substantial obstacle for GNNs. These diverse graphs necessitate models that can accurately represent a broad range of links and interactions. The Forman Ricci curvature can once again have a crucial influence in this context. By considering both the existence of edges and their arrangement, as well as the surrounding topology, the curvature provides a sophisticated method to directly integrate the graph's heterogeneity into the learning process.

The Forman Ricci curvature can be used to assign weights to vertices in a graph, which can help identify essential vertices and edges that contribute to the graph's distinct structure. This, in turn, guides the GNN to pay attention to noteworthy patterns and relationships. This approach inherently adjusts to the intricacy and variety of the graph's structure, augmenting the model's capacity to acquire knowledge from dissimilar data without necessitating substantial alterations to the network's design.

Another challenge in GNNs is capturing long-range dependencies within graphs. Traditional methods often rely on increasing the number of layers or utilizing sophisticated mechanisms like attention to extend the model's reach across the graph. However, this often comes with increased computational costs and complexity.

Vertex weighting using the Forman Ricci curvature offers an elegant solution. By encoding information about the graph's curvature at each vertex, the weights can implicitly reflect long-range topological features and dependencies. Vertices with specific curvature values may indicate critical bridges or paths within the graph that facilitate long-range interactions. Integrating these weights into the GNN's operations allows the model to naturally account for distant relationships without necessitating deeper or more complex network structures.

Ricci curvature in Riemannian geometry quantifies the extent to which the local geometry deviates from flatness at a given position. The Forman Ricci curvature, initially proposed by Robin Forman as a discrete counterpart in [55] and applied extensively in other disciplines [56 - 60], applies this notion to graphs, which can be regarded as discrete geometric entities. In the context of weighted graphs, the weights assigned to edges and vertices can reflect various attributes such as lengths, capacities, or probabilities. This introduces an additional level of intricacy to the calculation of curvature.

For a given vertex $v$ in a weighted graph, the Forman Ricci curvature $F(v)$ is defined in a way that accounts for the weighted structure of the graph. The definition involves both the edges incident to $v$ and the neighboring vertices. Let's consider a weighted graph G=($V,E,W$), where $v$ is the set of vertices, $E$ is the set of edges, and $w:V \cup E \rightarrow$ R+ is a weight function that assigns positive weights to both vertices and edges.

To construct a graph neural network (GNN) for networks that are both vertex and edge-weighted and incorporate Forman Ricci curvature for vertices, we need to consider how these elements interact within the framework of GNNs. The Forman Ricci curvature, adapted from its geometric origins to graph theory by Forman, provides a measure of curvature for edges and can also be extended to vertices in the graph. This extension allows for a nuanced understanding of the graph's topology, which can be particularly useful in the design and operation of GNNs for complex networks.

In a vertex and edge-weighted graph, each vertex $v$ and each edge $e$ is assigned a weight, denoted as $w(v)$ for vertices and $w(e)$ for edges. These weights can represent various attributes or capacities depending on the specific application, such as the strength of connections (edges) or the importance/attributes of entities (vertices). The Forman Ricci curvature for an edge $e$ connecting vertices $u$ and $v$ in a weighted graph is given by:

$$F(e) = w(e) \left( \frac{w(u) + w(v)}{w(e)^2} - \sum_{e' \in E(u) \cap E(v)} \frac{1}{w(e')} \right), \qquad (4)$$

where $w(u)$ and $w(v)$ are the weights of the vertices u and v at either end of e, E(u) and E(v) are the sets of edges incident to vertices u and v, respectively, e' represents the edges adjacent to e through a common vertex (either u or v).

This formula integrates the edge weights and vertex weights into the calculation of curvature, reflecting the "cost" of moving across edges relative to the "importance" of the vertices they connect.

To extend the Forman Ricci curvature to vertices, one approach is to aggregate the curvature of edges incident to a vertex, reflecting the overall curvature of the local topology around the vertex. A formulation for the curvature at a vertex v is

$$F(v) = \sum_{e \in E(v)} F(e),$$ (5)

where E(v) denotes the set of edges incident to vertex v. This summation aggregates the Forman Ricci curvature of all edges connected to v, providing a measure of the vertex's topological and geometrical significance within the graph.

In a GNN, the Forman Ricci curvature of vertices and edges can be utilized as features or weights in the message-passing mechanism. Specifically, the curvature could influence:

The aggregation step can be weighted by the curvature, where messages from neighbors connected by edges with higher curvature are given more significance. This allows the GNN to prioritize information from topologically significant regions of the graph. The vertex update mechanism can incorporate the vertex's curvature as an additional feature, enabling the model to adjust its representation based on the vertex's topological context.

The mathematical integration of Forman Ricci curvature into a GNN's operations can be formulated as follows: For a vertex v, the aggregated message $m_v^{(l)}$ at layer l could be modified to include curvature weights:

$$m_v^{(l)} = AGGREGATE^{(l)} \left( \left\{ h_u^{(l-1)} \cdot F(u,v) : u \in \mathcal{N}(v) \right\} \right).$$ (6)

Here, F(u, v) represents the Forman Ricci curvature between vertices u and v, influencing the aggregation of neighbor messages based on the topological significance of their connections.

Curvature-Enhanced Update Function: The update function for vertex v's representation can explicitly incorporate the vertex's own curvature:

$$h_v^{(l)} = UPDATE^{(l)} \left( h_v^{(l-1)}, m_v^{(l)}, F(v) \right).$$ (7)

This formulation allows the vertex representation to evolve based on both the aggregated messages and the vertex's geometric significance within the graph.

# RESULTS

## Data Set

The data included in this study comes from a point cloud dataset that was created using laser scanning of plants [61, 62]. The dataset documented the growth of three distinct crops in various habitats: tomato, tobacco, and sorghum. Over the course of 30 days, every crop was scanned numerous times. The dataset's scanning error is kept within ±25μm. The collection includes 546 distinct point clouds, 312 of which are tomatoes, 105 of which are tobacco, and 129 of which are sorghum. The smallest point cloud has approximately 10,000 points, while the biggest one has over 8,000,000 points. As reported in detail in the research [63 - 65], we used Semantic Segmentation Editor (SSE) [66] to annotate the stem and leaves with semantic labels and the instance label for each individually. For the purpose of semantic annotation, we specifically divide the stem system and leaves of various species into separate semantic groups. Thus, our dataset has six semantic categories.

Due to the heavy dependence of our methods on the graph structures that arise from point clouds, crop models including more than 10,000 points would result in computational inefficiencies. Therefore, we apply the Voxelized Farthest Point Sampling (VFPS) down sampling strategy, as suggested in [64], to each point cloud model. For all data entering, we consistently set the down-sampled size to 4096.

## Graph Weighting

Incorporating mathematical formulations for edge and vertex weighting in graphs derived from point clouds of plants, using methodologies such as β-Skeleton, Gabriel Graph, Delaunay Graph, and Relative Neighbor Graph, provides a sophisticated framework for capturing the intricate structural and spatial relationships within botanical structures. This approach is grounded in the principles of spatial interaction and graph centrality, offering a nuanced perspective on plant morphology and connectivity.

The decision to weight edges by the exponential function of their negative distance, mathematically represented as

$$w(e) = \exp(-\|p_u - p_v\|_2), \tag{8}$$

is the weight of an edge e connecting vertices u and v having 3D coordinates $p_u$ and $p_v$, respectively, and $\|.\|_2$ is the spatial distance between these vertices, inspired by physical and biological phenomena where interactions between entities decay exponentially with distance. This weighting scheme ensures that edges connecting closer vertices in the point cloud, which likely represent more immediate or significant botanical relationships (*e.g.,* adjacent leaves or branches), are given higher importance. Such an approach mirrors the efficiency of biological processes that tend to be more effective over shorter distances, effectively capturing the underlying functional connectivity within the plant structure.

Vertex weighting based on centrality metrics introduces an additional dimension of structural significance. Centrality, in a graph-theoretical context, can be quantified through various measures such as degree centrality, closeness centrality, and betweenness centrality, among others. Each of these measures offers a different lens through which the importance of a vertex v in the graph can be assessed—whether it be through the number of direct connections (degree), the average distance to other vertices (closeness), or the frequency with which a vertex acts as a bridge along the shortest path between two other vertices (betweenness). Since our aim is to classify crop organs in this study, we prefer to use closeness centrality defined with

$$C_C(v) = \frac{1}{\sum_{u \in V \setminus \{v\}} d(v, u)},$$ 
(9)

where d(v,u) is the shortest path distance between vertices v and u, considering the weights of the edges along the path. By assigning weights to vertices based on their centrality, the graph model emphasizes vertices that play pivotal roles in the connectivity and topology of the plant structure. This could, for instance, spotlight major vertices of nutrient distribution or signal transmission within the plant, aligning with key biological functions and structural dependencies.

Examples of graph representations using vertex and edge weight distributions from the dataset we utilize are shown in Fig. (**2**) for a tobacco shoot. We display the distributions in normalized form in subfigures to emphasize them. The Forman Ricci curvatures of the edges and vertices are likewise included in subfigures.

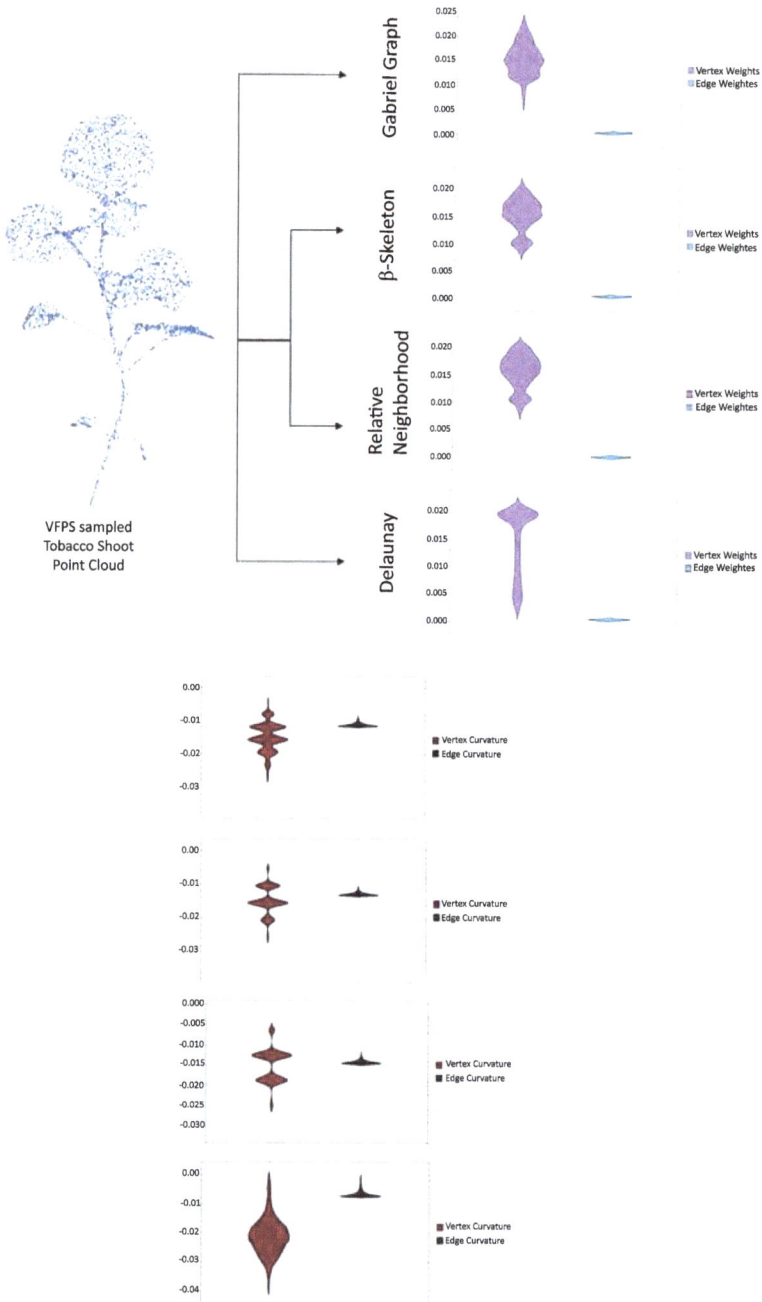

**Fig. (2).** Graph theoretic measures on a point cloud from a tobacco shoot.

The mathematical integration of these weighting schemes enriches the graph representation of plant point clouds by prioritizing spatial proximity and topological prominence. For edge weights, the exponential decay function ensures that the graph reflects realistic biological interactions, emphasizing closer spatial connections. For vertex weights, centrality measures highlight the nodes of greatest significance in the plant's structure and function. Together, these mathematical strategies enable more accurate and biologically relevant analyses of plant morphology, support complex graph-based machine learning models like GNNs, and facilitate a deeper understanding of plant architecture and its functional implications.

This mathematical framework for edge and vertex weighting thus establishes a robust foundation for exploring the structural and functional complexity of plants through graph-based models, leveraging the spatial and topological nuances inherent in botanical point clouds to drive forward research in plant science, ecology, and beyond.

## Graph Neural Network (GNN) Classification

Dataset division and extension are necessary steps in getting the data ready for network training and testing. We start by splitting the initial dataset in half, creating a training set and a test set with a ratio of 2:1. We utilized the VFPS approach, which was presented in a study [64], to down sample each of the 546-point clouds in the original dataset to a smaller one. We then repeated this process ten times, each time using a randomly picked beginning point in the last phase of VFPS, in order to replenish the dataset. Data augmentation is inherently unpredictable due to discrepancies in the starting point of FPS iterations following voxelization in VFPS. Consequently, upon augmentation, every point cloud in the dataset now comprises 4096 points. Interestingly, even though the enhanced 10 clouds were all created from the same original point cloud, the distributions of the local points were significantly different despite their outward similarities. Finally, we take 3640-point clouds for the training set and 1820 for the test set.

Designing a graph neural network (GNN) architecture for classifying plant organs, such as leaves and stems, from a point cloud involves creating a network capable of processing graph-structured data. The initial step in designing such an architecture involves preparing the input layer, which consists of the graph representation where nodes correspond to parts of the plant and edges reflect the connections based on the constructed graph. Nodes carry features derived from closeness centrality and Forman Ricci curvature, while edges are weighted by the exponential of the negative distance between points.

The core of the GNN architecture comprises several graph convolutional layers. The first of these layers processes the input feature matrix, which includes node features, and a weighted adjacency matrix that represents the graph structure, possibly augmented with self-connections to include the node's own features in the aggregation process. This layer aims to capture the local neighborhood information by transforming node features through a combination of the adjacency matrix and the feature matrix, applying a weight matrix and a non-linear activation function ReLU. A global pooling layer is incorporated to aggregate information across the entire graph, producing a single vector that represents the whole graph. This is particularly useful for classification tasks, where a global understanding of the graph is necessary to make a decision. Following the convolutional and optional pooling layers, the architecture typically includes one fully connected (dense) layer that leads to the final classification output. The last of these layers employs a softmax activation function to provide probabilities for each class, indicating whether a given segment of the plant is a leaf or a stem.

To optimize the network, a loss function cross-entropy is used, which is suitable for classification tasks. Adam optimizer updates the network weights to minimize the loss function during training. Additionally, to prevent overfitting, regularization techniques such as L2 regularization and dropout are applied in the fully connected layers.

This GNN architecture leverages both the spatial and topological information encoded in the graph, enabling the network to distinguish between different plant organs based on the unique structure and features of the graph representing the plant shoot.

In Tables **1 - 4**, we present the classification performance metrics for GNNs with Gabriel graphs, β-skeletons, RNGs, and Delaunay graphs.

**Table 1. Classification results for the GNN having Gabriel graph inputs.**

| - | Tobacco | | Tomato | | Sorghum | |
|---|---|---|---|---|---|---|
| - | Stem | Leaf | Stem | Leaf | Stem | Leaf |
| Precision | 0.9130 | 0.9478 | 0.9403 | 0.9585 | 0.9694 | 0.9594 |
| Recall | 0.95 | 0.9095 | 0.9594 | 0.9391 | 0.9590 | 0.9698 |
| F1 | 0.9311 | 0.9283 | 0.9497 | 0.9487 | 0.9642 | 0.9646 |
| AUC | 0.8640 | 0.8640 | 0.9009 | 0.9009 | 0.9301 | 0.9301 |
| MCC | 0.8602 | 0.8602 | 0.8986 | 0.8986 | 0.9289 | 0.9289 |

**Table 2. Classification results for the GNN having β-skeleton inputs.**

| - | Tobacco | | Tomato | | Sorghum | |
|---|---|---|---|---|---|---|
| - | **Stem** | **Leaf** | **Stem** | **Leaf** | **Stem** | **Leaf** |
| Precision | 0.9484 | 0.9214 | 0.9677 | 0.9321 | 0.9401 | 0.9478 |
| Recall | 0.9190 | 0.95 | 0.9294 | 0.9690 | 0.9482 | 0.9396 |
| F1 | 0.9334 | 0.9355 | 0.9482 | 0.9502 | 0.9442 | 0.9437 |
| AUC | 0.8731 | 0.8731 | 0.9006 | 0.9006 | 0.8910 | 0.8910 |
| MCC | 0.8694 | 0.8694 | 0.8992 | 0.8992 | 0.8879 | 0.8879 |

**Table 3. Classification results for the GNN having RNG inputs.**

| - | Tobacco | | Tomato | | Sorghum | |
|---|---|---|---|---|---|---|
| - | **Stem** | **Leaf** | **Stem** | **Leaf** | **Stem** | **Leaf** |
| Precision | 0.9478 | 0.9103 | 0.9684 | 0.9507 | 0.9044 | 0.9564 |
| Recall | 0.9095 | 0.95 | 0.9497 | 0.9690 | 0.9590 | 0.8987 |
| F1 | 0.9283 | 0.9311 | 0.9590 | 0.9597 | 0.9309 | 0.9266 |
| AUC | 0.8640 | 0.8640 | 0.9203 | 0.9203 | 0.8619 | 0.8619 |
| MCC | 0.8602 | 0.8602 | 0.9189 | 0.9189 | 0.8593 | 0.8593 |

**Table 4. Classification results for the GNN having Delaunay graph inputs.**

| - | Tobacco | | Tomato | | Sorghum | |
|---|---|---|---|---|---|---|
| - | **Stem** | **Leaf** | **Stem** | **Leaf** | **Stem** | **Leaf** |
| Precision | 0.9214 | 0.9484 | 0.9492 | 0.9397 | 0.9006 | 0.9164 |
| Recall | 0.95 | 0.9190 | 0.9391 | 0.9497 | 0.9181 | 0.8987 |
| F1 | 0.9355 | 0.9334 | 0.9441 | 0.9447 | 0.9092 | 0.9075 |
| AUC | 0.8731 | 0.8731 | 0.8919 | 0.8919 | 0.8251 | 0.8251 |
| MCC | 0.8694 | 0.8694 | 0.8889 | 0.8889 | 0.8169 | 0.8169 |

Precision is the ratio of correctly predicted positive observations to the total predicted positives. It is a measure of the accuracy of the positive predictions made by the model and is defined as follows:

$$Precision = \frac{TP}{TP + FP}, \tag{10}$$

where TP is the number of true positives, and FP is the number of false positives.

Recall is the ratio of correctly predicted positive observations to all the observations in the actual class. It measures the model's ability to capture all relevant instances defined with

$$Recall = \frac{TP}{TP + FN} \, , \tag{11}$$

where TP is the number of true positives, and FN is the number of false negatives.

The F1 score is the harmonic mean of precision and recall. It is a balance between the precision and recall metrics, providing a single metric to assess the model's performance when both false positives and false negatives are important and defined as follows:

$$F1 = 2 \cdot \frac{Precision \times Recall}{Precision + Recall}. \tag{12}$$

The ROC curve is a graphical representation of the true positive rate (recall) against the false positive rate (FPR) at various threshold settings. The AUC measures the entire two-dimensional area underneath the entire ROC curve from (0,0) to (1,1) and provides an aggregate measure of performance across all possible classification thresholds. A model whose predictions are 100% wrong has an AUC of 0.0; one whose predictions are 100% correct has an AUC of 1.0. The MCC is a measure of the quality of binary classifications. It takes into account true and false positives and negatives and is generally regarded as a balanced measure which can be used even if the classes are of very different sizes and defined with

$$MCC = \frac{(TP \times TN) - (FP \times FN)}{\sqrt{(TP + FP)(TP + FN)(TN + FP)(TN + FN)}}, \tag{13}$$

where TN is the number of true negatives. The MCC is in the range of [-1, 1]. A coefficient of +1 represents a perfect prediction, 0 is no better than a random prediction, and -1 indicates total disagreement between prediction and observation.

## Comparative Analysis

Using the same plant dataset, our GNN is compared to multiple popular point cloud segmentation networks in this section. Some of these networks can only perform semantic segmentation, such as PointNet [24], PointNet++ [25], and DGCNN [67]. Using the same set of semantic and instance labels for training, ASIS [68] and PlantNet [64] may also perform instance and semantic segmentation tasks simultaneously, similar to a study [69]. We utilized the suggested parameter settings for the comparison networks outlined in their respective original publications.

On the semantic segmentation test, Tables **5** - **7** display the quantitative comparisons with respect to Precision, Recall, and F1 scores among six other networks.

**Table 5. Comparison of Precision values (%).**

| - | Tobacco | | Tomato | | Sorghum | | Mean |
|---|---|---|---|---|---|---|---|
| - | Stem | Leaf | Stem | Leaf | Stem | Leaf | |
| PointNet | 77.15 | 94.02 | 93.99 | 96.71 | 77.87 | 95.37 | 89.19 |
| PointNet++ | 87.78 | 95.62 | 93.65 | 96.80 | 78.01 | 98.33 | 91.70 |
| ASIS | 91.65 | 91.94 | 93.55 | 97.14 | 85.47 | 95.17 | 92.49 |
| PlantNet | 89.45 | 96.80 | 95.90 | 96.30 | 89.07 | 97.43 | 94.16 |
| DGCNN | 90.55 | 96.42 | 95.24 | 97.86 | 83.95 | 97.37 | 93.57 |
| PSegNet | 92.71 | 96.76 | 96.36 | 97.98 | 89.54 | 98.04 | 95.23 |
| GNN+Gabriel | 91.3 | 94.78 | 94.03 | 95.85 | 96.94 | 95.94 | 94.80 |
| GNN+β-Skeleton | 94.84 | 92.14 | 96.77 | 93.21 | 94.01 | 94.78 | 94.27 |
| GNN+RNG | 94.78 | 91.30 | 96.84 | 95.07 | 90.44 | 95.64 | 94.01 |
| GNN+Delaunay | 92.14 | 94.84 | 94.92 | 93.97 | 90.06 | 91.64 | 92.93 |

**Table 6. Comparison of Recall values (%).**

| - | Tobacco | | Tomato | | Sorghum | | Mean |
|---|---|---|---|---|---|---|---|
| - | Stem | Leaf | Stem | Leaf | Stem | Leaf | |
| PointNet | 79.20 | 93.31 | 91.85 | 97.61 | 61.45 | 97.85 | 86.88 |
| PointNet++ | 90.83 | 94.05 | 92.45 | 97.33 | 78.66 | 98.27 | 91.93 |
| ASIS | 83.85 | 96.11 | 92.87 | 95.51 | 81.65 | 97.88 | 91.31 |
| PlantNet | 86.12 | 92.97 | 95.24 | 98.23 | 86.06 | 98.07 | 92.78 |
| DGCNN | 85.55 | 97.76 | 94.15 | 98.27 | 78.05 | 98.20 | 92.00 |
| PSegNet | 87.42 | 97.73 | 95.02 | 98.59 | 85.69 | 98.63 | 93.85 |
| GNN+Gabriel | 95.00 | 90.95 | 95.94 | 93.91 | 95.90 | 96.98 | 94.78 |
| GNN+β-Skeleton | 91.90 | 95.00 | 92.94 | 96.90 | 94.82 | 93.96 | 94.25 |
| GNN+RNG | 90.95 | 95.00 | 94.97 | 96.90 | 95.90 | 89.87 | 93.93 |
| GNN+Delaunay | 95.00 | 91.90 | 93.91 | 94.97 | 91.81 | 89.87 | 92.91 |

**Table 7. Comparison of F1 Score values (%).**

| - | Tobacco | | Tomato | | Sorghum | | Mean |
|---|---|---|---|---|---|---|---|
| - | Stem | Leaf | Stem | Leaf | Stem | Leaf | |
| PointNet | 78.16 | 93.66 | 92.91 | 97.16 | 68.69 | 96.59 | 87.86 |
| PointNet++ | 89.28 | 94.83 | 93.05 | 97.07 | 78.33 | 98.30 | 91.81 |
| ASIS | 87.58 | 93.98 | 93.21 | 96.32 | 83.52 | 96.51 | 91.85 |
| PlantNet | 87.75 | 94.85 | 95.56 | 97.26 | 87.54 | 97.75 | 93.45 |
| DGCNN | 87.98 | 97.09 | 94.69 | 98.07 | 80.89 | 97.78 | 92.75 |
| PSegNet | 89.99 | 97.24 | 95.68 | 98.29 | 87.57 | 98.33 | 94.52 |
| GNN+Gabriel | 93.11 | 92.83 | 94.97 | 94.87 | 96.42 | 96.46 | 94.77 |
| GNN+β-Skeleton | 93.34 | 93.55 | 94.82 | 95.02 | 94.42 | 94.37 | 94.25 |
| GNN+RNG | 92.83 | 93.11 | 95.90 | 95.97 | 93.09 | 92.66 | 93.92 |
| GNN+Delaunay | 93.55 | 93.34 | 94.41 | 94.47 | 90.92 | 90.75 | 92.91 |

## DISCUSSION

This study intended to enhance the categorization of plant organs, such as leaves and stems, by utilizing Graph Neural Networks (GNN) on point cloud data of plant shoots. Our research involved creating different graph representations such as the Gabriel graph, β-Skeleton, Relative Neighbor Graph, and Delaunay graph to understand the intricate patterns of plant shoots. The graphs were carefully analyzed using exponential functions based on point distances and vertex closeness centrality, establishing a new standard for representing spatial interactions in a GNN framework. The inclusion of Forman Ricci curvatures for vertices and edges is a groundbreaking advancement in combining topological and geometric characteristics in the classification procedure.

Our method stands out by using a variety of network structures and edge and vertex weights based on proximity centrality and distance metrics. This method highlights the ability of graph neural networks (GNNs) to depict the complex characteristics of plant organ structures using advanced graph models. We manually labeled leaf and stem segmentations and integrated them into our graph neural network (GNN) model, paving the way for automatic plant organ categorization.

In Table **1**, precision remains consistently high across the three species for both leaf and stem classifications, with scores ranging from 0.9130 for Tobacco stems to 0.9694 for Sorghum stems. This indicates that the model is quite successful in accurately recognizing positive samples within the anticipated classes.

The recall rate is consistently high in all categories, demonstrating the model's ability to accurately detect the majority of leaf and stem segments. The algorithm demonstrates its robustness in identifying relevant samples across several plant species, with scores ranging from 0.9095 for Tobacco leaves to 0.9698 for Sorghum leaves. The F1 values in this table closely align with the precision and recall scores for each category, indicating that the model achieves a balanced performance in properly detecting positive occurrences and capturing the majority of relevant cases. The scores vary from 0.9283 for Tobacco leaves to 0.9646 for Sorghum leaves, showing a consistently high performance for both plant parts and species. The AUC ratings vary from 0.8640 for Tobacco to 0.9301 for Sorghum, indicating that the model demonstrates significant discriminative capability among various plant species, especially in Sorghum. The model's MCC scores are high, ranging from 0.8602 for Tobacco to 0.9289 for Sorghum, indicating outstanding performance in categorizing leaf and stem segments across the studied species. The results show that the GNN model, using Gabriel graph inputs and derived features, excels at categorizing plant parts as either leaves or stems across various species.

Table **2** displays high precision values for all categories and species, ranging from 0.9214 for Tomato leaves to 0.9677 for Tomato stems, indicating the accuracy of positive predictions. The high precision scores indicate that the model excels at accurately detecting positive instances of both stems and leaves, with a little edge in identifying stems, notably in Tomato. The recall metric assesses the model's ability to correctly identify all relevant instances within a specific class. The scores range from 0.9190 for Tobacco stems to 0.9690 for Tomato leaves, demonstrating remarkable performance. The model demonstrates a high capability in accurately identifying the truest positive cases in various plant parts and species, particularly excelling in detecting leaf segments in Tomato. The F1 scores range from 0.9334 for Tobacco stems to 0.9502 for Tomato leaves, showcasing the model's balanced accuracy in classification tasks. The model excels in classifying Tomato leaves, demonstrated by the highest F1 score, which shows a good balance between precision and recall. The AUC scores, which indicate the model's ability to differentiate between classes at different threshold settings, are consistently high for all species, ranging from 0.8731 for Tobacco to 0.9006 for Tomato. The scores indicate that the model has a strong ability to differentiate between stem and leaf segments in various species, with the best performance observed in Tomato. The Matthews Correlation Coefficient, a thorough metric that accounts for true and false positives and negatives, displays high values ranging from 0.8694 for Tobacco to 0.8992 for Tomato. The MCC scores demonstrate that the model excels at classifying leaf and stem segments, with the most accurate predictions seen in Tomato. The comprehensive findings from Table **2** show that the GNN model, utilizing β-skeleton inputs and advanced feature calculations, is highly effective in categorizing plant parts as either leaves or stems in different species. The model demonstrates accuracy, dependability, and robustness in distinguishing between the intricate geometries of leaves and stems in point cloud models, as seen by its consistently high scores in precision, recall, F1, AUC, and MCC metrics.

Table 3 displays the precision values for different categories and species. Tomato stems have the highest precision at 0.9684, while Sorghum stems have the lowest at 0.9044. The results indicate that the model is mostly precise in its positive identifications, especially in recognizing Tomato stems. The recall scores, which indicate the model's capability to detect all true positives, are notably high, especially for Tomato leaves and Sorghum stems, both above 0.95. The model demonstrates the ability to accurately identify a large proportion of relevant cases within the specified categories, with a somewhat lower recall rate of 0.8987 for Sorghum leaves. The F1 values are similar among many plant parts and species, with Tomato leaves achieving the highest score of 0.9597 and Sorghum leaves the lowest at 0.9266. The data highlights the model's well-rounded performance in classification tasks, showing a modest advantage in processing Tomato plant segments. The AUC scores are identical for Tobacco and Sorghum but significantly higher for Tomato at 0.9203. This indicates that the model excels at differentiating between leaf and stem segments in Tomato, in contrast to Tobacco and Sorghum. Finally, the MCC values correspond to the AUC patterns. Tomato had the highest score of 0.9189, showing exceptional categorization quality, particularly for this species. The GNN model, incorporating RNG inputs and computed features, demonstrates high performance in categorizing plant organs as either leaves or stems, as shown in Table 3. The model demonstrates exceptional precision and discriminative capability for Tomato, indicating possible species-specific benefits of the RNG-based graph representations.

Table 4 displays impressive precision scores for all categories and species. Tomato stems have a precision of 0.9492, indicating strong accuracy in identifying stem segments, whereas Sorghum stems have a precision of 0.9006, suggesting somewhat less accuracy but still a robust capacity to classify stem segments. Recall values assess the model's ability to correctly identify all true positives in a given category. Tobacco stems have the highest recall rate of 0.95, demonstrating the model's ability to accurately identify most real stem segments in Tobacco. Sorghum leaves have the lowest recall rate of 0.8987, suggesting difficulty in accurately distinguishing all genuine leaf segments in Sorghum. F1 scores show consistent performance across all plant parts and species, with Tomato showing somewhat higher scores, suggesting a balanced precision and recall ability, particularly for Tomato leaves at 0.9447. The AUC ratings are consistently high for Tobacco and Tomato but significantly lower for

Sorghum at 0.8251. This suggests a generally strong capacity to discriminate, yet there is a noticeable decline in performance, specifically for Sorghum. The MCC values offer a comprehensive assessment of the model's performance in all quadrants of the confusion matrix and are similar to the AUC scores. High MCC values are found for Tobacco and Tomato, indicating outstanding classification performance. Sorghum, on the other hand, has a lower MCC score of 0.8169, indicating a minor decrease in classification quality for this species. The results in Table **4** show that the GNN model, using Delaunay graph inputs and advanced feature calculations, accurately classifies plant parts into leaf and stem categories for the species studied. The model demonstrates exceptional performance in distinguishing Tomato stems and leaves, as seen by its high precision, recall, and F1 scores. This suggests a distinct advantage of the Delaunay graph-based representations for Tomato species.

Comparing the classification results in Tables 1 to 4, which show the performance of a GNN using different graph inputs to classify plant organs into leaf and stem segments across Tobacco, Tomato, and Sorghum species, reveals insights about the effectiveness and flexibility of the GNN model with different graph structures. The precision and recall metrics demonstrate the GNN's consistent high accuracy in identifying and capturing the truest positive leaf and stem segments across the three species. The precision values are typically high, suggesting that the model is successful in accurately predicting the class labels when it makes a positive classification. The recall scores are strong, indicating that the model can accurately identify a large fraction of the true positive examples in each class. Tomato frequently has somewhat greater precision and recall rates, suggesting that the model may be well-suited or optimized for this species. The F1 scores, which consider both precision and recall, consistently demonstrate high performance for all types of graph input and plant species, indicating the model's accuracy and efficiency in classifying plant organs. It is essential for practical use that the model maintains a balance between precision and recall without favoring one over the other. AUC and MCC metrics offer a comprehensive evaluation of the model's performance, taking into account its capacity to differentiate between classes and its overall quality in all elements of the classification task. Although AUC and MCC scores are often high, suggesting good classification ability and quality, there are modest variances according to the graph input type and plant species. These differences indicate that specific network structures may be more compatible with the natural geometric and topological characteristics of particular plant species, impacting the model's capacity to differentiate between leaves and stems. The performance metrics for various graph input types demonstrate the subtle influence of graph architecture on the classification ability of the Graph Neural Network (GNN). However, all graph forms exhibit strong classification performance, modest variations in precision, recall, F1, AUC, and MCC scores indicate that particular graph designs may more effectively represent the intricate spatial relationships in plant shoot point clouds for specific species.

The comparison of these tables shows that the GNN model demonstrates strong and flexible performance with modest variances that could guide future improvements, spanning various graph input types and plant species. The constant high-performance metrics of various graph types and species highlight the promise of GNNs and advanced graph-based inputs in improving classification tasks in botanical research and plant phenotyping. These findings support additional investigations into the applications of graph neural networks, highlighting the significance of selecting suitable graph structures that align with the unique features and needs of the classification task.

Table **5** displays a detailed comparison of precision values for categorizing stem and leaf segments in point cloud models of Tobacco, Tomato, and Sorghum using various models. The precision metric, quantified as a percentage, evaluates the correctness of positive predictions in each category. PointNet++ shows considerable advancements over PointNet, consistently surpassing it in all aspects. This indicates the effectiveness of hierarchical feature learning in PointNet++ for handling point cloud data, though there's still room for improvement, especially in accurately classifying stems in Tobacco and Sorghum. ASIS demonstrates a well-rounded performance in both leaf and stem categories, excelling particularly in Tomato leaf classification. Combining instance and semantic segmentation tasks enhances precision in point cloud categorization. PlantNet and DGCNN demonstrate high precision values, especially when categorizing leaf segments. PlantNet performs exceptionally well across all species. Both models excel at identifying intricate geometric elements in plant point clouds, with PlantNet showing somewhat higher precision overall. PSegNet demonstrates the best average precision among all models, showcasing its capability in accurately categorizing point cloud segments. PSegNet's method, which may utilize sophisticated segmentation techniques and deep learning characteristics, appears to be highly successful at categorizing plant organs. The GNN+Gabriel Graph demonstrates high efficacy in categorizing stem segments in Sorghum, indicating that the Gabriel graph's method of capturing spatial relationships is especially beneficial for specific species. The GNN+β-Skeleton method demonstrates superior accuracy in classifying stems in Tobacco and Tomato and also delivers strong performance in other categories. This suggests that the β-skeleton graph input is adept at capturing the inherent geometric characteristics of plant structures. GNN+RNG demonstrates strong performance, especially in stem classifications for Tobacco and Tomato, highlighting the effectiveness of RNG in capturing crucial spatial correlations for classification purposes. The GNN+Delaunay Graph shows strong performance across all species; however, it has a slightly lower mean precision compared to other GNN variations. Delaunay graphs are useful for these tasks, although alternative graph inputs may better represent the complexities of plant organ geometries. Overall, the comparison highlights the advantages of several methods for categorizing plant organs in point cloud data. GNN models, utilizing diverse graph inputs, demonstrate competitive and sometimes greater precision values when compared to existing models such as PointNet, PointNet++, and DGCNN. GNN variants that use β-skeleton and RNG inputs demonstrate significant precision benefits, emphasizing the necessity of selecting the appropriate graph input to represent the intricate spatial relationships found in plant point clouds. The consistently high precision values of these models highlight the progress in point cloud processing approaches, offering useful insights for creating more precise and dependable classification models in botanical research.

Table **6** provides a comprehensive comparison of recall values across different models used to categorize stem and leaf segments in point cloud models of Tobacco, Tomato, and Sorghum plants. Recall, represented as a percentage, gauges the model's capacity to detect all true positives inside each category, offering insight into each model's sensitivity and efficiency in identifying pertinent events. PointNet and PointNet++ demonstrate a significant enhancement in recall from PointNet to PointNet++, with PointNet++ providing superior sensitivity in most categories. Both models have weaknesses, with PointNet showing much worse recall for Sorghum stems, indicating difficulties in catching all relevant instances in this category. ASIS demonstrates strong performance, particularly in accurately classifying leaves of all species, showcasing its successful implementation of semantic segmentation to enhance model sensitivity. Nevertheless, the recall rates for stems, particularly in Tobacco and Sorghum, indicate the need for enhancement. PlantNet and DGCNN have significant recall values, especially in leaf segmentation. DGCNN demonstrates remarkably high memory rates in Tomato and Sorghum. Both models are effective in recognizing relevant leaf segments, with PlantNet showing a more balanced performance across different organ types. PSegNet demonstrates exceptional performance with a high mean recall value, showcasing its remarkable capacity to detect almost all true positive events comprehensively. The model's excellent segmentation skills are highlighted by its impressive performance in leaf categorization. The GNN+Gabriel Graph has the highest average recall compared to other GNN variations, especially in stem classification for all three species. This indicates that the spatial representation of the Gabriel graph accurately captures the pertinent cases for stem categorization. The GNN+β-Skeleton algorithm demonstrates a consistent recall rate for both stems and leaves, with significantly elevated values for leaf segments in Tomato and Sorghum, showcasing its precision in identifying leaf structures. GNN+RNG demonstrates high recall, particularly for Sorghum stems, showcasing RNG's ability to accurately recognize stem segments in this species. The model's recall for Sorghum leaves is the lowest among the GNN variations, indicating a potential selectivity in its sensitivity to various organ types. The GNN+Delaunay Graph model exhibits strong memory rates for stems in Tobacco and Tomato but experiences a drop in recall for Sorghum leaves. This pattern indicates that the Delaunay graph input is efficient at representing stem instances but may not be as responsive to leaf structures in specific species. This comparison demonstrates the nuanced performance of several models in categorizing plant organs using point cloud data, with each model showing strengths in specific areas. The GNN models, especially those using the Gabriel graph and RNG inputs, demonstrate high recall values, showing their usefulness in catching important instances across various plant species and organ kinds. PSegNet has remarkable segmentation capabilities and high recall rates for both stems and leaves. The fluctuation in recall rates for classic models such as PointNet and ASIS highlights the persistent difficulties and potential for enhancement in point cloud categorization jobs.

Table 7 presents a detailed comparison of F1 scores among several models used to classify stem and leaf segments in point cloud models of Tobacco, Tomato, and Sorghum plants. The F1 score, a metric derived from the harmonic mean of precision and recall, provides a balanced evaluation of a model's accuracy by considering both its precision in positive predictions and its capability to identify all pertinent instances. PointNet and PointNet++ provide strong performance, with PointNet++ showing a significant enhancement compared to PointNet. PointNet++ demonstrates improved performance in balancing precision and recall but encounters difficulties in reliably classifying Sorghum stems. ASIS provides a competitive F1 score, which is almost equivalent to PointNet++, indicating that its effective incorporation of semantic segmentation enhances its overall accuracy. It falls slightly short compared to more advanced models in attaining the optimal balance between precision and recall. PlantNet and DGCNN both provide high F1 scores, with PlantNet displaying a well-balanced performance across all species and organ types. DGCNN demonstrates superior performance in leaf classification, as evidenced by its high F1 scores for leaf segments, indicating its proficiency in recognizing intricate geometric characteristics in leaves. PSegNet excels with a high mean F1 score, showcasing its exceptional capacity to maintain a balance between precision and recall comprehensively. The software excels in leaf classification due to its excellent segmentation skills and deep learning characteristics. The GNN+Gabriel Graph model demonstrates remarkable F1 scores, particularly in Sorghum, surpassing other models. This indicates that the spatial representation obtained from the Gabriel graph is highly efficient for this species. The GNN+β-Skeleton method demonstrates well-balanced F1 scores in all categories, showing its ability to effectively capture the inherent geometric characteristics of plant structures for stem and leaf categorization. GNN+RNG demonstrates high F1 scores, particularly excelling in Tomato classification, highlighting RNG's efficacy in capturing crucial spatial correlations for classification purposes. The GNN+Delaunay Graph exhibits good F1 scores, especially for stem classification in Tobacco and Tomato, with a little drop observed for Sorghum leaves. This pattern indicates that the Delaunay graph, albeit efficient, may not be as responsive to leaf structures in Sorghum compared to alternative graph inputs. This comparison demonstrates the detailed performance of several models in categorizing plant organs using point cloud data, with GNN models displaying notably high F1 scores. The differences in F1 scores among models and graph inputs emphasize the significance of choosing suitable graph structures and deep learning architectures to optimize both precision and recall. The GNN models, particularly those using Gabriel graph and β-skeleton inputs, have attained high F1 scores, demonstrating their potential for accurate and reliable classification in botanical research and plant phenotyping.

## CONCLUSION

Utilizing graph neural networks (GNN) to categorize plant organs, such as stems and leaves, from point cloud models of plant shoots marks a significant advancement in the domains of plant phenotyping and computational botany. This methodology, especially with the incorporation of Forman curvature characteristics for vertices and edges, underscores a refined approach to dissecting and leveraging the intricate spatial and geometric connections inherent in plant formations. The application of GNNs, enriched with Forman curvature features, signifies a pivotal shift toward more sophisticated, geometry-aware machine learning models in botanical studies, highlighting the method's novelty and potential for transformative insights in the field.

Forman curvature, a topological measure that captures the curvature of edges and vertices within a graph, enhances the model's capability to intricately represent the geometric nuances of plant organs. This innovative addition not only elevates the accuracy of classification tasks but also deepens our understanding of the complex architectures of plant shoots, facilitating a granular analysis of plant morphology through computational lenses. The adoption of such advanced features introduces a novel dimension to the analysis, enabling researchers to explore the morphological complexities of plants in unprecedented detail.

The detailed examination of Tables **1** through **4** reveals the nuanced efficacy of different graph inputs in capturing the spatial relationships critical for accurate organ classification. Table **1**, showcasing Gabriel graph inputs, reveals exceptional precision and recall across all examined species, particularly shining in the classification of Sorghum. This outcome illustrates the robustness of combining Gabriel graph inputs with Forman curvature features in delineating the unique spatial interactions that define plant structures. Table **2** further accentuates the potency of the β-skeleton in delivering high precision in stem classification, suggesting its unparalleled capacity for conducting in-depth morphological analyses, a testament to the method's precision in differentiating between stems and leaves.

Similarly, Table **3** highlights the competitive edge of the relative neighbor graph (RNG) in precisely classifying Tomato stems and leaves, underscoring the RNG's adeptness at emphasizing critical spatial connections essential for organ categorization. This underscores the importance of selecting graph structures tailored to the specific needs of the species and organ types under investigation. Meanwhile, Table **4** points to the solid classification accuracy achieved through Delaunay graph inputs, albeit with slightly lower mean precision and recall compared to other graph inputs, suggesting a nuanced alignment between graph selection and the structural characteristics of the target species.

The comparative analysis provided in Tables **5** through **7** further illuminates the superior performance of GNN models, particularly those utilizing β-skeleton and Gabriel graph inputs, against conventional models like PointNet and DGCNN. This precision and recall advantage underline the GNN models' capacity to accurately classify plant organs by tapping into the intricate geometric information encapsulated by Forman curvature features. Such findings not only validate the approach's effectiveness but also spotlight the potential for GNNs to set new benchmarks in the classification accuracy of plant phenotyping tasks.

Despite the promising outcomes, the approach harbors inherent limitations, notably the reliance on manually tagged segmentation for model training and validation, which may introduce biases and variability. Additionally, the computational demands associated with calculating Forman curvature and generating various graph inputs pose challenges to scalability and real-time application feasibility.

Future research directions should aim at automating segmentation processes to mitigate manual biases and enhance dataset preparation efficiency. Exploring the scalability and optimization of GNN models to accommodate larger datasets and real-time applications could significantly extend the methodology's applicability. Expanding the classification scope to include a broader array of plant species and organ types, as well as additional structural features, can vastly enrich the methodology's utility in plant science. Comparative analyses with other cutting-edge machine learning and deep learning models will further delineate the relative merits and limitations of GNNs in plant organ classification, paving the way for refined models and methodologies.

In conclusion, the deployment of GNNs equipped with Forman curvature features for the classification of plant organs heralds a promising avenue in computational botany, offering intricate insights into plant morphology through advanced computational techniques. While challenges persist, the pathway toward future enhancements and applications of this methodology in plant science and beyond brims with potential, heralding a new era of precision and insight in botanical research.

# REFERENCES

[1]     Luo L, Jiang X, Yang Y, Samy ER, Lefsrud M, Hoyos-Villegas V, Sun S. Eff-3DPSeg: 3D organ-level plant shoot segmentation using annotation-efficient deep learning. Plant Phenomics. 2023; 5: 0080.

[2]     Li Y, Wen W, Miao T, *et al.* Automatic organ-level point cloud segmentation of maize shoots by integrating high-throughput data acquisition and deep learning. Comput Electron Agric 2022; 193: 106702.
[http://dx.doi.org/10.1016/j.compag.2022.106702]

[3]     Wahabzada M, Paulus S, Kersting K, Mahlein AK. Automated interpretation of 3D laserscanned point clouds for plant organ segmentation. BMC Bioinformatics 2015; 16(1): 248.
[http://dx.doi.org/10.1186/s12859-015-0665-2] [PMID: 26253564]

[4]     Herrero M, Thornton PK, Mason-D'Croz D, *et al.* Innovation can accelerate the transition towards a sustainable food system. Nat Food 2020; 1(5): 266-72.
[http://dx.doi.org/10.1038/s43016-020-0074-1]

[5]     Bode L, Weinmann M, Klein R. BoundED: Neural boundary and edge detection in 3D point clouds *via* local neighborhood statistics. ISPRS J Photogramm Remote Sens 2023; 205: 334-51.
[http://dx.doi.org/10.1016/j.isprsjprs.2023.09.023]

[6]     Rivera G, Porras R, Florencia R, Sánchez-Solís JP. LiDAR applications in precision agriculture for cultivating crops: A review of recent advances. Comput Electron Agric 2023; 207: 107737.
[http://dx.doi.org/10.1016/j.compag.2023.107737]

[7]     Harandi N, Vandenberghe B, Vankerschaver J, Depuydt S, Van Messem A. How to make sense of 3D representations for plant phenotyping: a compendium of processing and analysis techniques. Plant Methods 2023; 19(1): 60.
[http://dx.doi.org/10.1186/s13007-023-01031-z] [PMID: 37353846]

[8]     Cai S, Gou W, Wen W, Lu X, Fan J, Guo X. Design and development of a low-cost UGV 3D phenotyping platform with integrated LiDAR and electric slide rail. Plants 2023; 12(3): 483.
[http://dx.doi.org/10.3390/plants12030483] [PMID: 36771568]

[9]     Gill T, Gill SK, Saini DK, Chopra Y, de Koff JP, Sandhu KS. A comprehensive review of high throughput phenotyping and machine learning for plant stress phenotyping. Phenomics 2022; 2(3): 156-83.
[http://dx.doi.org/10.1007/s43657-022-00048-z] [PMID: 36939773]

[10]    Mansoor S, Karunathilake EMBM, Tuan TT, Chung YS. Genomics, phenomics, and machine learning in transforming plant research: advancements and challenges. Hortic Plant J 2024; 13
[http://dx.doi.org/10.1016/j.hpj.2023.09.005]

[11]    Stock M, Pieters O, De Swaef T, wyffels F. Plant science in the age of simulation intelligence. Front Plant Sci 2024; 14: 1299208.

[http://dx.doi.org/10.3389/fpls.2023.1299208] [PMID: 38293629]

[12]  Sarkar S, Ganapathysubramanian B, Singh A, *et al.* Cyber-agricultural systems for crop breeding and sustainable production. Trends Plant Sci 2023; 28
[http://dx.doi.org/10.1016/j.tplants.2023.08.001] [PMID: 37648631]

[13]  Duan T, Chapman SC, Holland E, Rebetzke GJ, Guo Y, Zheng B. Dynamic quantification of canopy structure to characterize early plant vigour in wheat genotypes. J Exp Bot 2016; 67(15): 4523-34.
[http://dx.doi.org/10.1093/jxb/erw227] [PMID: 27312669]

[14]  Du R, Ma Z, Xie P, He Y, Cen H. PST: Plant segmentation transformer for 3D point clouds of rapeseed plants at the podding stage. ISPRS J Photogramm Remote Sens 2023; 195: 380-92.
[http://dx.doi.org/10.1016/j.isprsjprs.2022.11.022]

[15]  Debnath S, Paul M, Debnath T. Applications of LiDAR in agriculture and future research directions. J Imaging 2023; 9(3): 57.
[http://dx.doi.org/10.3390/jimaging9030057] [PMID: 36976108]

[16]  Mkaouar A, Kallel A. Leaf properties estimation enhancement over heterogeneous vegetation by correcting for terrestrial laser scanning beam divergence effect. Remote Sens Environ 2024; 302: 113959.
[http://dx.doi.org/10.1016/j.rse.2023.113959]

[17]  Zhu JJ, Yang M, Ren ZJ. Machine learning in environmental research: common pitfalls and best practices. Environ Sci Technol 2023; 57(46): 17671-89.
[http://dx.doi.org/10.1021/acs.est.3c00026] [PMID: 37384597]

[18]  Guo Q, Jin S, Li M, *et al.* Application of deep learning in ecological resource research: Theories, methods, and challenges. Sci China Earth Sci 2020; 63(10): 1457-74.
[http://dx.doi.org/10.1007/s11430-019-9584-9]

[19]  Qi CR, Su H, Nießner M, Dai A, Yan M, Guibas LJ. Volumetric and Multi-View CNNs for Object Classification on 3D data. Proceedings of the IEEE Conference on Computer Vision and Pattern Recognition. USA: IEEE Computer Society 2016; pp. 5648-56. Available from: http://www.computer.org/csdl/proceedings/cvpr/index.html
[http://dx.doi.org/10.1109/CVPR.2016.609]

[20]  Song R, Zhang W, Zhao Y, Liu Y. Unsupervised multi-view CNN for salient view selection and 3D interest point detection. Int J Comput Vis 2022; 130(5): 1210-27.
[http://dx.doi.org/10.1007/s11263-022-01592-x]

[21]  Peng Y, Lin S, Wu H, Cao G. Point cloud registration based on fast point feature histogram descriptors for 3D reconstruction of trees. Remote Sens (Basel) 2023; 15(15): 3775.
[http://dx.doi.org/10.3390/rs15153775]

[22]  Li S, Corney J. Multi-view expressive graph neural networks for 3D CAD model classification. Comput Ind 2023; 151: 103993.
[http://dx.doi.org/10.1016/j.compind.2023.103993]

[23]  Jin S, Su Y, Gao S, *et al.* Separating the structural components of maize for field phenotyping using terrestrial LiDAR data and deep convolutional neural networks. IEEE Trans Geosci Remote Sens 2020; 58(4): 2644-58.
[http://dx.doi.org/10.1109/TGRS.2019.2953092]

[24]  Qi CR, Su H, Mo K, Guibas LJ. Pointnet: deep learning on point sets for 3d classification and segmentation. Proceedings of the IEEE Conference on Computer Vision and Pattern Recognition 2017; 652-60.

[25]  Qi CR, Yi L, Su H, Guibas LJ. Pointnet++: deep hierarchical feature learning on point sets in a metric space. Adv Neural Inf Process Syst 2017; 30.

[26]  Wang W, Yu R, Huang Q, Neumann U. Sgpn: similarity group proposal network for 3D point cloud instance segmentation. Proceedings of the IEEE Conference on Computer Vision and Pattern

Recognition 2018; 2569-78.
[http://dx.doi.org/10.1109/CVPR.2018.00272]

[27]    Fan H, Yang Y. PointRNN: point recurrent neural network for moving point cloud processing. 2019.

[28]    Xiao A, Huang J, Guan D, Zhang X, Lu S, Shao L. Unsupervised point cloud representation learning with deep neural networks: a survey. IEEE Trans Pattern Anal Mach Intell 2023; 45(9): 11321-39.
[http://dx.doi.org/10.1109/TPAMI.2023.3262786] [PMID: 37030870]

[29]    Ye X, Li J, Huang H, Du L, Zhang X. 3D recurrent neural networks with context fusion for point cloud semantic segmentation. Proceedings of the European Conference on Computer Vision (ECCV) 2018; 403-17.
[http://dx.doi.org/10.1007/978-3-030-01234-2_25]

[30]    Yang F, Davoine F, Wang H, Jin Z. Continuous conditional random field convolution for point cloud segmentation. Pattern Recognit 2022; 122: 108357.
[http://dx.doi.org/10.1016/j.patcog.2021.108357]

[31]    Chen Z, Ledoux H, Khademi S, Nan L. Reconstructing compact building models from point clouds using deep implicit fields. ISPRS J Photogramm Remote Sens 2022; 194: 58-73.
[http://dx.doi.org/10.1016/j.isprsjprs.2022.09.017]

[32]    Kragh M, Underwood J. Multimodal obstacle detection in unstructured environments with conditional random fields. J Field Robot 2020; 37(1): 53-72.
[http://dx.doi.org/10.1002/rob.21866]

[33]    Chou G, Bahat Y, Heide F. Diffusion-SDF: conditional generative modeling of signed distance functions. Proceedings of the IEEE/CVF International Conference on Computer Vision 2023; 2262-72.
[http://dx.doi.org/10.1109/ICCV51070.2023.00215]

[34]    Arora M, Wiesmann L, Chen X, Stachniss C. Static map generation from 3D LiDAR point clouds exploiting ground segmentation. Robot Auton Syst 2023; 159: 104287.
[http://dx.doi.org/10.1016/j.robot.2022.104287]

[35]    Fischer K, Simon M, Olsner F, Milz S, Gross HM, Mader P. Stickypillars: robust and efficient feature matching on point clouds using graph neural networks. Proceedings of the IEEE/CVF Conference on Computer Vision and Pattern Recognition 2021; 313-23.
[http://dx.doi.org/10.1109/CVPR46437.2021.00038]

[36]    Yin J, Shen J, Gao X, Crandall D, Yang R. Graph neural network and spatiotemporal transformer attention for 3D video object detection from point clouds. IEEE Trans Pattern Anal Mach Intell 2021; 9
[http://dx.doi.org/10.1109/TPAMI.2021.3125981] [PMID: 34752380]

[37]    Chen C, Fragonara LZ, Tsourdos A. GAPointNet: Graph attention based point neural network for exploiting local feature of point cloud. Neurocomputing 2021; 438: 122-32.
[http://dx.doi.org/10.1016/j.neucom.2021.01.095]

[38]    Chen J, Lei B, Song Q, Ying H, Chen DZ, Wu J. A hierarchical graph network for 3d object detection on point clouds. Proceedings of the IEEE/CVF Conference on Computer Vision and Pattern Recognition 2020; 392-401.
[http://dx.doi.org/10.1109/CVPR42600.2020.00047]

[39]    Xiong S, Li B, Zhu S. DCGNN: a single-stage 3D object detection network based on density clustering and graph neural network. Complex & Intelligent Systems 2023; 9(3): 3399-408.
[http://dx.doi.org/10.1007/s40747-022-00926-z]

[40]    Wen C, Li X, Yao X, Peng L, Chi T. Airborne LiDAR point cloud classification with global-local graph attention convolution neural network. ISPRS J Photogramm Remote Sens 2021; 173: 181-94.
[http://dx.doi.org/10.1016/j.isprsjprs.2021.01.007]

[41]    Li M, Duncan K, Topp CN, Chitwood DH. Persistent homology and the branching topologies of

plants. Am J Bot 2017; 104(3): 349-53. Available from: https://www.jstor.org/stable/26410931
[http://dx.doi.org/10.3732/ajb.1700046] [PMID: 28341629]

[42]   Li M, Liu Z, Jiang N, *et al.* Topological data analysis expands the genotype to phenotype map for 3D maize root system architecture. Front Plant Sci 2024; 14: 1260005.
[http://dx.doi.org/10.3389/fpls.2023.1260005] [PMID: 38288407]

[43]   Xiang S, Li D. Research on plant growth tracking based on point cloud segmentation and registration. International Conference on Image Processing, Computer Vision and Machine Learning (ICICML) 2022; 469-78.
[http://dx.doi.org/10.1109/ICICML57342.2022.10009765]

[44]   Ziamtsov I, Navlakha S. Plant 3D (P3D): a plant phenotyping toolkit for 3D point clouds. Bioinformatics 2020; 36(12): 3949-50.
[http://dx.doi.org/10.1093/bioinformatics/btaa220] [PMID: 32232439]

[45]   Kumari P, Gangwar H, Kumar V, Jaiswal V, Gahlaut V. Crop Phenomics and High-Throughput Phenotyping. In: Priyadarshan PM, Jain SM, Penna S, Al-Khayri JM, Eds. Digital Agriculture: A Solution for Sustainable Food and Nutritional Security. Cham: Springer 2024; pp. 391-423.
[http://dx.doi.org/10.1007/978-3-031-43548-5_13]

[46]   Das G, Joseph D. Which triangulations approximate the complete graph? In: Djidjev H, Ed. Optimal Algorithms. Berlin, Heidelberg: Springer 1989; pp. 168-92.
[http://dx.doi.org/10.1007/3-540-51859-2_15]

[47]   Musin OR. Properties of the delaunay triangulation. Proceedings of the Thirteenth Annual Symposium on Computational Geometry 1997; 424-6.
[http://dx.doi.org/10.1145/262839.263061]

[48]   Lingas A. A linear-time construction of the relative neighborhood graph from the *Delaunay triangulation.* Comput Geom 1994; 4(4): 199-208.
[http://dx.doi.org/10.1016/0925-7721(94)90018-3]

[49]   Han Q. TrustGNN: enhancing GNN *via* multi-similarity neighbors identifying for social recommendation. IEEE Conference on Telecommunications, Optics and Computer Science (TOCS) 2022; 749-55.
[http://dx.doi.org/10.1109/TOCS56154.2022.10015957]

[50]   La Rosa M, Fiannaca A, La Paglia L, Urso A. A Graph neural network approach for the snalysis of siRNA-target biological networks. Int J Mol Sci 2022; 23(22): 14211.
[http://dx.doi.org/10.3390/ijms232214211] [PMID: 36430688]

[51]   Ju H, Kim K, Kim BI, Woo SK. Graph neural network model for prediction of non-small cell lung cancer lymph node metastasis using protein–protein interaction network and 18F-FDG PET/CT radiomics. Int J Mol Sci 2024; 25(2): 698.
[http://dx.doi.org/10.3390/ijms25020698] [PMID: 38255770]

[52]   Li XS, Liu X, Lu L, Hua XS, Chi Y, Xia K. Multiphysical graph neural network (MP-GNN) for COVID-19 drug design. Brief Bioinform 2022; 23(4): bbac231.
[http://dx.doi.org/10.1093/bib/bbac231] [PMID: 35696650]

[53]   He T, Zhou H, Ong YS. Cong G. Not all neighbors are worth attending to: graph selective attention networks for semi-supervised learning. 2022.

[54]   Sun Y, Ma H, Ko YB, Wang J. LNGAT: local neighborhood graph attention network. J. Electron. Imaging. 2022; 31(5): 053034.
[http://dx.doi.org/10.1117/1.JEI.31.5.053034]

[55]   Forman R. Forman. Bochner's method for cell complexes and combinatorial *Ricci curvature.* Discrete Comput Geom 2003; 29(3): 323-74.
[http://dx.doi.org/10.1007/s00454-002-0743-x]

[56]   Anand DV, Xu Q, Wee J, Xia K, Sum TC. Topological feature engineering for machine learning based

halide perovskite materials design. npj Computational Materials 2022; 8(1): 203.
[http://dx.doi.org/10.1038/s41524-022-00883-8]

[57]    Elumalai P, Yadav Y, Williams N, Saucan E, Jost J, Samal A. Graph Ricci curvatures reveal atypical functional connectivity in autism spectrum disorder. Sci Rep 2022; 12(1): 8295.
[http://dx.doi.org/10.1038/s41598-022-12171-y] [PMID: 35585156]

[58]    Han X, Zhu G, Zhao L, *et al.* Ollivier–Ricci curvature based spatio-temporal graph neural networks for traffic flow forecasting. Symmetry (Basel) 2023; 15(5): 995.
[http://dx.doi.org/10.3390/sym15050995]

[59]    Wee J, Xia K. Ollivier persistent Ricci curvature-based machine learning for the protein–ligand binding affinity prediction. J Chem Inf Model 2021; 61(4): 1617-26.
[http://dx.doi.org/10.1021/acs.jcim.0c01415] [PMID: 33724038]

[60]    Santos FAN, Raposo EP, Coutinho-Filho MD, Copelli M, Stam CJ, Douw L. Topological phase transitions in functional brain networks. Phys Rev E 2019; 100(3): 032414.
[http://dx.doi.org/10.1103/PhysRevE.100.032414] [PMID: 31640025]

[61]    Conn A, Pedmale UV, Chory J, Navlakha S. High-resolution laser scanning reveals plant architectures that reflect universal network design principles. Cell Syst 2017; 5(1): 53-62.e3.
[http://dx.doi.org/10.1016/j.cels.2017.06.017] [PMID: 28750198]

[62]    Conn A, Pedmale UV, Chory J, Stevens CF, Navlakha S. A statistical description of plant shoot architecture. Curr Biol 2017; 27(14): 2078-2088.e3.
[http://dx.doi.org/10.1016/j.cub.2017.06.009] [PMID: 28690115]

[63]    Li D, Li J, Xiang S, Pan A. PSegNet: simultaneous semantic and instance segmentation for point clouds of plants. Plant Phenomics 2022; 2022/9787643.
[http://dx.doi.org/10.34133/2022/9787643] [PMID: 35693119]

[64]    Li D, Shi G, Li J, *et al.* PlantNet: A dual-function point cloud segmentation network for multiple plant species. ISPRS J Photogramm Remote Sens 2022; 184: 243-63.
[http://dx.doi.org/10.1016/j.isprsjprs.2022.01.007]

[65]    Li D, Wei Y, Zhu R. A comparative study on point cloud down-sampling strategies for deep learning-based crop organ segmentation. Plant Methods 2023; 19(1): 124.
[http://dx.doi.org/10.1186/s13007-023-01099-7] [PMID: 37951912]

[66]    Schindelin L, Kusztos F, Schmid B. Segmentation editor. ImageJ Documentation Wiki. 2007.

[67]    Phan AV, Nguyen ML, Nguyen YLH, Bui LT. DGCNN: A convolutional neural network over large-scale labeled graphs. Neural Netw 2018; 108: 533-43.
[http://dx.doi.org/10.1016/j.neunet.2018.09.001] [PMID: 30458952]

[68]    Wang X, Liu S, Shen X, Shen C, Jia J. Associatively Segmenting Instances and Semantics in Point Clouds. Proceedings of the IEEE/CVF Conference on Computer Vision and Pattern Recognition. 4096-105.
[http://dx.doi.org/10.1109/CVPR.2019.00422]

[69]    Mirande K, Godin C, Tisserand M, *et al.* A graph-based approach for simultaneous semantic and instance segmentation of plant 3D point clouds. Front Plant Sci 2022; 13: 1012669.
[http://dx.doi.org/10.3389/fpls.2022.1012669]

# SUBJECT INDEX